DISCRETE MATHEMATICS
Methods and Applications

Third Edition

ABDUL-MAJID WAZWAZ
Saint Xavier University

Stipes Publishing L.L.C.

204 W. University Avenue

Champaign, IL 61820

Published by

Stipes Publishing L.L.C.

204 W. University Avenue

Champaign, IL 61820

THIS BOOK IS DEDICATED TO

My wife, our son and our three daughters
for supporting me in all my endeavors

Contents

Preface

This book has evolved over many years of teaching from lecture notes that were used successfully to reinforce the concepts of discrete mathematics. This text is designed for an introductory level course to those areas of discrete mathematics used by students of mathematics and computer science. *Discrete Mathematics: Methods and Applications* is especially designed for those who wish to understand discrete mathematics without having the extensive mathematical background. In this fashion, this text leaves out the difficult approaches and emphasizes the material in an accessible way through worked examples and applications.

From my experience in teaching, I have found that the material can indeed be taught in an accessible manner. Students have shown both a lot of motivation and capability to grasp the subject once the material introduced in a clear manner. I have translated my means of introducing and fully teaching this subject into this text so that the intended user can take full advantage of the easily presented and explained material. The text books used in discrete mathematics courses differ widely in contents, approach and applications.

The emphasis in the second edition of this book is on introducing the material in a manner to build the student's skill, supported by numerous worked examples and applications with full explanations. In addition, new material and concepts with lot of useful applications have been added to increase the knowledge and the skills that students need. New illustrative examples and updated exercises have been added to reinforce the relevance of the text.

Two primary objectives are the main concern of this text:

1. to introduce the fundamental concepts of discrete mathematics topic, starting from number theory to algorithm analysis along with matrices and determinants, and

2. to introduce and reinforce the proof-writing skills through a variety of methods of proof.

The second edition of this book consists of thirteen chapters, each being divided into sections. Three Appendices are also added. Algorithms and their analysis will be found in this text, but the emphasis is on mathematics.

Chapter 1 introduces a practical approach to the number systems. The primary focus of this chapter is on decimal, binary, octal and hexadecimal systems. Converting from binary to decimal by using hexadecimal concepts is emphasized. Equations involving different number systems is handled in a useful way.

Chapter 2 provides the reader with a comprehensive discussion of the literature related to primes and Euclidean algorithm. Functions generating primes and Mersenne primes are examined.

Chapter 3 introduces topics that include specific cases of linear congruences, set of linear congruences and Diophantine equations. The Chinese remainder theorem is presented with illustrative examples and exercises.

Chapter 4 provides a concise discussion of the methods of proofs, namely, mathematical induction, finite differences method, partial fraction method, direct proof method and proof by contradiction. This chapter emphasizes the powerful method of mathematical induction, and the other alternative methods. The methods provide supportive techniques to build up the proof-writing skills.

Chapter 5 presents a reliable study on set theory. The important set theory arise in logic, switching circuits, Boolean algebra and probability areas.

Chapter 6 provides the reader with a comprehensive study on logic, switching circuits, and Boolean algebra. These concepts became recently the focus of study in discrete mathematics due to its applications in computers and computer information systems.

Chapter 7 covers the counting techniques and the combinatorics. The chapter covers permutations, combinations and the binomial theorem. The concepts lead naturally to probability theory.

Chapter 8 provides the reader with an introductory study on probability. The chapter handles the finite space probability, the conditional probability, and the repeated experiments with an effective use of set theory discussed in Chapter 5.

Chapter 9 provides a concise discussion of functions and the basic definitions of limits. Algorithms and the Big-Oh notation are discussed supported by an extensive set of worked examples that lay the foundation for handling the rate of growth.

Chapter 10 presents in details the recursion concept. Fibonacci sequence and Lucas sequence are covered in an easy-to-read style that will build up the student's ability for their needs in other courses. Identities of both sequences are presented and proved by mathematical induction to increase the students proof-writing skills

Chapter 11 provides a concise discussion of matrices. The classic problems of these concepts have been switched to equations that combine all related operations on matrices.

Chapter 12 provides a concise discussion of determinants. Applications to the inverse matrix and solving system of equations are introduced. The *Q-matrix*, the *M-Matrix*, and the *R-Matrix* are presented, where mathematical induction is used to handle applications to Fibonacci and Lucas identities.

In Chapter 13, relations, digraphs and binary trees are introduced in a useful way that will build the students experience for a data structure course. A fairly detailed treatment of the rooted binary tree and tree searching topics is discussed, supported by many practical examples.

Throughout the text, numerous examples are provided with a substantial amount of explanation to introduce the material in a clear fashion. Examples ranging in level from easy to reasonable, are given in each section to give the students the knowledge, the practice and the skill they are seeking.

Finally, this book is suitable for a one semester course in discrete mathematics, depending on the extent of coverage required.

I am indebted and grateful to my wife, my son and my daughters who provided me with their encouragement, patience and constant support. I would like to make a special mention and dedication to my daughter, Mai Wazwaz. This book could not have been made possible without her natural artistic skills, creative suggestions, and most of all, her encouragement, and help. Mai, thank you for being such a wonderful daughter.

The author would highly appreciate any note concerning any error found and for any constructive suggestion.

Chicago, IL 60655
August 1, 2013

Abdul-Majid Wazwaz
Web: http://web.sxu.edu/aw1
e-mail:wazwaz@sxu.edu

Preface to the Third Edition

While the first two editions of this Book were certainly excellent, I feel strongly that the new edition of "Discrete Mathematics: Methods and Applications" is far superior to its two predecessors. Since I began writing the first edition in 1999 and the second edition in 2004, much has happened that makes a third edition of the book timely necessary. In the last few years since the second edition of this book was published, I have also built up a lot of ideas based on my own experiences in teaching this course for several years and on the countless suggestions from many colleagues. Many specific useful changes and improvements are added in this edition. Updated examples are used throughout the third edition, and new concepts and developments were also added that make this edition reflect the current and future needs for discrete mathematics. The text has been updated in several places to improve its readability and pedagogy. The Solution Manual providing solutions to the chapter-end exercises given in this book is available.

Chicago, IL 60655
August 1, 2013

Abdul-Majid Wazwaz
Web: http://web.sxu.edu/aw1
e-mail:wazwaz@sxu.edu

Chapter 1

Number Systems

1.1 Introduction

A number system is the set of symbols, characters, or digits used to express data and to perform computational works. We are familiar with the decimal system, where the prefix "deci" stands for ten. However, other number systems are used in the computer field. In digital computers we use the very efficient binary number system, where the prefix "bi" stands for two, to express data and to perform arithmetic computations. The hexadecimal number system (with base 16) provides a useful method of working with binary numbers. The octal number system (with base 8) is the less commonly used system.

The aforementioned number systems are positional number systems. In any system, the rightmost symbol or digit of any number has a place value of one, and every other symbol or digit has a place value b times that of the place value of the symbol or digit to its right, where b is the base of the system used.

The number of symbols or digits used in a positional number system depends on its base. The base or radix is usually the number of unique symbols or digits, including zero. The highest numerical symbol or digit always has a value of $b - 1$, i.e. one less than the base b of the system. The highest numerical symbols for the decimal system, octal system, and the binary system are 9, 7, and 1 respectively.

In this chapter, we will study the binary numerical system, the octal numerical system, and the hexadecimal numerical system. Arithmetic operations will also be studied using the binary numerical system.

1.2   The Decimal System

The decimal system, or the base 10 system, is the number system that we use in arithmetic calculations. The decimal system employs 10 distinct digits to express any number by combining these digits. The 10 digits or symbols that we use to write decimal numbers are $0, 1, 2, 3, 4, 5, 6, 7, 8$ and 9. The decimal system or the base ten system is a place-value system. This means that the place or the position of a digit in a base ten number determines its corresponding numerical value that is a power of 10. Each digit represents a quantity depending on the position of the digit in the number. By considering the number 22, the right-most digit 2 represents the number of units in this number, while the next digit 2 represents the number of tens.

The place values in decimal system increase from right to left. In any decimal number without any decimal point, the first place or the right-most place is the one's place, the second to the left is the ten's place, followed by the hundred's place, and so on. The following table shows the place value for each digit's position of any number. This table was established by placing 1 at the right-most place, then multiply by 10 from right to left to define the values of the other places.

multiply by 10 from right to left
$\longleftarrow$

Power of the Base	10^3	10^2	10^1	10^0
Decimal Value	1000	100	10	1

Table 1.1

Using Table 1.1, we can write each digit of the number 1356 in its proper position as shown by the following table

$\cdots$	$10^3 = 1000$	$10^2 = 100$	$10^1 = 10$	$10^0 = 1$
$\cdots$	1	3	5	6

Table 1.2

Accordingly, using the weighted value for each position, the decimal number 1356 can be written in an expanded notation by

$$1356 \ = \ 1(10^3) + 3(10^2) + 5(10^1) + 6(10^0)$$

Based on the principle of the place values of the digit's position, we can construct several number systems, each with a specified base. Generally speaking, if a number system with base b is to be established, then b distinct digits or symbols, including zero, should be used to represent any number in that system. The digits are $0, 1, 2, 3, \cdots, (b-1)$, where b is an integer. The place values of the digit's position follow the table:

multiply by b from right to left

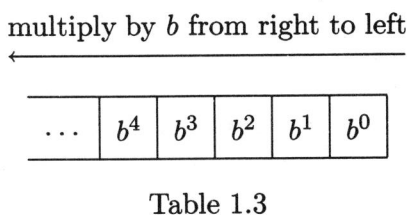

Table 1.3

It is to be noted that for systems with base $b > 10$, we usually use distinct alphabetical characters to represent the digits that are greater than 9. This will be explained in details in Section 1.5. In the coming sections, we will focus our study on three well-known number systems, namely, the binary system, the octal system and the hexadecimal system. Other systems can be constructed and will be subjected to the same rules that we will discuss.

1.3 The Binary System

The binary system, or the base 2 system, is the number system used by computers and calculators to store decimal numbers, alphabetical letter, and a single character. In this system, two digits, namely, 0 and 1 are used. A **binary digit** is called a **bit**, which is an abbreviation for binary digit. Therefore, the symbols 0 and 1 are called **bits**, short for binary digits. A **byte** is a unit of memory containing 8 bits. Large number of bytes are usually expressed by **kilobyte** (KB) and **megabyte** (MB). The value of one kilobyte (1KB) is $2^{10} = 1024$ bytes. Normally, we use kilo to refer to 1000, but in computer concepts, kilobyte is 1024 bytes. Likewise, one megabyte refers to $2^{10} \times 2^{10} = 1024$KB $= 1048576$ bytes. Normally, we use mega to refer to a million whereas in computer concepts the mega is 48576 more than a million.

Setting $b = 2$ in Table 1.3, the binary place values representation is given in Table 1.4.

multiply by 2 from right to left
←——————————————————

Power of the Base	2^6	2^5	2^4	2^3	2^2	2^1	2^0
Decimal Value	64	32	16	8	4	2	1

Table 1.4

The weighted values for each position in binary system increase from right to left. It is worth noting that a binary number should be recognized by writing a subscript 2 to the lower right of that number. For example, the binary number $11_{(2)}$ is different than the decimal number 11. The binary number $11_{(2)}$ is equivalent to the decimal number 3. In this section we will convert a binary number to a decimal number, a decimal number to binary number, and will study addition and multiplication of binary numbers.

1.3.1   Converting from Binary to Decimal

To convert a binary number to a decimal number, we use the place value table 1.4. We then multiply each binary digit by the related decimal value that corresponds to the position of that digit. Adding the obtained results gives the equivalent decimal number. This is explained by the following illustrative examples.

Example 1. Convert the binary number $1011_{(2)}$ to a decimal number.

Solution. Writing the digits of the binary number $1011_{(2)}$ at the proper places of the place value table we obtain

multiply by 2 from right to left
←——————————————————

Decimal Value	16	8	4	2	1
Binary Number	$\cdots$	1	0	1	1

Hence, we write

$$
\begin{aligned}
1011_{(2)} &= 1(8) + 0(4) + 1(2) + 1(1) \\
&= 8 + 0 + 2 + 1 \\
&= 11
\end{aligned}
$$

Example 2. Convert the binary number $110101_{(2)}$ to a decimal number.

Solution. Writing the digits of the binary number $110101_{(2)}$ at the proper places, we obtain

multiply by 2 from right to left

←

Decimal Value	32	16	8	4	2	1
Binary Number	1	1	0	1	0	1

Hence, we write

$$
\begin{aligned}
110101_{(2)} &= 1(32) + 1(16) + 0(8) + 1(4) + 0(2) + 1(1) \\
&= 32 + 16 + 0 + 4 + 0 + 1 \\
&= 53
\end{aligned}
$$

Example 3. Convert the binary number $1101_{(2)}$ to a decimal number.

Solution. Writing the digits of the binary number $1101_{(2)}$ at the proper places, we obtain

multiply by 2 from right to left

←

Decimal Value	8	4	2	1
Binary Number	1	1	0	1

Therefore, we write

$$
\begin{aligned}
1101_{(2)} &= 1(8) + 1(4) + 0(2) + 1(1) \\
&= 8 + 4 + 0 + 1 \\
&= 13
\end{aligned}
$$

1.3.2   Converting from Decimal to Binary

It remains now to show how we can convert any decimal number to its equivalent number in the binary system. To convert a decimal number to a binary number, we divide the given decimal number by the base 2, write the quotient and the remainder in this step. Repeat this step by dividing the quotient obtained in the first step by the base 2, and write the quotient and the remainder of this step. Division by 2 is repeated until the quotient is 0. Notice that the last remainder, when the quotient is 0, should be listed. The binary number is consequently obtained by reading the remainders from the bottom to the top in the inverse order. The conversion rule will be illustrated by the following examples.

Example 4. Convert the decimal number 26 to a binary number.

Solution. Using the successive divisions we obtain

base 2	decimal number	
2	26	remainder in binary digits
2	13	0
2	6	1
2	3	0
2	1	1
	0	1

Reading the remainders upwards gives

$$26 \ = \ 11010_{(2)}$$

The result can be justified by converting the obtained binary number $11010_{(2)}$ to its equivalent decimal number.

Example 5. Convert the decimal number 43 to a binary number.

Solution. Using the repeated division process we obtain

base 2	decimal number	
2	43	remainder in binary digits
2	21	1
2	10	1
2	5	0
2	2	1
2	1	0
	0	1

Reading the remainders upward gives

$$43 = 101011_{(2)}$$

Example 6. Convert the decimal number 53 to a binary number.

Solution. Following the procedure used before, we obtain

base 2	decimal number	
2	53	remainder in binary digits
2	26	1
2	13	0
2	6	1
2	3	0
2	1	1
	0	1

Hence, we write

$$53 = 110101_{(2)}$$

Example 7. Find x by solving the equation:

$$11x_{(2)} \;=\; 7.$$

Solution. Notice that the left side is a binary number, and the right side is a decimal number. Converting the binary number to a decimal number we find

$$
\begin{aligned}
1(4) + 1(2) + x(1) &= 7 \\
6 + x &= 7 \\
x &= 1
\end{aligned}
$$

1.3.3 Binary Addition

The following table shows the first few decimal numbers with their equivalent binary numbers:

Decimal	Binary Equivalent
0	0
1	1
2	10
3	11
4	100
5	101
6	110
7	111
8	1000
9	1001
10	1010

To add two binary numbers, such as $11101_{(2)} + 10101_{(2)}$, we begin from right as in decimal addition, noting that $1 + 1 = 10_{(2)}$; thus we write 0 and carry 1 to the next column to the left. We continue in this way using the table presented above, arriving at the answer shown below.

$$11101_{(2)}$$

$$+10101_{(2)}$$

$$110010_{(2)}$$

1.3.4 Binary Subtraction

To subtract two binary numbers, such as $11101_{(2)} - 10111_{(2)}$, we begin from right as in decimal subtraction, noting that $1-1 = 0_{(2)}$. However, to subtract 1 from 9, we borrow from the next digit noting that the 1 we borrow equals 2 in the decimal, hence $0-1 = 1$ and we carry the remaining 1. We continue in this way, arriving at the answer shown below.

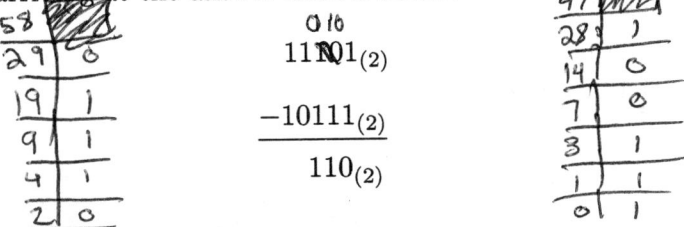

$$11\mathbf{N}01_{(2)}$$

$$-10111_{(2)}$$

$$110_{(2)}$$

1.3.5 Binary Multiplication

The method we usually use for decimal multiplication is used for binary multiplication. For example, we find that $111_{(2)} \times 101_{(2)} = 100011_{(2)}$.

Exercises 1.3

1. Convert each decimal number to a binary number
   (a) 11 (b) 21 (c) 113 (d) 171

2. Convert each binary number to a decimal number
   (a) $1101_{(2)}$ (b) $10101_{(2)}$ (c) $1011_{(2)}$ (d) $11011_{(2)}$

3. Convert each decimal number to a binary number
   (a) 47 (b) 58 (c) 85 (d) 91

4. Convert each decimal number to a binary number

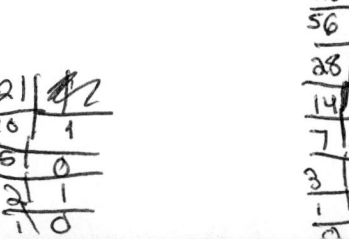

    (a) 97   (b) 89   (c) 117   (d) 202

5. Solve the following equations:

    (a) $11x1_{(2)} = 15$ (b) $1x11_{(2)} = 11$ (c) $110_{(x)} = 6$ (d) $111_{(x)} = 7$

6. Add the following binary numbers:

    (a) $1111_{(2)} + 1100_{(2)}$ (b) $11011_{(2)} + 10111_{(2)}$ (c) $1001_{(2)} + 1101_{(2)}$

7. Multiply the following binary numbers:

    (a) $1111_{(2)} \times 1010_{(2)}$ (b) $1110_{(2)} \times 1011_{(2)}$

8. Subtract the following binary numbers:

    (a) $1111_{(2)} - 1100_{(2)}$ (b) $11011_{(2)} - 10111_{(2)}$ (c) $11010101_{(2)} - 1001011_{(2)}$

1.4 The Octal System

In the octal system, the system of base 8, eight distinct digits, namely $0, 1, 2, 3, 4, 5, 6$, and 7 are used to represent any octal number. The octal number is distinguished from the decimal number by using a subscript (8) at the lower right of that number. For example, the octal number $347_{(8)}$ is different than the decimal number 347. It will be discussed later that $347_{(8)}$ is equivalent to the decimal number 231.

As discussed before, each digit in any number system represents a quantity depending on the position of that digit in the number. Setting $b = 8$ in Table 1.3, the octal place values representation is given by the following table:

multiply by 8 from right to left
←

Power of the Base	8^4	8^3	8^2	8^1	8^0
Decimal value	4096	512	64	8	1

Table 1.5

1.4.1 Converting from Octal to Decimal

To convert any octal number to an equivalent decimal number, we simply use the place value representation shown in Table 1.5. In this case, we follow the same process we used in converting a binary number to a decimal number. We then multiply each digit of the octal number by its equivalent decimal value that corresponds to the position of that digit. Adding the results of these products gives the equivalent decimal number. The technique will be illustrated by discussing the following examples.

Example 1. Convert $534_{(8)}$ to a decimal number.

Solution. Writing the digits of the octal number $534_{(8)}$ at the proper places, we obtain

multiply by 8 from right to left

←—————————————

Decimal value	512	64	8	1
Octal Number	$\cdots$	5	3	4

Hence, we find

$$\begin{aligned} 534_{(8)} &= 5(64) + 3(8) + 4(1) \\ &= 320 + 24 + 4 \\ &= 348 \end{aligned}$$

Example 2. Convert the octal number $1743_{(8)}$ to a decimal number.

Solution. Writing the digits of the octal number $1743_{(8)}$ at the proper places, we obtain

multiply by 8 from right to left

←—————————————

Decimal value	4096	512	64	8	1
Octal Number	$\cdots$	1	7	4	3

Using the table shown above, we obtain

$$1743_{(8)} = 1(512) + 7(64) + 4(8) + 3(1)$$
$$= 512 + 448 + 32 + 3 = 995$$

1.4.2 Converting from Decimal to Octal

In this part we will study the process of converting any decimal number to it equivalent octal number. The process is straightforward and identical to that used in converting a decimal number to a binary number that was discussed in the preceding section. The repeated division process consists of dividing the decimal number by the base 8, repeat the division process as many times until the quotient is zero, and list the remainders in a separate column as discussed before. The equivalent octal number is obtained by writing the remainders upwards, i.e from the bottom to the top. Notice that the remainders will be any digit from 0 to 7.

The following examples illustrate the procedure that was discussed.

Example 3. Convert 679 to an octal number.

Solution. Using the repeated division by 8 we obtain

base 8	decimal number	
8	679	remainder in octal digits
8	84	7
8	10	4
8	1	2
	0	1

Hence, we obtain

$$679 = 1247_{(8)}$$

As stated before, each digit in the resulting octal number is less than the base 8.

Example 4. Convert the decimal number 3675 to an octal number.

Solution. Using the repeated division we obtain the following:

base 8	decimal number	
8	3675	remainder in octal digits
8	459	3
8	57	3
8	7	1
	0	7

Consequently, we obtain

$$3675 = 7133_{(8)}$$

Each digit in the resulting octal number is less than the base 8.

Example 5. Convert the decimal number 62467 to an octal number.

Solution. Proceeding as before we obtain

base 8	decimal number	
8	62467	remainder in octal digits
8	7808	3
8	976	0
8	122	0
8	15	2
8	1	7
	0	1

Consequently, we obtain

$$62467 = 172003_{(8)}$$

We close this section by discussing the following equations related to the number systems.

Example 6. Find y by solving the equation:

$$1y7_{(8)} \ = \ 103$$

Solution. We first note that the left hand side is an octal number, therefore y should be any digit from 0 to 7. Also, notice that the right hand side is a decimal number. We therefore set

$$
\begin{aligned}
1y7_{(8)} &= 103 \\
1(8^2) + y(8^1) + 7(8^0) &= 103 \\
64 + 8y + 7 &= 103 \\
8y &= 32 \\
y &= 4
\end{aligned}
$$

Example 7. Find y by solving the equation:

$$y36_{(8)} \ = \ 360 - 1010_{(2)}$$

Solution. It is clear that the left hand side is an octal number, therefore y should be any digit from 0 to 7. Also, notice that the right hand side includes a decimal number and a binary number. To determine y, we convert the octal and the binary numbers to decimal numbers and then proceed to solve a standard equation. Accordingly, we find

$$
\begin{aligned}
y36_{(8)} &= 360 - 1010_{(2)} \\
y(8^2) + 3(8^1) + 6(1) &= 360 - \left(1(2^3) + 0(2^2) + 1(2^1) + 0(2^0)\right) \\
64y + 30 &= 360 - 10 \\
64y &= 320 \\
y &= 5
\end{aligned}
$$

Example 8. Find x by solving the equation:

$$123_{(x)} \ = \ 70 + 1101_{(2)}$$

Solution. It is clear that the left hand side is expressed in a system where the base x is unknown. Also, notice that the right hand side includes a decimal number and a binary number. To determine x, we convert all numbers to decimal numbers and then proceed to solve the resulting standard equation. Accordingly, we find

$$\begin{aligned}
123_x &= 70 + 1101_{(2)} \\
1(x^2) + 2(x^1) + 3(x^0) &= 70 + \left(1(2^3) + 1(2^2) + 0(2^1) + 1(2^0)\right) \\
x^2 + 2x + 3 &= 70 + 13 \\
x^2 + 2x - 80 &= 0 \\
(x - 8)(x + 10) &= 0 \\
x &= 8
\end{aligned}$$

Notice that the base of any system should be a positive integer, and for this reason the negative value of x is not valid in the number system of any base. Therefore $x = -10$ is not possible here.

Exercises 1.4

1. Convert the following decimal numbers to octal numbers:

   (a) 231   (b) 314   (c) 561   (d) 723

2. Convert the following octal numbers to decimal numbers:

   (a) $136_{(8)}$ (b) $517_{(8)}$ (c) $716_{(8)}$ (d) $777_{(8)}$

3. Convert the following decimal numbers to octal numbers:

   (a) 63   (b) 97   (c) 111   (d) 157

4. Convert the following octal numbers to decimal numbers:

   (a) $43_{(8)}$ (b) $76_{(8)}$ (c) $123_{(8)}$ (d) $327_{(8)}$

5. Solve the following equations:
   (a) $1y3_{(8)} = 99$ (b) $y56_{(8)} = 200 - 11010_{(2)}$

   (c) $147_{(x)} = 103$ (d) $216_{(x)} = 132 + 1010_{(2)}$

6. Show that the following statements are true:
   (a) $1101_{(2)} = 15_{(8)}$ (b) $1011_{(2)} < 101_{(8)}$ (c) $1111_{(2)} > 11_{(8)}$

7. Solve the following equations:

   (a) $23x_{(8)} = 156$ (b) $5x7_{(8)} = 375$

   (c) $173_{(x)} = 123$ (d) $473_{(x)} = 390$

1.4.3   Converting from Octal to Binary Notation

To convert any octal number directly to a binary number, a table that relates octal number to its 3-bit binary equivalent is normally used. The following table shows the octal digits together with their decimal equivalents and their 3-bit binary equivalents:

Decimal	3-Bit Binary Equivalent	Octal
0	000	0
1	001	1
2	010	2
3	011	3
4	100	4
5	101	5
6	110	6
7	111	7

Table 1.6

The process to convert an integer from octal to binary is as follows:
1. We first write each octal digit of the given numeral in fixed 3-bit binary notation by using Table 1.6.
2. We then group the resulting binary digits.

The method presented above can be illustrated by discussing the following examples.

Example 10. Convert the octal number $751_{(8)}$ to a binary number.

Solution. We first write each octal digit of the given numeral $751_{(8)}$ in fixed 3-bit binary notation by using Table 1.6, therefore we find:
$7 = 111_{(2)}$, $5 = 101_{(2)}$, $1 = 001_{(2)}$. In other words we can write

$$7 \quad 5 \quad 1$$
$$\updownarrow \quad \updownarrow \quad \updownarrow$$
$$111 \quad 101 \quad 001$$

By grouping together the resulting binary digits, the answer is $111101001_{(2)}$.

Example 11. Convert the octal number $362_{(8)}$ to a binary number.

Solution. We first write each octal digit of the given numeral $362_{(8)}$ in fixed 3-bit binary notation by using Table 1.6, therefore we find:
$3 = 011_{(2)}$, $6 = 110_{(2)}$, $2 = 010_{(2)}$. In other words we can write

$$3 \quad 6 \quad 2$$
$$\updownarrow \quad \updownarrow \quad \updownarrow$$
$$011 \quad 110 \quad 010$$

By grouping together the resulting binary digits, the answer is $11110010_{(2)}$.

Example 12. Convert the octal number $1504_{(8)}$ to a binary number.

Solution. We first write each octal digit of the given numeral $1504_{(8)}$ in fixed 3-bit binary notation by using Table 1.6, therefore we find:
$1 = 001_{(2)}$, $5 = 101_{(2)}$, $0 = 000_{(2)}$, $4 = 100_{(2)}$. In other words we can write

$$1 \quad 5 \quad 0 \quad 4$$
$$\updownarrow \quad \updownarrow \quad \updownarrow \quad \updownarrow$$
$$001 \quad 101 \quad 000 \quad 100$$

By grouping the resulting binary digits, the answer is $1101000100_{(2)}$.

1.4.4   Converting from Binary to Octal Notation

To convert any binary number directly to an octal number, Table 1.6 that relates octal number to its 3-bit binary equivalent will be used. The conversion is the reverse of the process followed above. Consequently, the conversion process to convert an integer from binary to an octal number is as follows:

1. We first group the digits of the given binary number into sets of 3-bits, starting from the right of the binary number to the left, adding leading zeros as needed, usually one or two zeros.

2. Using Table 1.6, we find the equivalent octal digit to each set of 3-binary bits.

3. We then group the resulting octal digits, using 8 as a subscript to the resulting octal number.

The process presented above can be illustrated by discussing the following examples.

Example 13. Convert the binary number $110011101_{(2)}$ to an octal number.

Solution. Starting from right to left, we first group the binary number $110011101_{(2)}$ into sets of fixed 3-bit binary digits, therefore we set: 110 011 101. Using Table 1.6, we find

$$\underbrace{110} \quad \underbrace{011} \quad \underbrace{101}$$

$$\updownarrow \qquad \updownarrow \qquad \updownarrow$$

$$6 \qquad \quad 3 \qquad \quad 5$$

By grouping together the resulting octal digits, the answer is $635_{(8)}$.

Example 14. Convert the binary number $10111011_{(2)}$ to an octal number.

Solution. Starting from right to left, we first group the binary number $10111011_{(2)}$ into sets of fixed 3-bit binary digits, therefore we set: 010 111 011, by adding one leading zero. Using Table 1.6 we find

$$\underbrace{010} \quad \underbrace{111} \quad \underbrace{011}$$

$$\updownarrow \qquad \updownarrow \qquad \updownarrow$$

$$2 \qquad \quad 7 \qquad \quad 3$$

By grouping together the resulting octal digits, the answer is $273_{(8)}$.

Example 15. Convert the binary number $1110100_{(2)}$ to an octal number.

Solution. Starting from right to left, we first group the binary number $1110100_{(2)}$ into sets of fixed 3-bit binary digits, therefore we set: 001 110 100, by adding two leading zeros. Using Table 1.6, we find the equivalent octal digit to each set, hence we find

$$\underbrace{001}\quad \underbrace{110}\quad \underbrace{100}$$

$$\updownarrow\quad \updownarrow\quad \updownarrow$$

$$1\quad\quad 6\quad\quad 4$$

By grouping together the resulting octal digits, the answer is $164_{(8)}$.

Exercises 1.4.4

1. Convert the following octal numbers to binary numbers:

   (a) $543_{(8)}$ (b) $345_{(8)}$ (c) $176_{(8)}$ (d) $261_{(8)}$

2. Convert the following octal numbers to binary numbers:

   (a) $1234_{(8)}$ (b) $5210_{(8)}$ (c) $2137_{(8)}$ (d) $7171_{(8)}$

3. Convert the following binary numbers to octal numbers:

   (a) $111011001_{(2)}$ (b) $101101011_{(2)}$

   (c) $10111101_{(2)}$ (d) $1011010_{(2)}$

4. Convert the following binary numbers to octal numbers:

   (a) $111101111110_{(2)}$ (b) $1110101100_{(2)}$

   (c) $10111011010_{(2)}$ (d) $11111110101_{(2)}$

5. Find x of the following equations (x can be one or more digits):

   (a) $3x1_{(8)} = 11100001_{(2)}$ (b) $761_{(8)} = 111x001_{(2)}$

   (c) $1x24_{(8)} = 1011010100_{(2)}$ (d) $2565_{(8)} = 10x110101_{(2)}$

1.5   The Hexadecimal System

In this section, we will discuss the hexadecimal system that gained its importance in computer science. The hexadecimal system is a positional number system as are the decimal number system, the octal system, and the binary number system. The system is called the hexadecimal system, and for this reason the base of this system is 16. This means that sixteen distinct symbols are needed to represent any hexadecimal number. Recall that one and only one symbol must be used to represent a value in any position of the number. There are ten well-known standard digits we normally use in the decimal system. The hexadecimal number system uses not only the numerals 0 through 9, but also uses the letters A, B, C, D, E, and F to represent the equivalent of 10 through 15, respectively. This means that 6 distinct symbols, such as the first six alphabetical letters A, B, C, D, E and F are used, in addition to the ten digits, to form the base of this system. In other words, for the hexadecimal system we use the symbols $0, 1, 2, 3, 4, 5, 6, 7, 8, 9, A, B, C, D, E$ and F. It is to be noted here that the hexadecimal symbols A, B, C, D, E and F correspond to the decimal integers 10, 11, 12, 13, 14 and 15 respectively.

It is interesting to note that the letters A through F are selected instead of the integers 10 through 15, because each place in any system should be represented only by a single symbol, as A or B to fill only one position in the table of the place values, whereas using 10 or 11 means that two places are filled instead of one. As stated before, each symbol of a hexadecimal number is weighted differently depending on its position or its location. The hexadecimal number is distinguished from the decimal number by writing a subscript (16) to the lower right of the number. Because the base of the hexadecimal number system is 16, so each position of the hexadecimal number represents a successive power of 16 as shown below by Table 1.7 obtained by setting $b = 16$ in Table 1.3.

multiply by 16 from right to left

←————————————

Power of the Base	16^4	16^3	16^2	16^1	16^0
Decimal Value	65536	4096	256	16	1

Table 1.7

1.5.1 Converting from Hexadecimal to Decimal

The conversion rules for converting a number from hexadecimal to decimal are identical to the rules discussed earlier in previous sections, hence we skip details. The following illustrative examples will be used to explain the conversion rules.

Example 1. Convert the hexadecimal number $1AB_{(16)}$ to a decimal number.

Solution. We first write the digits of the hexadecimal number $1AB_{(16)}$ at the proper places in the table of the place values representation given above. It should be noted that in converting to decimal number, we should substitute $A = 10$ and $b = 11$. Consequently, we find

Decimal Value	256	16	1
Hexadecimal Number	1	A	B
	↕	↕	↕
Decimal Equivalent	1	10	11

Hence, we find

$$\begin{aligned} 1AB_{(16)} &= 1(256) + 10(16) + 11(1) \\ &= 256 + 160 + 11 \\ &= 427 \end{aligned}$$

Example 2. Convert the hexadecimal number $49E(16)$ to a decimal number.

Solution. Writing the digits of the octal number $49E_{(16)}$ at the proper places, we obtain

Decimal Value	256	16	1
Hexadecimal Number	4	9	E
	↕	↕	↕
Decimal Equivalent	4	9	14

Using the table shown above, we obtain

$$49E_{(16)} \ = \ 4(256) + 9(16) + 14(1)$$
$$= \ 1024 + 144 + 14 = 1182$$

1.5.2   Converting from Decimal to Hexadecimal

To convert any decimal number to its equivalent hexadecimal number, we follow the same technique used in the binary system and the octal system. The repeated division by 16 is the technique that we should use, noting that the remainder should be listed as a one distinct symbol. For example, if the remainder is 14 or 15, it should be written as E or F in the remainder column.

The following two examples will illustrate the procedure of the converting from a decimal number to a hexadecimal number.

Example 3. Convert 2748 to a hexadecimal number.

Solution. We divide 2748 successively by 16 and list the remainder at each step as done before. Attention here must be drawn that the resulting remainder at each step must be written as a single digit or a symbol from 0 to F. For example, if the remainder is 10, then the letter A should be written in the right column of remainders in order to fill one place only. Proceeding as before we obtain the following table:

base 16	decimal number	
16	2748	remainder in hexadecimal digits
16	171	C
16	10	B
	0	A

Hence, we obtain

$$2748 = ABC_{(16)}$$

Example 4. Convert the decimal number 44015 to a hexadecimal number.

Solution. Proceeding as before we find

base 16	decimal number	
16	44015	remainder in hexadecimal digits
16	2750	F
16	171	E
16	10	B
	0	A

Consequently, we obtain

$$44015 = ABEF_{(16)}$$

1.5.3 Number Systems Equations

In the following examples, equations that include numbers expressed in different number systems will be discussed. It is recommended that each number involved be converted to the decimal system, then we proceed to solve the resulting standard equation.

Example 5. Find x that satisfies the equation

$$xAD_{(16)} \ = \ 125_{(8)} + 600$$

Solution. We first note that the left hand side is a hexadecimal number, therefore x should be any digit from 0 to F. Also, notice that the right hand side includes a decimal number and an octal number. As stated before, we first convert the hexadecimal and the octal numbers to decimal numbers. Hence, we find

$$\begin{aligned} xAD_{(16)} &= x(256) + A(16) + D(1) \\ &= 256x + 160 + 13 \\ &= 256x + 173 \end{aligned}$$

and

$$\begin{aligned} 125_{(8)} &= 1(64) + 2(8) + 5(1) \\ &= 64 + 16 + 5 \\ &= 85 \end{aligned}$$

Substituting in the equation gives

$$\begin{aligned} 256x + 173 &= 600 + 85 \\ 256x &= 512 \\ x &= 2 \end{aligned}$$

Example 6. Find x that satisfies the equation

$$xxF_{(16)} = 700 + 620_{(8)} + 11_{(2)}$$

Solution. Proceeding as before, and converting non decimal numbers to decimals, we find

$$\begin{aligned} xxF_{(16)} &= 700 + 620_{(8)} + 11_{(2)} \\ x(256) + x(16) + F(1) &= 700 + 6(64) + 2(8) + 0(1) + 1(2) + 1(1) \\ 256x + 16x + 15 &= 700 + 400 + 3 \\ 272x &= 1088 \\ x &= 4 \end{aligned}$$

Example 7. Find x that satisfies the equation

$$345_{(x)} = 25x_{(8)} + 101_{(2)}$$

Solution. Proceeding as before, and converting non decimal numbers to decimals, we find

$$\begin{aligned} 345_{(x)} &= 25x_{(8)} + 101_{(2)} \\ 3(x^2) + 4(x) + 5(x^0) &= 2(64) + 5(8) + x(1) + 1(4) + 0(2) + 1(1) \\ 3x^2 + 4x + 5 &= 128 + 40 + x + 5 \\ 3x^2 + 4x + 5 &= 173 + x \\ x^2 + x - 56 &= 0 \\ (x - 7)(x + 8) &= 0 \\ x &= 7 \end{aligned}$$

Notice that the base of any system should be a positive integer, and for this reason the negative value of x is not valid in the number system of any base.

Exercises 1.5

1. Convert the following decimal numbers to hexadecimal numbers:

   (a) 651   (b)3258   (c) 3567   (d) 52129

2. Convert the following hexadecimal numbers to decimal numbers:

   (a) $BCD_{(16)}$ (b) $9AB_{(16)}$ (c) $123A_{(16)}$ (d) $F09_{(16)}$

3. Convert the following decimal numbers to hexadecimal numbers:

   (a) 2330   (b) 2475   (c) 897   (d) 2635

4. Convert the following hexadecimal numbers to decimal numbers:

   (a) $381_{(16)}$ (b) $9AB_{(16)}$ (c) $DEF_{(16)}$ (d) $123A_{(16)}$

5. Classify the following as True or False. Justify your answer by converting to decimal numbers:

   (a) $12_{(16)} > 12_{(8)}$ (b) $12_{(5)} = 21_{(3)}$

   (c) $37_{(8)} < 41_{(7)}$ (d) $AB_{(16)} = BA_{(12)}$

6. Find the largest number from each group of different number systems. Justify your answer:

   (a) $6_{(7)}$, $12_{(4)}$, $110_{(2)}$

   (b) $431_{(5)}$, $312_{(6)}$, $224_{(7)}$

   (c) $AB_{(16)}$, $BA_{(12)}$, $234_{(8)}$

   (d) $134_{(5)}$, $125_{(6)}$, $87_{(9)}$

7.   Change each of the following decimal numbers to the indicated base:

(a) 194 to the base-five number

(b) 310 to the base-six number

(c) 466 to the base-seven number

(d) 668 to the base-eight number

8. Find x if $411_{(x)} = 304_{(7)}$

9. Find x if $225_{(x)} = A7_{(11)}$

10. Find x if $110_{(x)} = 183_{(9)}$

11. Find x if $x2_{(9)} = 132_{(x)}$

12. Find x if $1x1_{(8)} = 49_{(16)}$

13. Find x if $x23_{(8)} = 1x3_{(16)}$

14. Find x if $1x1_{(8)} = 1100001_{(2)}$

15. Find x if $123_{(x)} = 53_{(16)}$

16. Find x if $346_{(x)} = 1506_{(8)}$

17. Find x if $603_{(x)} = 1x3_{(16)}$

18. Find x if $x56_{(8)} = AE_{(16)}$

19. Find x if $114_{(x)} = 424_{(8)}$

20. Find x if $ABC_{(13)} = 111_{(x)} + 1814$

21. Find x if $234_{(x)} = 1x2_{(8)} - 202_{(3)}$

22. Find x if $142_{(x)} = A2_{(16)} - 101011_{(2)}$

23. Find x if $AFB4_{(x)} = 44980$

24. Find x if $13AB_{(x)} = 4211$

25. Find x if $5A2E_{(16)} = 23086_{(x)}$

26. Find x if $2x48 = 11004_{(x)}$

27. Find x if $xxx_{(5)} = 1330_{(x)}$

1.5.4 Converting from Hexadecimal to Binary Notation

To convert any hexadecimal number directly to a binary number, we use the following table that shows the hexadecimal digits together with their decimal equivalents and their 4-bit binary equivalents:

Decimal	4-Bit Binary Equivalent	Hexadecimal
0	0000	0
1	0001	1
2	0010	2
3	0011	3
4	0100	4
5	0101	5
6	0110	6
7	0111	7
8	1000	8
9	1001	9
10	1010	A
11	1011	B
12	1100	C
13	1101	D
14	1110	E
15	1111	F

Table 1.8

The process to convert an integer from hexadecimal to binary is as follows:
1. We first write each hexadecimal digit of the given numeral in fixed 4-bit binary notation by using Table 1.8.
2. We then combine together the resulting binary digits by removing the spaces.

The process presented above can be illustrated by discussing the following examples.

Example 9. Convert the hexadecimal number $ABC_{(16)}$ to a binary number.

Solution. We first write each hexadecimal digit of the given numeral $ABC_{(16)}$ in fixed 4-bit binary notation by using Table 1.8, therefore we find:

$$A = 1010_{(2)}, \ B = 1011_{(2)}, \ C = 1100_{(2)}$$

In other words we can write

$$A \quad B \quad C$$
$$\updownarrow \quad \updownarrow \quad \updownarrow$$
$$1010 \quad 1011 \quad 1100$$

By grouping together the binary digits, the answer is $101010111100_{(2)}$.

Example 10. Convert the hexadecimal number $F71_{(16)}$ to a binary number.

Solution. We first write each hexadecimal digit of the given numeral $F71_{(16)}$ in fixed 4-bit binary notation by using Table 1.8, hence we obtain:

$$F = 1111_{(2)}, \ 7 = 0111_{(2)}, \ 1 = 0001_{(2)}$$

In other words we can write

$$F \quad 7 \quad 1$$
$$\updownarrow \quad \updownarrow \quad \updownarrow$$
$$1111 \quad 0111 \quad 0001$$

By grouping together the binary digits, the answer is $111101110001_{(2)}$.

Example 11. Convert the hexadecimal number $D4E6_{(16)}$ to a binary number.

Solution. We first write each hexadecimal digit of the given numeral $D4E6_{(16)}$ in fixed 4-bit binary notation by using Table 1.8 to find that:

$$D = 1101_{(2)}, \ 4 = 0100_{(2)}, \ E = 1110_{(2)}, \ 6 = 0110_{(2)}$$

In other words we can write

$$D \quad 4 \quad E \quad 6$$
$$\updownarrow \quad \updownarrow \quad \updownarrow \quad \updownarrow$$
$$1101 \quad 0100 \quad 1110 \quad 0110$$

Grouping the binary digits, the answer is $1101010011100110_{(2)}$.

Example 12. Convert the hexadecimal number $ABC_{(16)}$ to an octal number by converting first to binary number then to octal.

Solution. We first write each hexadecimal digit of the given numeral $ABC_{(16)}$ in fixed 4-bit binary notation by using Table 1.8 to find that:

$$A = 1010_{(2)}, \ B = 1011_{(2)}, \ C = 1100_{(2)}$$

In other words we can write

$$A \quad B \quad C$$
$$\updownarrow \quad \updownarrow \quad \updownarrow$$
$$1010 \quad 1011 \quad 1100$$

The resulting binary number is $101010111100_{(2)}$. We then group this binary number into sets of fixed 3-bit binary digits and using table 1.6 we obtain

$$101 \quad 010 \quad 111 \quad 100$$
$$\updownarrow \quad \updownarrow \quad \updownarrow \quad \updownarrow$$
$$5 \quad 2 \quad 7 \quad 4$$

By grouping together the resulting octal digits, the answer is $5274_{(8)}$.

1.5.5   Converting from Binary to Hexadecimal Notation

To convert any binary number directly to a hexadecimal number, Table 1.8 that relates each hexadecimal number to its 4-bit binary equivalent should be used. The conversion in this case is the reverse of the process followed above from hexadecimal to binary. The process to convert an integer from hexadecimal number to a binary number is as follows:

1. We first group the digits of the given binary number into sets of **4-bits**, starting from the right of the binary digits to the left, adding leading zeros as needed, normally we add one, two or three zeros.

2. Using Table 1.8, we find the equivalent hexadecimal digit to each set of 4-binary bits.

3. We then group the resulting hexadecimal digits, using 16 as a subscript to the resulting hexadecimal number.

The process presented above can be illustrated by discussing the following examples.

Example 13. Convert the binary number $110110011111_{(2)}$ to a hexadecimal number.

Solution. Starting from right to left, we first group the binary number $110110011111_{(2)}$ into sets of fixed 4-bit binary digits, therefore we set: 1101 1001 1111.

Using Table 1.8, we find the equivalent hexadecimal digit to each set, hence we find

$$
\begin{array}{ccc}
\underbrace{1101} & \underbrace{1001} & \underbrace{1111} \\
\updownarrow & \updownarrow & \updownarrow \\
D & 9 & F
\end{array}
$$

By grouping together the resulting hexadecimal digits, the answer is $D9F_{(16)}$.

Example 14. Convert the binary number $1110001011_{(2)}$ to a hexadecimal number.

Solution. Starting from right to left, we first group the binary number $1110001011_{(2)}$ into sets of fixed 4-bits, therefore we set: 0011 1000 1011, by adding two leading zeros.

Using Table 1.8, and proceeding as before we find the equivalent octal digit to each set, therefore we set

$$\underbrace{0011}_{} \quad \underbrace{1000}_{} \quad \underbrace{1011}_{}$$

$$\updownarrow \qquad \updownarrow \qquad \updownarrow$$

$$3 \qquad 8 \qquad B$$

By grouping together the resulting hexadecimal digits, the answer is $38B_{(16)}$.

Example 15. Convert the binary number $1110001110100_{(2)}$ to a hexadecimal number.

Solution. Starting from right to left, we first group the binary number $1110001110100_{(2)}$ into sets of fixed 4-bit binary digits, therefore we set: $0001\,1100\,0111\,0100$, by adding three leading zeros. Using Table 1.8, we find the equivalent hexadecimal digit to each set, hence we find

$$\underbrace{0001}_{} \quad \underbrace{1100}_{} \quad \underbrace{0111}_{} \quad \underbrace{0100}_{}$$

$$\updownarrow \qquad \updownarrow \qquad \updownarrow \qquad \updownarrow$$

$$1 \qquad C \qquad 7 \qquad 4$$

By grouping together the resulting octal digits, the answer is $1C74_{(16)}$.

Exercises 1.5.5

1. Convert the following hexadecimal numbers to binary numbers:

(a) $34A_{(16)}$ (b) $DE1_{(16)}$ (c) $23C_{(16)}$

2. Convert the following hexadecimal numbers to binary numbers:

(a) $1234_{(16)}$ (b) $2BCD_{(16)}$ (c) $4C1F_{(16)}$

3. Convert the following binary numbers to hexadecimal numbers:

(a) $110101101100_{(2)}$ (b) $111111000001_{(2)}$

(c) $11010101101011_{(2)}$ (d) $100000011100_{(2)}$

4. Convert the following hexadecimal numbers to octal numbers by converting first to binary numbers:

(a) $81C_{(16)}$ (b) $BCD_{(16)}$ (c) $C19E_{(16)}$ (d) $A01F_{(16)}$

5. Find x of the following equations (x can be one or more digits):

    (a) $8x9_{(16)} = 100010111001_{(2)}$

    (b) $ACF_{(16)} = 1010x1111_{(2)}$

    (c) $xBCD_{(16)} = 1010101111001101_{(2)}$

    (d) $FED0_{(16)} = x111011010000_{(2)}$

6. Convert the following octal numbers to hexadecimal numbers by converting first to binary numbers:

    (a) $347_{(8)}$ (b) $462_{(8)}$ (c) $1256_{(8)}$ (d) $5274_{(8)}$

7. Add the following hexadecimal numbers:

    (a) $ABC_{(16)} + 347_{(16)}$ (b) $89A_{(16)} + 78B_{(16)}$

    (c) $ABC_{(16)} + DEF_{(16)}$ (d) $CBA_{(16)} + FED_{(16)}$

Chapter 2

Prime Numbers and Euclidean Algorithm

2.1 Prime Numbers

Consider two positive integers a and b. The integer a is a factor of b, or a is a divisor of b, if a divides b, and this is denoted by $a \mid b$. In this case, there is an integer k such that $b = ak$.

It is to be noted that there are integers whose factors are 1 and the integer itself only. Such integers are called *primes*. We define a prime number as an integer greater than 1 and divisible by one and by itself only. In other words, $p > 1$ is a prime number that has exactly two positive factors, 1 and itself. An integer $n > 1$ that is not a prime is termed *composite*. The number 6 is composite, because it has 1,2,3,6 as factors. The definition of a prime does not allow 1 to be a prime, simply 1 has only one factor, 1 itself. The number 1 is not considered neither prime nor composite.

The following are the first few prime numbers:

$$2, 3, 5, 7, 11, 13, 17, 19, 23, 29, 31, 37$$

$$41, 43, 47, 53, 59, 61, 67, 71, 73, 79, 83, 89$$

$$97, 101, 103, 107, 109, 113, 127, 131, 137, 139, 149, 151$$

$$157, 163, 167, 173, 179, 181, 191, 193, 197, 199, 211, 223, 227, 229$$

More primes are listed in Appendix A.1. It is worth noting that the integer 2 is the smallest prime and the only even prime. The remaining primes are odd integers. Prime numbers have been the focus of study for many mathematicians, and enormous amounts of time and work have been spent for computing large primes. We point out here that other numbers, that are not primes, can be factored into two or more factors. Such numbers are usually called **composite** numbers. For example, the numbers 6 and 10 are composite numbers because $6 = 2 \times 3$ and $10 = 2 \times 5$.

Excluding 2 and 3, all primes are of the form $p = 6n \pm 1, n \geq 1$. This means that if we divide any prime by 6, the remainder is 1 or 5. There is an infinite number of primes. The work to find ever-larger primes is going on and attracts a considerable amount of efforts. A very large prime discovered recently is $2^{37156667} - 1$. Another large known prime number as of (2009) is $p = 2^{42643801} - 1$ found by Odd Magnar Strindmo. The largest known prime as of (2008) is $p = 2^{43112609} - 1$ found by Edson Smith.

How can we decide whether a number is a prime or not? For small numbers, this can be easily done by examining if the number has factors other than 1 or not. By using the factor command of any calculator, we can define if the number has factors or not. If the given number has factors then it is composite, otherwise it is a prime number.

However, for large numbers, several methods have been developed to achieve this goal. In this section, a commonly used technique will be applied. Another old technique, called **Sieve of Eratosthenes** will be discussed later.

It is now useful to explain the method mostly used to examine if a number is a prime or not. For a given number n, we need only to examine if it is divisible by any prime number less than or equal to $\sqrt{n}$. In other words, we have to examine if any prime number that is less than or equal to $\sqrt{n}$ is a factor of n. If such a factor exists, then n is not a prime number, otherwise, n is a prime number. Our study will be focused on odd numbers greater than 2. The technique described above can be explained by discussing the following illustrative examples.

Example 1.   Check if 143 is a prime number or not.

Solution. We first find

$$\sqrt{143} \approx 11$$

We now examine if any of the primes $2, 3, 5, 7,$ and 11 divides 143. It is clear that 11 divides 143, therefore we write

$$143 = 11 \times 13$$

This result can also be obtained by using the calculator and use the command factor to obtain $143 = 11 \cdot 13$. This shows that 143 is not a prime number.

Example 2. Check if 877 is a prime number or not.

Solution. We first find

$$\sqrt{877} \approx 29$$

We now examine if any of the primes $2, 3, 5, 7, 11, 13, 17, 19, 23$, or 29 divides the number 877. It is obvious that none of these primes divides 877. Also by using the calculator we factor 877 to find that $877 = 1 \cdot 877$. As a result, we conclude that 877 is a prime number.

It should be noted that the divisibility process can be carried out by a simple calculator or a simple computer program. In the case of large numbers, tables of primes or computer programs are usually used to examine if a number is a prime or not.

2.1.1 Sieve of Eratosthenes

This is an ancient method that determines prime numbers. To find prime numbers from $1, 2, 3, 4, 5, \cdots, 100$ we make a chart of these numbers as follows:

1	2	3	4	5	6	7	8	9	10
11	12	13	14	15	16	17	18	19	20
21	22	23	24	25	26	27	28	29	30
31	32	33	34	35	36	37	38	39	40
41	42	43	44	45	46	47	48	49	50
51	52	53	54	55	56	57	58	59	60
61	62	63	64	65	66	67	68	69	70
71	72	73	74	75	76	77	78	79	80
81	82	83	84	85	86	87	88	89	90
91	92	93	94	95	96	97	98	99	100

The Sieve of Eratosthenes consists of the following steps:
1. Cross out 1 because it is not a prime.
2. Circle 2, then cross out all multiples of 2. This means that we should cross out every second number.
3. Circle 3, then cross out all multiples of 3. This means that we should cross out every third number.
4. Circle 5, then cross out all multiples of 5. This means that we should cross out every fifth number.
5. We continue in this manner.
6. The remaining numbers that are circled, i.e. not crossed out, are the prime numbers from 2 to 100.

The sieve is like a strainer that drains out the composite numbers and leaves the prime numbers behind.

2.1.2   Functions Generating Primes

We point out that mathematicians tried to develop formulas that may generate all prime numbers. However, their works to establish such formulas have failed. Several formulas have been established that work for specific cases, but none of these formulas can generate all prime numbers. We can easily show that each developed formula can provide some primes, but fails to produce others. For example, consider the formula

$$x^2 + 2$$

for the following values of x given by

$$x = 0, 1, 2, 3, \cdots, 10$$

We can easily observe that this formula generates the primes $2, 3, 11$, and 83 for $x = 0, 1, 3$, and 9. Other primes are not obtainable by this formula such as $5, 7, 13, 17, 19, 23$, and 29. Euler noticed that $f(n) = n^2 + n + 41$ is prime for $n = 0, 1, \cdots, 39$. It is clear that $f(40) = 1681$ and $f(41) = 1763$ are divisible by 41, hence $f(40)$ and $f(41)$ are not primes. The concept of formulas that may generate some prime numbers will be examined by the following examples.

Example 3. Write all primes that are generated from the formula

$$x^2 + x + 3$$

for

$$x = 0, 1, 2, 3, 4, 5, 6, 7$$

Solution. The primes $3, 5, 23$ and 59 are the only primes that this formula provides for $x = 0, 1, 4,$ and 7. Other values of x will generate composite numbers.

Example 4. Write all primes that are generated from the formula

$$p^2 + p + 1$$

where p is a prime number, $p \leq 11$.

Solution. The primes $7, 13$ and 31 are the only primes generated from the formula for $p = 2, 3$ and 5. However, the composite numbers 57 and 133 are generated for $p = 7$ and $p = 11$.

2.1.3 Mersenne Prime Numbers

As stated before, no formula can generate all prime numbers. Attention has turned to find a specific formula that can generate prime numbers. One formula that has attracted mathematicians is that of *Mersenne primes*, named for the French monk Marin Mersenne who studied prime numbers in the early 1600's. Mersenne introduced a formula of the form

$$M_n = 2^n - 1$$

where n is a positive integer. For $n = 1, 2, 3, \cdots, 8$, we obtain the Mersenne numbers $1, 3, 7, 15, 31, 63, 127, 255$. It is interesting to point out that the Mersenne numbers can be expressed by binary numbers where only the 1 digit is used, namely $1_{(2)}, 11_{(2)}, 111_{(2)}, 1111_{(2)}, 11111_{(2)}, \cdots$.

However, it was found that for p a prime number, the formula

$$M_p = 2^p - 1$$

provide prime numbers, but not all prime numbers. This formula allows us to define a Mersenne prime as a prime number that is one less than a prime power of two. For $p = 2, 3$ we obtain the prime numbers $3, 7$. Notice that the Mersenne formula does not provide primes for all prime numbers p. For example, for $p = 11$, $2^{11} - 1 = 2047 = 23 \times 89$ is not a prime. Moreover, for $p = 23$, $2^{23} - 1 = 8388607 = 47 \times 178481$ is not a prime.

In (2009) and (2008), it was discovered two Mersenne prime numbers, 46th and 47th, that are given by

$$p = 2^{42,643,801} - 1 \text{ and } p = 2^{43,112,609} - 1$$

that contain $12,837,064$ and $12,978,189$ decimal digits respectively. Recently, in (2013) the 48th Mersenne prime number was discovered that contains $17,425,170$ decimal digits. Work is still going on, and more large Mersenne prime numbers are expected to be discovered. Mersenne primes are listed in Appendix A.2.

Example 5. Write all primes that are generated from the formula

$$F = 3^{p-1} + 2^{p-1}$$

where p is all prime numbers < 10.

Solution. Prime numbers < 10 are 2, 3, 5, and 7. Substituting these values of p in the given formula we find

$$
\begin{aligned}
F_1 &= 3^1 + 2^1 = 5 \\
F_2 &= 3^2 + 2^2 = 13 \\
F_3 &= 3^4 + 2^4 = 97 \\
F_4 &= 3^6 + 2^6 = 793 = 13 \cdot 61
\end{aligned}
$$

The numbers 5, 13, and 97 are the only primes we obtained from this formula for $p < 10$.

Example 6. Write all primes that are generated from the formula

$$G = \frac{3^{p-3} + 1}{2}$$

where p is all odd prime numbers ≤ 11.

Solution. The odd prime numbers ≤ 11 are 3, 5, 7, and 11. Substituting these values of p in the given formula we find

$$
\begin{aligned}
G_1 &= \frac{3^{3-3} + 1}{2} = 1 \\
G_2 &= \frac{3^{5-3} + 1}{2} = 5 \\
G_3 &= \frac{3^{7-3} + 1}{2} = 41 \\
G_4 &= \frac{3^{11-3} + 1}{2} = 3281 = 17 \cdot 193
\end{aligned}
$$

The numbers 5 and 41 are the only primes we obtained from this formula for p odd and $p \leq 11$.

Exercises 2.1

1. List all factors of 48, 57.

2. Which of the following numbers are prime numbers:

(a) 71   (b) 91   (c) 137   (d) 147   (e) 487

3. Mathematicians were not able to establish a formula that gives all primes. Some formulas produce some prime numbers for specific values of x. Use the formula $x^2 + 1$ for $x = 0, 1, 2, 3, \ldots, 10$ and list the primes that may result from using this formula. How many resulting primes you obtained by using this formula.

4. Repeat the idea of Example 4 for the formula $x^2 - x + 41$ for $x = 0, 1, 2, 3, \cdots, 15$.

5. The formula $x^2 + x + 17$ can produce sixteen primes for specific values of x. Write the first ten primes and the values of x that produce these primes.

6. **Twin primes.** Pairs of primes with difference 2 are called twin primes. For example, 3 and 5, 5 and 7, and 11 and 13 are twin primes. Write three more twin primes.

7. **Sieve of Eratosthenes.** Make a chart of the numbers $1, 2, 3, \cdots, 100$. Use Sieve of Eratosthenes to find all prime numbers between 50 1nd 100.

8. Find three consecutive odd integers where all these integers are primes.

9. Show that the formula

$$p^2 + 2$$

generates composite numbers for the primes

$$p = 5, 7, 11, 13, 17$$

10. Given the formulas

$$6k - 1 \quad \text{and} \quad 6k + 1$$

Find four values of $k \geq 1$ that will generate four twin primes from these formulas.

11. Given the formulas

$$8k + 1 \quad \text{and} \quad 8k + 3$$

Find two values of $k \geq 1$ that will generate two twin primes from these formulas.

12. **Perfect numbers** Given a number n. Add the divisors of the number n other than the number itself. If the resulting sum is equal to the number itself, then the number n is called a perfect number. For example, the number 6 has divisors 1,2,3. The sum of the divisors $1+2+3 = 6$. Therefore, 6 is called a perfect number.

If $2^p - 1$ is a prime number, use all primes $p \leq 7$ to determine the first four perfect numbers by using the formula

$$P = 2^{p-1}(2^p - 1)$$

Check your answer from Appendix A.3.

13. Given the formula
$$4k + 1, k \geq 0$$

Find six prime numbers that can be generated from this formula.

14. Given the formula
$$4k - 1, k \geq 0$$

Find six prime numbers that can be generated from this formula.

15. Select five even numbers > 2. Show that each even number > 2 is the sum of two primes.

16. Use $p = 5, 7$ and find two Mersenne primes.

17. Show that for $p = 29$, the Mersenne number is composite.

18. Use the formula $M = \frac{2^p + 1}{3}$ to generate three prime numbers for three prime values of p.

19. Write all primes that are generated from the formula

$$M = 3^{p-1} + 2$$

where p is all prime numbers < 10.

20. Write all primes that are generated from the formula

$$N = 3^{n-1} + 2^{n-1}$$

where n is any integer such that $1 \leq n < 6$.

21. Write all primes that are generated from the formula

$$F = \frac{5^p - 2^p}{3}$$

where p is all prime numbers < 10.

22. Write all primes that are generated from the formula

$$G = 3^{p-1} - 2$$

where p is all prime numbers < 10.

2.2 Euclidean Algorithm

In Section 2.1, we discussed the factors or divisors of any positive integer. It is clear that we can factor the integers 30 and 42 as follows:

$$30 = 2 \times 3 \times 5$$

$$42 = 2 \times 3 \times 7$$

We can easily observe that 2 is a common factor of 30 and 42. Similarly, the number 3 is a common factor of both integers 30 and 42 as well. This leads to the question of the **greatest common factor** or the **greatest common divisor** of any two integers a and b denoted by $\gcd(a, b)$. In this particular example that we are discussing, it is obvious that 6 is a common factor, or a common divisor, of 30 and 42, and in addition it is the largest factor of these two integers compared to the other common factors 2 and 3. Therefore, we call 6 the greatest common factor, or the greatest common divisor, of the two integers 30 and 42, where we denote this by $\gcd(42, 30) = 6$.

Based on the discussion made above, we define a divisor d as the greatest common divisor of the integers a and b, denoted by $\gcd(a, b)$, if the following conditions are satisfied:

(i) $d \mid a$ and $d \mid b$.

(ii) d is largest common divisor of the given integers a and b.

We point out that it is easy to find the greatest common divisor gcd for two small numbers as in the example discussed above. This can simply be done by factoring each integer and the product of the common factors will give the greatest common divisor.

However, for large numbers, where usually factoring is not always easy, the factoring procedure is not normally used. For large integers, we usually use a practical technique called the **Euclidean Algorithm**. It is named after the Greek mathematician Euclid. Although this algorithm is an ancient one, it is the most popular and the effective technique to handle this concept. The Euclidean Algorithm will be explained as follows.

2.2.1 The Division Algorithm

Given any two integers a and b, where $a > b$, then there are unique integers q and r such that

$$a = qb + r, 0 \leq r < b$$

where q and r are called the quotient and the remainder respectively.

Example 1. Find the quotient q and the remainder r when dividing 1157 by 219.

Solution. Dividing 1157 by 219, the quotient $q = 5$ and the remainder $r = 62$. This can be expressed as

$$1157 = 5(219) + 62$$

Example 2. Find the quotient q and the remainder r when dividing 27911 by 113.

Solution. Dividing 27911 by 113, the quotient $q = 247$ and the remainder $r = 0$. This can be expressed as

$$27911 = 247(113) + 0$$

2.2.2 The Concept of the Euclidean Algorithm

To determine the $\gcd(a, b)$ where $a > b$, we divide a by b and find the remainder r_1 of this division. If $r_1 = 0$, then b is the greatest common divisor of (a, b). If not, then we express the outcome of this division by

$$a = q_1 b + r_1, \ 0 < r_1 < a$$

In the next step, we find the remainder r_2 when b is divided by r_1. Continuing the steps stated above until we obtain a zero remainder. In this case the **last nonzero remainder** r_n is the required greatest common divisor of (a, b). It is useful to note that the remainders we will get through the division technique will decrease, and the division process will end by obtaining a zero remainder as stated above. The division procedures and the remainders will be expressed by the following scheme

$$
\begin{aligned}
a &= q_1 b + r_1 \\
b &= q_2 r_1 + r_2 \\
r_1 &= q_3 r_2 + r_3 \\
r_2 &= q_4 r_3 + r_4 \\
&\vdots \\
r_{n-2} &= q_n r_{n-1} + \underline{r_n} \\
r_{n-1} &= q_{n+1} r_n + 0
\end{aligned}
$$

The Euclid's algorithm can be described by the following properties:

$$
\begin{aligned}
\gcd(a, 0) &= a \\
\gcd(a, a) &= a \\
\gcd(a, b) &= \gcd(a - b, b), \text{ if } b < a \\
\gcd(a, b) &= \gcd(a, b - a), \text{ if } b > a
\end{aligned}
$$

The Euclidean Algorithm by using the long division or by using the aforementioned properties, will be illustrated by discussing the following examples.

Example 3. Find gcd(1767, 1519).

Solution. Dividing 1767 by 1519, the quotient is 1 and the remainder is 248. We next divide 1519 by 248 to obtain 6 with a remainder given by 31. We then divide 248 by 31 to obtain 6 with a zero remainder. The process used above can summarized by

$$
\begin{aligned}
1767 &= 1 \times 1519 + 248 \\
1519 &= 6 \times 248 + \underline{31} \\
248 &= 8 \times 31 + 0
\end{aligned}
$$

It is obvious that the last nonzero remainder is 31. Hence, gcd(1767, 1519)=31.

It is useful to note that the Euclidean method can also be viewed by using the long division process. We divide 1767 by 1519 by using the long division where the remainder is listed at the bottom as usual. We next divide 1519 by the remainder and find the new remainder of this step. The last nonzero remainder will be the greatest common divisor. This can be illustrated by the following figure.

$$
\begin{array}{r}
1 \\
1519\overline{)1767} \\
1519 \\
\hline 248
\end{array}
\quad\longrightarrow\quad
\begin{array}{r}
6 \\
248\overline{)1519} \\
1488 \\
\hline \mathbf{31}
\end{array}
\quad\longrightarrow\quad
\begin{array}{r}
8 \\
31\overline{)248} \\
248 \\
\hline 0
\end{array}
$$

We can also use the properties of the gcd listed before to obtain the gcd$(1767, 1519)$ as follows.

$$
\begin{aligned}
\gcd(1767, 1519) &= \gcd(1767 - 1519, 1519) = \gcd(248, 1519) \\
\gcd(248, 1519) &= \gcd(248, 1519 - 6 \times 248) = \gcd(248, 31) \\
\gcd(248, 31) &= \gcd(248 - 8 \times 31, 31) = \gcd(0, 31) = 31
\end{aligned}
$$

It is clear that the last nonzero remainder by using the long division is 31. Moreover, we also found that gcd$(1767, 1519) = 31$ by using the properties of the Euclid's algorithm. This means we have

$$
\begin{aligned}
1767 &= 31 \times 57 \\
1519 &= 31 \times 49
\end{aligned}
$$

Example 4. Find gcd $(3337, 2769)$.

Solution. We follow the procedure used in the previous example, by using the successive division and we stop when we obtain a zero remainder. Accordingly, we find

$$
\begin{aligned}
3337 &= 1 \times 2769 + 568 \\
2769 &= 4 \times 568 + 497 \\
568 &= 1 \times 497 + \underline{71} \\
497 &= 7 \times 71 + 0
\end{aligned}
$$

It is obvious that the last nonzero remainder is 71. From this, we find that the gcd(3337, 2769) =71. This also means

$$3337 = 47 \times 71 \text{ and } 2769 = 39 \times 71$$

On the other hand, we can use the long division process introduced above to find that the gcd(3337, 2769) = 71 as shown by the long division figure given below.

We can also use the properties of the gcd listed before to obtain the gcd(1767, 1519) as follows.

$$
\begin{aligned}
\gcd(3337, 2769) &= \gcd(3337 - 2769, 2769) = \gcd(568, 2769) \\
\gcd(568, 2769) &= \gcd(568, 2769 - 4 \times 568) = \gcd(568, 497) \\
\gcd(568, 497) &= \gcd(568 - 497, 497) = \gcd(71, 497) \\
\gcd(71, 497) &= \gcd(71, 497 - 7 \times 71) = \gcd(71, 0) = 71
\end{aligned}
$$

It is clear that the last nonzero remainder by using the long division is 71. Moreover, we also found that gcd(3337, 2769) = 71 by using the properties of the Euclid's algorithm. This means we have

$$
\begin{aligned}
3337 &= 71 \times 47 \\
2769 &= 71 \times 39
\end{aligned}
$$

Definition Two integers a and b, not both of which are zero, are said to be *relatively prime* whenever gcd(a,b)=1. It is not necessary for both a and b to be primes in order a and b are relatively prime suchh as 6 and 7.

Exercises 2.2

1. Find the greatest common divisor for each pair of integers by using the Euclidean Algorithm:

(a) (481, 377)   (b) (1681, 1517)   (c) (1189, 1073)

2. Find the greatest common divisor for each pair of integers by using the Euclidean Algorithm:

(a) (1921, 1243)   (b) (1927, 1739)   (c) (647, 563)

3. Find the greatest common divisor for each pair of integers by using the Euclidean Algorithm:

(a) (3599, 3233)   (b) (2294, 1798)   (c) (997, 947)

4. **Relatively prime.** If the gcd of two integers a and b is 1, then a and b are called relatively prime, such as (3, 5), (4, 11) and (8, 39). Show that the following pairs are relatively primes:

(a) (34,43)   (b) (37,73)   (c) (86,111)

5. Show that the following pairs are relatively primes:

(a) (703,611)   (b) (899,407)   (c) (412,389)

6. Use the greatest common divisor idea to factor each integer of the following pairs:

(a) (377,319)   (b) (703,437)   (c) (749,535)

7. Answer the following:

(a) Find gcd (247, 221)

(b) Find gcd (247 ± 221, 221)

(c) Write a conclusion from the answers obtained in (a), (b) and (c)

8. Use the conclusion you made from Exercise 7 to evaluate the following for $a > b$:

(a) If gcd $(a, b) = c$, find gcd $(a + b, b)$

(b) If gcd $(a, b) = c$, find gcd $(a - b, b)$

(c) If gcd $(a, b) = c$, find gcd $(a + 2b, b)$

9. Find gcd $(1363, 1739)$. Use your answer to factor each number.

10. Find gcd $(2868, 2257)$. Use your answer to factor each number.

11. Find gcd $(16531, 18157)$. Use your answer to factor each number.

Chapter 3

Linear Congruences and Diophantine Equations

3.1 Linear Congruences

In this section, we will discuss the useful concept of congruences which is of great importance in number theory. Consider two integers a and b. If a and b give the same remainder when divided by a natural number m, then we say that a is **congruent** to b (modulo m), and this is usually expressed by

$$a \equiv b \bmod m$$

where m is called the **modulus** of the congruence. For example,

$$43 \equiv 13 \bmod 5$$

which indicates that the integers 43 and 13 leave the same remainder 3 when divided by 5. In other words, when we say that a is congruent to b modulo m it means that the difference $a - b$ is divisible by m. In the example

$$43 \equiv 13 \bmod 5$$

we notice that $5 \mid (43 - 13)$.

In the congruence $a \equiv b \bmod m$, the least positive integer $b < m$ is called the least remainder or the least residue obtained upon dividing a by m. The notion of the least positive remainder will be explained by the following example.

Example 1. Find the least remainder of 75 modulo 4 and write the result in a congruence formulation.

Solution. Dividing 75 by 4 gives a 3 as a remainder. Hence we write

$$75 \equiv 3 \bmod 4$$

It is useful to note that any arithmetic sequence can be expressed in a congruence notation. In an increasing arithmetic sequence, each term exceeds the preceding term by a fixed number called the difference between any two consecutive terms. An example of arithmetic sequences is given by

$$2, 7, 12, 17, 22, 27, 32, \cdots$$

where the difference between any two consecutive terms is 5. A second arithmetic sequence is given by

$$4, 7, 10, 13, 16, 19, \cdots$$

where the difference between any two consecutive terms is 3.

An important application of the congruence concept is the arithmetic sequence. Recall that an increasing arithmetic sequence of positive numbers consists of a sequence of terms, where each term exceeds the preceding term by a fixed number called the difference. To express an arithmetic sequence in a congruence formulation, the first term of the sequence will be considered the least remainder b, and the difference between any two consecutive terms will be m in the notation modulo m. Expressing an arithmetic sequence in a congruence form and expressing a congruence in a sequence form will be illustrated by the following examples.

Example 2. Write the following arithmetic sequence

$$3, 8, 13, 18, 23, 28, \cdots$$

in a congruence form.

Solution. In this example we set $b = 3$ and $m = 5$ obtained by finding the difference between any two consecutive terms. Hence, we write

$$x \equiv 3 \bmod 5$$

where x can be any term of the arithmetic sequence given above.

Example 3. Write the following arithmetic sequence

$$2, 8, 14, 20, 26, 32, \cdots$$

in a congruence form.

Solution. In this example we set $b = 2$ and $m = 6$ obtained by finding the difference between any two consecutive terms. Hence, we write

$$x \equiv 2 \bmod 6$$

where x can be any term of the arithmetic sequence given above.

In a parallel manner, we can list all positive numbers that satisfy a congruence by an arithmetic sequence. This can be done by setting r as the first term of the sequence. Other terms can be obtained by adding m to the first term and to the resulting terms obtained.

In what follows, we study two illustrative examples to explain the technique to convert a linear congruence to its equivalent sequence.

Example 4. Express the following congruence

$$x \equiv 2 \bmod 7$$

in a sequence form.

Solution. We set the first term of the equivalent sequence by 2. We then add the increment 7 (of the mod 7) to the first term and to all resulting terms. Accordingly, the arithmetic sequence is given by

$$2, 9, 16, 23, 30, 37, \cdots$$

Example 5. Express the following congruence

$$x \equiv 3 \bmod 11$$

in a sequence form.

Solution. We set the first term of the equivalent sequence by 3. We then add the increment 11 to the first term and to all resulting terms. Accordingly, the sequence is given by

$$3, 14, 25, 36, 47, 58, \cdots$$

3.1.1   Properties of Congruences

P1. If $a \equiv b \bmod m$, then for any k, $(a + k) \equiv (b + k) \bmod m$.

P2. If $a \equiv b \bmod m$, then for any k, $(a - k) \equiv (b - k) \bmod m$.

P3. If $ac \equiv bc \bmod m$, c and m have no common factor other than 1, then $a \equiv b \bmod m$. In other words, this property means that if the only common factor between c and m is 1, then we can divide ac and bc by c to get an equivalent congruence as shown above.

P4. If $a \equiv b \bmod m_1$, and $a \equiv b \bmod m_2$, then $a \equiv b \bmod (m_1 m_2)$.

P5. If $a \equiv b \bmod m$, then $a^k \equiv b^k \bmod m$.

3.1.2   Division and Congruences

The last property **P5** provides a helpful tool in obtaining the remainder when division process is used, especially if large numbers are used. This will be illustrated by the following examples:

Example 6. Find the remainder when 6^{1849} is divided by 37.

Solution. We can easily observe that

$$6^2 \equiv -1 \bmod 37$$

Using **P5**, we set

$$(6^2)^{924} \equiv (-1)^{924} \bmod 37$$

This is equivalent to

$$(6)^{1848} \equiv 1 \bmod 37$$

Multiplying both sides by 6 we get

$$6 \times 6^{1848} \equiv 6 \times 1 \bmod 37$$

This means that the remainder is 6 when 6^{1849} is divided by 37.

Example 7. Find the remainder when 3^{51} is divided by 8.

Solution. We can easily observe that

$$3^2 \equiv 1 \bmod 8$$

Using **P5**, we set
$$(3^2)^{25} \equiv (1)^{25} \bmod 8$$
This is equivalent to
$$(3)^{50} \equiv 1 \bmod 8$$
Multiplying both sides by 3 we get
$$3 \times (3^{50}) \equiv 3 \times 1 \bmod 8$$
This shows that the remainder is 3 when 3^{51} is divided by 8.

Example 8. Find the remainder when $25 \times (34)^4 + 9 \times 8^{51}$ is divided by 7.

Solution. We can easily observe that
$$25 \equiv 4 \bmod 7$$
$$34 \equiv -1 \bmod 7$$
$$34^4 \equiv (-1)^4 \bmod 7$$
$$9 \equiv 2 \bmod 7$$
$$8 \equiv 1 \bmod 7$$
$$8^{51} \equiv 1^{51} \bmod 7$$
This in turn gives
$$25 \times (34)^4 + 9 \times 8^{51} \equiv ((4)(1) + (2)(1)) \bmod 7$$
or equivalently
$$25 \times (34)^4 + 9 \times 8^{51} \equiv 6 \bmod 7$$

This means that the remainder is 6 when dividing the given expression by 7.

Example 9. Find the digit in the one's place of the number 3^{401}.

Solution. The digit in the one's place is the remainder when 3^{401} is divided by 10. Therefore we find
$$3^4 \equiv 1 \bmod 10$$
$$(3^4)^{100} \equiv (1)^{100} \bmod 10$$
$$3 \times 3^{400} \equiv 3 \times 1 \bmod 10$$
This is equivalent to
$$(3)^{401} \equiv 3 \bmod 10$$
The digit in the one's place is 3.

Existence of Solution

The general form of the linear congruence is given by

$$ax \equiv b \bmod m$$

is solvable only when gcd(a,m) divides b.

The congruence $3x \equiv 19 \bmod 27$ does not have a solution for x, because gcd(3,27)=3 does not divide 19. However, the congruence $3x \equiv 2 \bmod 5$ has a solution, because gcd(3,5)=1 that divides 2.

Exercises 3.1

1. Write the sequence of values of x that satisfies each of the following congruences:

(a) $x \equiv 1 \bmod 3$ (b) $x \equiv 2 \bmod 5$

2. Write the sequence of values of x that satisfies each of the following congruences:

(a) $2x \equiv 4 \bmod 7$ (b) $3x \equiv 1 \bmod 5$

3. Write the following sequences of values of numbers in congruence notation that satisfies each of the following congruences:

(a) $2, 5, 8, 11, \cdots$ (b) $1, 6, 11, 16, 21, \cdots$

4. Write the following sequences of values of numbers in congruence notation that satisfies each of the following congruences:

(a) $3, 7, 11, 15, \cdots$ (b) $4, 9, 14, 19, \cdots$

5. Find the remainder when $132^5 - 17285^2$ is divided by 13

6. Find the remainder when $3^{105} + 4^{106}$ is divided by 7

7. Find the remainder when $2^{50} + 1$ is divided by 5

8. Find the remainder when 2^{501} is divided by 31

9. Find the remainder when $2^{43} \times 125 + 3^{21} \times 432$ is divided by 7

10. Find the remainder when $2^{101} \times 311^2 - 5891^{50}$ is divided by 31

11. Find the remainder when 3^{36} is divided by 7

12. Find the remainder when 8^{307} is divided by 65

13. Find the remainder when 2^{64} is divided by 13

14. Find the digit in the one's place of the number 3^{103}

15. Find the digit in the one's place of the number 7^{102}

16. Find the digits in the one's place and the ten's place of the number 2^{40}

17. Find the digits in the one's place and the ten's place of the number 3^{24}

18. Find the digits in the one's place place of the number $3^{41} + 7^{43}$

19. Check if the congruence $6x \equiv 4 \bmod 8$ has a solution or not

20. Check if the congruence $7x \equiv 9 \bmod 14$ has a solution or not

3.2   Linear Congruences

If a, b and m are integers, $a \not\equiv 0 \bmod$ m, then the congruence

$$ax \equiv b \bmod m$$

is called a **linear congruence** (mod m), where x is an unknown that will be determined.

Theorem The linear congruence $ax \equiv b \bmod m$ has a solution if and only if $d \mid b$, where $d = \gcd(a, m)$. If $d \mid b$, then the linear congruence has d solutions modulo m.

3.2.1   Existence of Solutions

The general form of the linear congruence is given by

$$ax \equiv b \bmod m$$

To determine if a solution exists for a linear congruence, we first find the greatest common divisor $d = \gcd(a, m)$. Consequently, three distinct cases will be examined as follows:

3.2.2   No solution

If d does not divide b, then there is no solution for the congruence in this case.

Example 1. Solve the following congruence

$$2x \equiv 1 \bmod 4$$

Solution. gcd(2,4)=2. However 2 does not divide 1, hence there is no solution for this congruence.

Example 2. Solve the following congruence

$$3x \equiv 2 \bmod 6$$

Solution. gcd(3,6)=3. However 3 does not divide 2, hence there is no solution for this congruence.

3.2.3 Unique Solution

If $d = gcd(a, m) = 1$, i.e. if a and m are relatively prime, then the linear congruence $ax \equiv b \bmod m$ has a unique solution in congruent form. In this case, the linear congruences can be solved in an identical way as algebraic equations. The rules we use in algebra hold for congruences. The technique we will use depends mainly on using the properties discussed above. This means that we can add or subtract the same number to both sides of the congruence as explained in P1 and P2. Also, we can divide both sides of any congruence by the same number, if the necessary condition is satisfied, as stated in P3. The operations of adding, subtracting or dividing will not be applied to m in $(\bmod m)$.

The technique solving linear congruences will be explained by the following illustrative examples.

Example 3. Solve the following congruence

$$2x - 1 \equiv 3 \bmod 7$$

Solution. Using P1, we add 1 to both sides of the congruence, hence we find

$$2x \equiv 4 \bmod 7$$

It is clear that gcd$(2, 7) = 1$, hence there is only one solution for x modulo 7. Using P3, we divide both sides by 2, hence the solution is given by

$$x \equiv 2 \bmod 7$$

Example 4. Solve the following congruence

$$3x + 1 \equiv 10 \bmod 11$$

Solution. Using P2, we subtract 1 from both sides of the congruence, hence we find

$$3x \equiv 9 \bmod 11$$

The gcd (3,11)=1, hence there is a unique solution Using P3, we divide both sides by 3, hence the solution is given by

$$x \equiv 3 \bmod 11$$

Example 5. Solve the following congruence

$$3x + 2 \equiv 3 \bmod 11$$

Solution. Using P2, we subtract 2 from both sides of the congruence, hence we find

$$3x \equiv 1 \bmod 11$$

We note here that $b = 1$ is not divisible by 3, therefore we add 11, or multiples of 11 to b, to get a number that is divisible by 3. In this case we add 11 to $b = 1$ to find

$$3x \equiv 12 \bmod 11$$

Using P3, we divide both sides by 3 so that

$$x \equiv 4 \bmod 11$$

3.2.4 Multiple Solutions

As stated before in the theorem, if $d = \gcd(a, m) > 1$, and if $d \mid b$, then the linear congruence $ax \equiv b \bmod m$ has d mutually incongruent solutions modulo m. In this case, we divide both sides of the congruence by d to find

$$\frac{a}{d}x \equiv \frac{b}{d} \bmod \frac{m}{d}$$

It is clear now that by solving this reduced congruence, we obtain one solution x_0. The d solutions of the given congruence have the form

$$x = x_0 + \frac{m}{d}t, t = 0, 1, 2, \cdots, (d-1)$$

The technique presented above will be illustrated by the following examples

Example 6. Solve the following linear congruence

$$3x \equiv 6 \bmod 12$$

Solution. The gcd(3,12)=3, and 3 | 6. This means that there are three mutually incongruent solutions modulo 12. Dividing a, b, and m by 3 we get

$$x \equiv 2 \bmod 4$$

with solution given by $x_0 = 2$. All three solutions are of the form

$$x = 2 + 4t, t = 0, 1, 2$$

Consequently, the three incongruent solutions are given by

$$x = 2, 6, 10 \bmod 12$$

This is equivalent to

$$x \equiv 2 \bmod 12, \quad x \equiv 6 \bmod 12, \quad x \equiv 10 \bmod 12$$

Example 7. Solve the following linear congruence

$$6x \equiv 4 \bmod 10$$

Solution. The gcd(6,10)=2, and 2 | 4. This means that there are two mutually incongruent solutions modulo 10. Dividing a, b, and m of the congruence by 2 we get

$$3x \equiv 2 \bmod 5$$

with solution given by $x_0 = 4$. The two solutions are of the form

$$x = 4 + 5t, t = 0, 1$$

Consequently, the two incongruent solutions are given by

$$x \equiv 4, 9 \bmod 10$$

This is equivalent to

$$x \equiv 4 \bmod 10, \quad x \equiv 9 \bmod 10$$

Example 8. Solve the following linear congruence

$$12x \equiv 6 \bmod 18$$

Solution. The gcd(12,18)=6, and 6 | 6. This means that there are six mutually incongruent solutions modulo 18. Dividing a, b, and m of the congruence by 6 we get

$$2x \equiv 1 \bmod 3$$

with solution given by $x_0 = 2$. The six solutions are of the form

$$x = 2 + 3t, t = 0, 1, \cdots, 5$$

Consequently, the six incongruent solutions are given by

$$x \equiv 2, 5, 8, 11, 14, 17 \bmod 18$$

This is equivalent to

$$x \equiv 2 \bmod 18, \quad x \equiv 5 \bmod 18, \quad x \equiv 8 \bmod 18,$$
$$x \equiv 11 \bmod 18 \quad x \equiv 14 \bmod 18 \quad x \equiv 17 \bmod 18$$

3.2.5 The Chinese Remainder Theorem

In this section we consider the case when we have multiple congruences. Let $m_1, m_2, \cdots, m_n$ be pairwise relatively prime integers; in other words, $\gcd(m_i, m_j) = 1$, whenever $1 \leq i < j \leq n$. The simultaneous congruences

$$
\begin{aligned}
x &\equiv a_1 \bmod m_1 \\
x &\equiv a_2 \bmod m_2 \\
&\vdots \\
x &\equiv a_n \bmod m_n
\end{aligned}
$$

has a unique solution modulo $m = m_1 m_2 \cdots m_n$.

As a special case of the Chinese remainder theorem, if

$$a_1 = a_2 = \cdots = a_n = a$$

then the unique solution is given by

$$x \equiv a \bmod (m_1 m_2 \cdots m_n)$$

In what follows, we study two illustrative examples to explain the special case where the least remainders of the simultaneous congruences are equivalent.

Example 9. Solve the following system of linear congruences

$$2x - 1 \equiv 2 \quad \bmod 7$$
$$3x + 2 \equiv 6 \quad \bmod 11$$

Solution. Adding 1 to both sides of the first congruence and subtracting 2 from both sides of the second congruence give

$$2x \equiv 3 \quad \bmod 7$$
$$3x \equiv 4 \quad \bmod 11$$

We now proceed as before and add 7 and 11 to the least remainder of both congruences respectively to obtain

$$2x \equiv 10 \quad \bmod 7$$
$$3x \equiv 15 \quad \bmod 11$$

Using P3, we divide both sides of the first congruence by 2 and both sides of the second congruence by 3 to obtain

$$x \equiv 5 \quad \bmod 7$$
$$x \equiv 5 \quad \bmod 11$$

The solution of the system of congruences can be obtained by using P4, hence we find

$$x \equiv 5 \bmod (77)$$

Example 10. Solve the following system of linear congruences

$$5x - 1 \equiv 2 \quad \bmod 4$$
$$3x + 10 \equiv 4 \quad \bmod 5$$
$$7x + 2 \equiv 1 \quad \bmod 11$$

Solution. Adding 1, subtracting 10 and subtracting 2 from both sides of the three congruences respectively gives

$$5x \equiv 3 \quad \mathrm{mod}\ 4$$
$$3x \equiv -6 \quad \mathrm{mod}\ 5$$
$$7x \equiv -1 \quad \mathrm{mod}\ 11$$

We now proceed as before and add 12 , 15 and 22 to r of all congruences respectively to obtain

$$5x \equiv 15 \quad \mathrm{mod}\ 4$$
$$3x \equiv 9 \quad \mathrm{mod}\ 5$$
$$7x \equiv 21 \quad \mathrm{mod}\ 11$$

Using P3, we divide both sides by 5, 3 and 7 respectively so that

$$x \equiv 3 \quad \mathrm{mod}\ 4$$
$$x \equiv 3 \quad \mathrm{mod}\ 5$$
$$x \equiv 3 \quad \mathrm{mod}\ 11$$

The solution of the system of congruences can be obtained by using P4, hence we find

$$x \equiv 3 \,\mathrm{mod}\ 220$$

General Case of the Chinese Remainder Theorem

We now consider the general case of the Chinese remainder theorem, where the least remainders are not equivalent. In what follows we present the algorithm needed to determine the unique solution by using the Chinese remainder theorem, therefore we skip the proof. For simplicity reason, we consider $n = 4$, therefore we have

$$x \equiv a_1 \,\mathrm{mod}\ m_1$$
$$x \equiv a_2 \,\mathrm{mod}\ m_2$$
$$x \equiv a_3 \,\mathrm{mod}\ m_3$$
$$x \equiv a_4 \,\mathrm{mod}\ m_4$$

Notice that the moduli are pairwise relatively prime. The following steps are usually used:

1. We first form the product $m = m_1 m_2 m_3 m_4$.
2. We then form the products: $M_1 = \frac{m}{m_1}, M_2 = \frac{m}{m_2}, M_3 = \frac{m}{m_3}, M_4 = \frac{m}{m_4}$.
3. The following linear congruences

$$M_1 x \equiv 1 \bmod m_1$$
$$M_2 x \equiv 1 \bmod m_2$$
$$M_3 x \equiv 1 \bmod m_3$$
$$M_4 x \equiv 1 \bmod m_4$$

can be easily solved to find the solutions x_1, x_2, x_3, and x_4 respectively.
4. Based on these calculations, the unique solution x is given by

$$x \equiv (a_1 M_1 x_1 + a_2 M_2 x_2 + a_3 M_3 x_3 + a_4 M_4 x_4) \bmod m$$

Notice that the least remainder should be used by deleting all multiples of m.

An alternative method will be introduced when we present the illustrative examples.

Example 11. Use the Chinese remainder theorem to solve the system of linear congruences

$$x \equiv 1 \bmod 3$$
$$x \equiv 2 \bmod 4$$
$$x \equiv 3 \bmod 5$$

Solution. Notice that the moduli are pairwise relatively prime. Following the steps presented above we find
1. $m = 3 \cdot 4 \cdot 5 = 60$.
2. $M_1 = 20, M_2 = 15, M_3 = 12$
3. We then solve the congruences

$$20x \equiv 1 \bmod 3$$
$$15x \equiv 1 \bmod 4$$
$$12x \equiv 1 \bmod 5$$

to find that the solutions are given by $x_1 = 2, x_2 = 3, x_3 = 3$. Using the equation given above we find

$$x \equiv (1 \cdot 20 \cdot 2 + 2 \cdot 15 \cdot 3 + 3 \cdot 12 \cdot 3) \bmod 60$$

Writing the least remainder, by subtracting all multiples of 60, we get

$$x \equiv 58 \bmod 60$$

An alternative method can be used in this case. We notice that the first two congruences can be written as

$$x \equiv 10 \bmod 3$$

$$x \equiv 10 \bmod 4$$

These two linear congruences can be written as

$$x \equiv 10 \bmod 12$$

This means that $x = 10 + 12t, t = 0, 1, \cdots$. Substituting this value of x in the third congruence we obtain

$$10 + 12t \equiv 3 \bmod 5$$

Solving this congruence for t we find

$$t \equiv 4 \bmod 5$$

This can be written as $t = 4 + 5r, r = 0, 1, 2, \cdots$. Substituting this value of t into the last result for x we find

$$x = 10 + 12(4 + 5r) = 58 + 60r, r = 0, 1, \cdots$$

This is equivalent to

$$x \equiv 58 \bmod 60$$

Example 12. Use the Chinese remainder theorem to solve the system of linear congruences

$$
\begin{aligned}
x &\equiv 1 \bmod 2 \\
x &\equiv 2 \bmod 3 \\
x &\equiv 3 \bmod 5
\end{aligned}
$$

Solution. Notice that the moduli are pairwise relatively prime. Following the steps presented above we find

1. $m = 2 \cdot 3 \cdot 5 = 30$.

2. $M_1 = 15, M_2 = 10, M_3 = 6$

3. We then solve the congruences

$$
\begin{aligned}
15x &\equiv 1 \bmod 2 \\
10x &\equiv 1 \bmod 3 \\
6x &\equiv 1 \bmod 5
\end{aligned}
$$

to find that the solutions are given by $x_1 = 1, x_2 = 1, x_3 = 1$. Using the equation given above we find

$$x \equiv (1 \cdot 15 \cdot 1 + 2 \cdot 10 \cdot 1 + 3 \cdot 6 \cdot 1) \bmod 30$$

Writing the least remainder we get

$$x \equiv 23 \bmod 30$$

Using the alternative method we introduced before, we first write the first two congruences as

$$x \equiv 11 \bmod 2$$

$$x \equiv 11 \bmod 3$$

These two linear congruences can be written as

$$x \equiv 11 \bmod 6$$

This means that $x = 11 + 6t, t = 0, 1, \cdots$. Substituting this value of x in the third congruence we obtain

$$11 + 6t \equiv 3 \bmod 5$$

Solving this congruence for t we find

$$t \equiv 2 \bmod 5$$

This can be written as $t = 2 + 5r, r = 0, 1, 2, \cdots$. Substituting this value of t into the last result for x we find

$$x = 11 + 6(2 + 5r) = 23 + 30r, r = 0, 1, \cdots$$

This is equivalent to

$$x \equiv 23 \bmod 30$$

Example 13. Use the Chinese remainder theorem to solve the system of linear congruences

$$2x \equiv 1 \bmod 3$$
$$3x \equiv 2 \bmod 5$$
$$4x \equiv 3 \bmod 7$$

Solution. The moduli are pairwise relatively prime. We first transform the congruences to

$$x \equiv 2 \bmod 3$$
$$x \equiv 4 \bmod 5$$
$$x \equiv 6 \bmod 7$$

Following the steps presented above we find
1. $m = 3 \cdot 5 \cdot 7 = 105$.
2. $M_1 = 35, M_2 = 21, M_3 = 15$
3. We then solve the congruences

$$35x \equiv 1 \bmod 3$$
$$21x \equiv 1 \bmod 5$$
$$15x \equiv 1 \bmod 7$$

to find that the solutions are given by $x_1 = 2, x_2 = 1, x_3 = 1$. Using the equation given above we find

$$x \equiv (2 \cdot 35 \cdot 2 + 4 \cdot 21 \cdot 1 + 6 \cdot 15 \cdot 1) \bmod 105$$

Writing the least remainder we get

$$x \equiv 104 \bmod 105$$

Using the alternative method we introduced before, we first write the first two congruences as

$$x \equiv 14 \bmod 3$$

$$x \equiv 14 \bmod 5$$

These two linear congruences can be written as

$$x \equiv 14 \bmod 15$$

This means that $x = 14 + 15t, t = 0, 1, \cdots$. Substituting this value of x in the third congruence we obtain

$$14 + 15t \equiv 6 \bmod 7$$

Solving this congruence for t we find

$$t \equiv 6 \bmod 7$$

This can be written as $t = 6 + 7r, r = 0, 1, 2, \cdots$. Substituting this value of t into the last result for x we find

$$x = 14 + 15(6 + 7r) = 104 + 105r, r = 0, 1, \cdots$$

This is equivalent to

$$x \equiv 104 \bmod 105$$

Exercises 3.2

1. Solve the following congruences and write the first few terms of the sequence of solution

(a) $3x \equiv 6 \bmod 5$ (b) $2x \equiv 4 \bmod 7$

2. Solve the following congruences and write the first few terms of the sequence of solution

(a) $4x \equiv 12 \bmod 11$ (b) $5x \equiv 10 \bmod 19$

3. Solve for x the following congruences :

(a) $3x \equiv 1 \bmod 7$ (b) $5x \equiv 6 \bmod 11$

4. Solve for x the following congruences :

(a) $2x \equiv 3 \bmod 5$ (b) $7x \equiv 10 \bmod 11$

5. Solve the following congruences:

(a) $3(2x + 1) + 4 \equiv 8 \bmod 11$

(b) $3(2x - 1) - 2 \equiv 6 \bmod 7$

6. Solve the following congruences:

(a) $3(3x - 4) \equiv x \bmod 5$

(b) $5(2x - 3) \equiv 4x \bmod 13$

7. Solve the following system of linear congruences

(a) $3x \equiv 6 \bmod 7$
$2x \equiv 4 \bmod 5$

(b) $5x - 1 \equiv 1 \bmod 3$
$3x + 4 \equiv 2 \bmod 5$

8. Solve the following system of linear congruences

(a) $7x + 4 \equiv 2 \bmod 3$
$3x + 6 \equiv 9 \bmod 11$

(b) $5x \equiv 1 \bmod 3$
$11x \equiv 1 \bmod 7$

9. Solve the following system of linear congruences

(a) $5x \equiv 1 \bmod 2$
$7x \equiv 1 \bmod 3$
$11x \equiv 1 \bmod 5$

(b) $5x \equiv 1 \bmod 3$
$3x \equiv 1 \bmod 5$
$19x \equiv 3 \bmod 7$

10. Solve the following system of linear congruences

(a) $3x + 1 \equiv 3 \bmod 7$
$4x - 1 \equiv 1 \bmod 5$
$7x + 3 \equiv 2 \bmod 11$

(b) $4x + 1 \equiv 4 \bmod 5$
$3x - 2 \equiv 4 \bmod 7$
$7x + 1 \equiv 4 \bmod 11$

11. Solve the following system of linear congruences

$2x - 1 \equiv 0 \bmod 3$
$5x + 1 \equiv 4 \bmod 7$
$4x - 5 \equiv 3 \bmod 11$

Check if $x = 22871$ is a possible value of x
Hint: Divide 22871 by the resulting modulus, and check if the remainder is

2 or not

12. Solve the following system of linear congruences

$x - 1 \equiv 0 \bmod 2$
$2x - 1 \equiv 1 \bmod 3$
$3x - 1 \equiv 2 \bmod 7$
Check if $x = 2081$ is a possible value of x

13. Solve the following system of linear congruences

$2x + 3 \equiv 6 \bmod 7$
$3x + 8 \equiv 1 \bmod 11$
$4x + 4 \equiv 11 \bmod 13$
Check if $x = 19019$ is a possible value of x

14. Use Chinese remainder theorem to solve the following system of linear congruences

$x \equiv 2 \bmod 3$
$x \equiv 3 \bmod 5$
$x \equiv 2 \bmod 7$

15. Use Chinese remainder theorem to solve the following system of linear congruences

$x \equiv 3 \bmod 5$
$x \equiv 2 \bmod 6$
$x \equiv 4 \bmod 7$

16. Use Chinese remainder theorem to solve the following system of linear congruences

$x \equiv 2 \bmod 6$
$x \equiv 3 \bmod 5$
$x \equiv 6 \bmod 7$

17. Use Chinese remainder theorem to solve the following system of linear congruences

$x \equiv 2 \bmod 5$
$x \equiv 0 \bmod 7$
$x \equiv 4 \bmod 11$

18. Use Chinese remainder theorem to solve the following system of linear congruences

$$x \equiv 1 \bmod 3$$
$$x \equiv 1 \bmod 4$$
$$x \equiv 6 \bmod 11$$

19. Use Chinese remainder theorem to solve the following system of linear congruences

$$x \equiv 1 \bmod 2$$
$$x \equiv 1 \bmod 5$$
$$x \equiv 5 \bmod 13$$

20. Use Chinese remainder theorem to solve the following system of linear congruences

$$x \equiv 1 \bmod 2$$
$$x \equiv 1 \bmod 3$$
$$x \equiv 3 \bmod 5$$
$$x \equiv 3 \bmod 7$$

21. Solve the following system of linear congruences

$$5(x - 1) \equiv (x - 2) \bmod 7$$
$$3(x + 2) \equiv (x - 4) \bmod 11$$

22. Solve the following system of linear congruences

$$2(2x - 3) \equiv (x - 2) \bmod 5$$
$$3(3x + 1) \equiv (x - 1) \bmod 7$$

23. Solve the following system of linear congruences

$$2(3x - 5) \equiv (x - 8) \bmod 11$$
$$3(2x + 1) \equiv 2(x - 4) \bmod 13$$

24. Solve the following system of linear congruences

$$3x - 5 \equiv (x - 9) \bmod 13$$
$$5x + 1 \equiv (2x - 17) \bmod 17$$

25. Solve the following system of linear congruences

$$2x + 3 \equiv 1 \bmod 7$$

$$3x + 8 \equiv (x - 2) \bmod 11$$
$$4x + 4 \equiv (x - 4) \bmod 13$$

26. Solve the following system of linear congruences

$$5x + 3 \equiv 2 \bmod 7$$
$$7x - 1 \equiv (x + 1) \bmod 11$$
$$11x - 2 \equiv (x - 1) \bmod 13$$

27. Solve the following system of linear congruences

$$3x + 4 \equiv x \bmod 7$$
$$5x + 8 \equiv (x - 5) \bmod 11$$
$$4x + 17 \equiv (x - 7) \bmod 13$$

28. Solve the following system of linear congruences

$$4x + 2 \equiv 1 \bmod 3$$
$$3x + 7 \equiv (x + 1) \bmod 5$$
$$5x + 1 \equiv (x + 2) \bmod 7$$

29. Solve the following system of linear congruences

$$3x + 1 \equiv (x - 3) \bmod 9$$
$$5x - 2 \equiv (7 - x) \bmod 11$$
$$3x + 9 \equiv (x - 11) \bmod 17$$

30. Solve the following system of linear congruences

$$3(x - 3) \equiv (x - 4) \bmod 11$$
$$3(2x - 1) \equiv (1 - x) \bmod 13$$
$$4(3x - 2) \equiv (x + 4) \bmod 19$$

3.3   Linear Diophantine Equations

A student graduated from a junior college with 64 credit hours. The transcript of the student shows that he studies x courses, each of 3 credit hours, and y courses each of 4 credit hours. From this we can set

$$3x + 4y = 64$$

From algebra, and by graphing this linear equation, the resulting graph is a line that indicates that there is an infinite number of solutions, such as

x	0	1	2	64/3	4	8	12
y	16	61/4	58/4	0	13	10	7

However, the unknowns x and y in this equation represent the number of 3-hours courses and 4-hours courses respectively. This means that x and y must be positive integers only, and therefore x and y cannot accept negative values or fractional values in this example.

Generally speaking, an equation in two or more unknown integers x and y, such as

$$ax + by = c$$

where a, b and c are integers, is called Diophantine equation. In other words, Diophantine equation is defined as an equation with integral coefficients a, b and c, from which integral solutions for x and y are required. Consequently, there is an infinite number of integers represented by points on the related line. Our concern in this section is to find the unknown integers x and y, each in a congruence form or also by a sequence of integers.

Assuming that a, b and c have no common factor, we then convert the equation

$$ax + by = c$$

to an equivalent congruence given by

$$ax \equiv c \bmod b$$

Accordingly, the unknown integer x is completely determined by a congruence form as given above. Solving the resulting congruence for x as discussed in the previous section, then x can be determined in a congruence form given by

$$x \equiv r \bmod b$$

or equivalently in an integer form expressed by

$$x = r + bt$$

where t is an integer. Substituting the integer result of x in the given Diophantine equation gives the integer values of y. It is worth noting that solving Diophantine equations is an important application of the congruence concept.

An important theorem will be stated here but its proof will not be given in this text. The theorem gives the necessary condition for an integer solutions to exist for a Diophantine equation. In practice, it was proved that not all Diophantine equations provide integer solutions. The theorem is stated as follows:

Theorem. The Diophantine equation

$$ax + by = c$$

has integer solutions if $\gcd(a, b) \mid c$, where a, b and c are integers.

The procedure discussed above will be explained by the following illustrative examples.

Example 1. Solve the following Diophantine equation:

$$6x + 7y = 11$$

Solution. This equation can be rewritten as

$$6x \equiv 11 \bmod 7$$

This congruence can be rewritten as

$$6x \equiv 18 \bmod 7$$

and hence we find

$$x \equiv 3 \bmod 7$$

This means that

$$x = 3 + 7t$$

where t is any positive or negative integer. Substituting this value of x in the Diophantine equation and solving for y we find

$$y = -1 - 6t$$

or

$$y \equiv -1 \bmod 6$$

It is clear that we obtained an infinite integer values for x and y. Setting $t = 0$ we find the pair $(10, -7)$. Similarly for $t = 1$ we obtain the pair $(3, -1)$, and so on.

3.3. Linear Diophantine Equations

Example 2. Solve the following Diophantine equation:

$$5x + 11y = 43$$

Solution. This equation can be rewritten as

$$5x \equiv 43 \bmod 11$$

This congruence can be rewritten as

$$5x \equiv 10 \bmod 11$$

by subtracting 33 from 43. Hence, we find

$$x \equiv 2 \bmod 11$$

This means that

$$x = 2 + 11t$$

where t is an integer. Substituting this value of x in the Diophantine equation and solving for y we find

$$y = 3 - 5t$$

or

$$y \equiv 3 \bmod 5$$

It is clear that we obtained an infinite integer values for x and y. Setting $t = 0$ we find the pair $(2, 3)$. Similarly for $t = 1$ we find the pair $(13, -2)$, and so on.

Example 3. Find a Diophantine equation which has the solutions given by

$$x = 5 + 3t$$

$$y = -2 - 4t$$

Solution. To determine the Diophantine equation that has the solutions given above, we should eliminate t from both equations. This can be easily done by multiplying the first equation by 4 and the second equation by 3 to find

$$4x = 20 + 12t$$

$$3y = -6 - 12t$$

Adding these two equations gives

$$4x + 3y = 14$$

Exercises 3.3

1. Use congruences to find all positive integers, if any, that are solutions of each of the following Diophantine equations:

(a)   5x+3y=23   (b)   7x+11y=47

2. Use congruences to find all positive integers, if any, that are solutions of each of the following Diophantine equations:

(a)   3x+5y=19   (b)   11x+5y=71

3. Use congruences to find all positive integers, if any, that are solutions of each of the following Diophantine equations:

(a)   7x+9y=23   (b)   11x+13y=41

4. Use congruences to find all positive integers, if any, that are solutions of each of the following Diophantine equations:

(a)   9x+5y=19   (b)   13x+17y=43

5. Write two pairs of solutions for each of the following Diophantine equations:

(a)   7x+3y=24   (b)   7x+12y=36

6. Write two pairs of solutions for each of the following Diophantine equations:

(a)   7x+12y=124   (b)   5x+7y=10

7. Find the Diophantine equation which has the solutions

$$x = 2 + 3t$$
$$y = 4 - 5t$$

8. Find the Diophantine equation which has the solutions

$$x = 3 + 7t$$
$$y = -1 - 5t$$

Chapter 4

Methods of Proof

4.1 Introduction

In mathematics, formulas, theorems, identities and other relationships are usually developed to be used in several applications. In computer science, programs are designed to be applied for many applications. The ability to prove that these formulas, identities and programs are correct is very important. Several methods of proofs have been established to justify the correctness of these mathematical and computer statements. In this text we will discuss five well–known methods of proof, namely:

1. **Mathematical Induction.**

2. **Finite Differences Method.**

3. **Partial Fractions Method.**

4. **Direct Proof Method.**

5. **Proof by Contradiction.**

4.2 Mathematical Induction

The method of mathematical induction is considered the most powerful technique to prove the validity of a formula, statement or a relationship. As will be discussed later, the proof will be concluded (induced) from the formula itself. It is useful to explain the method first by discussing the following illustrative example.

Example 1. Use mathematical induction to prove that the statement

$$1 + 2 + 3 + 4 + \cdots + n = \frac{n(n+1)}{2}$$

is true for all positive integers n, $n \geq 1$.

Solution. We first note that this statement consists of left hand side (LHS) and right hand side (RHS). For example, using the left hand side for $n = 4$ means that we must find the sum of the first four terms. If we let S_n denote the sum of the first n terms, then

$$S_4 = 1 + 2 + 3 + 4 = 10$$

On the other hand, using the right hand side means that the sum S_4 can also be obtained by substituting $n = 4$ in the formula given at the right hand side, i.e

$$S_4 = \frac{4(4+1)}{2} = 10$$

The explanation presented above suggests that we use the first four terms of the left hand side if the left hand side is to be used. However, we substitute $n = 4$ in the right hand side if the right hand side is to be used.

For n large, using the left hand side to evaluate S_n is not practical. However, it is more effective in this case to substitute the given value of n in the right hand side. Accordingly, it is necessary to prove that this statement is valid for all integer values of n, where $n \geq 1$.

To use the method of **mathematical induction**, we usually apply the following standard steps:

Step 1. We first prove that the statement is true for $n = 1$. Accordingly, for $n = 1$
Left Hand Side (LHS) = 1
Right Hand Side (RHS) = $\frac{1(1+1)}{2} = 1$
Hence the statement is true for $n = 1$.

Step 2. We next assume that the statement is true for $n = k$. In other words, we assume that

$$1 + 2 + 3 + 4 + \cdots + k = \frac{k(k+1)}{2}$$

is true. Notice that this assumption is obtained by using the first k terms of the left hand side, and by substituting $n = k$ in the right hand side as stated above.

Step 3. We then use the assumption of Step 2 to show that the statement is true for $n = k + 1$. This means that we should show that

$$1 + 2 + 3 + 4 + \cdots + k + (k+1) = \frac{(k+1)[(k+1)+1]}{2}$$

or equivalently,

$$1 + 2 + 3 + 4 + \cdots + k + (k+1) = \frac{(k+1)(k+2)}{2}$$

by using the assumption of equality made up in Step 2. Notice again that we wrote the first $(k+1)$ terms in the LHS of the original statement and we substituted $n = k + 1$ in the RHS of that statement.

To show that the last statement in Step 3 holds, two distinct approaches are usually used. We first start by the commonly used approach:
1. Using the LHS of the statement in Step 3 above we find

$$
\begin{aligned}
\text{LHS} &= 1 + 2 + 3 + 4 + \cdots + k + (k+1) \\
&= \underbrace{1 + 2 + 3 + 4 + \cdots + k}_{\text{from assumption 2}} + (k+1) \\
&= \underbrace{\frac{k(k+1)}{2}}_{\text{from assumption 2}} + (k+1) \\
&= \frac{k(k+1) + 2(k+1)}{2} = \frac{(k+1)(k+2)}{2} = \text{RHS}
\end{aligned}
$$

This shows that the statement is true for $n = k + 1$. In other words, validity of the statement for $n = k$ implies its validity for $n = k + 1$. Accordingly, the statement under discussion is true for every positive integer n, $n \geq 1$.
2. In the following we explain an alternative approach to that discussed before. The statement in Step 3 can also be obtained by adding $(k + 1)$ to both sides of the equality assumed in Step 2, where we obtain

$$
\begin{aligned}
\underbrace{1 + 2 + 3 + \cdots + k}_{\text{from assumption 2}} + (k+1) &= \underbrace{\frac{k(k+1)}{2}}_{\text{assumption 2}} + (k+1) \\
&= \frac{k(k+1) + 2(k+1)}{2} \\
&= \frac{(k+1)(k+2)}{2}
\end{aligned}
$$

This shows that the statement is true for $n = k + 1$. Accordingly, the statement under discussion is true for every positive integers n, $n \geq 1$.

In the following we list the necessary steps of the **principle of mathematical induction**. To prove that a statement S_n is true for all positive integers n, we should:

1. Show that S_1 is true, i.e show that the statement is true for $n = 1$.
2. Assume that the statement is true for $n = k$.
3. Use the assumption made in 2 for $n = k$ to show that the statement is true for $n = k + 1$.

Generally speaking, given any statement $S(n)$ about a natural number n. If $S(n_0)$ is true, and if we can show that if $S(k)$ is true then $S(k + 1)$ is also true, then $S(n)$ is true for every n.

The three essential steps of the method of mathematical induction will be illustrated by discussing the following examples.

Example 2. Use mathematical induction to prove that the statement

$$3 + 7 + 11 + 15 + \cdots + (4n - 1) = n(2n + 1)$$

is true for all positive integers n.

Solution.

Step 1. For $n = 1$
Left Hand Side (LHS) $= 3$
Right Hand Side (RHS) $= 1(2 + 1) = 3$
Hence, the statement is true for $n = 1$.
Step 2. Assume that the statement is true for $n = k$, i.e assume that

$$3 + 7 + 11 + 15 + \cdots + (4k - 1) = k(2k + 1)$$

Step 3. We next show that S_n is true for $n = k + 1$, i.e we should show that

$$3 + 7 + 11 + 15 + \cdots + (4k - 1) + [4(k + 1) - 1] = (k + 1)[2(k + 1) + 1]$$

or equivalently,

$$3 + 7 + 11 + 15 + \cdots + (4k - 1) + (4k + 3) = (k + 1)(2k + 3)$$

This can be proved by setting

$$\text{LHS} \quad = \quad 3 + 7 + 11 + 15 + \cdots + (4k - 1) + (4k + 3)$$

$$= \underbrace{3 + 7 + 11 + 15 + \cdots + (4k - 1)}_{\text{from assumption 2}} + (4k + 3)$$

$$= \underbrace{k(2k + 1)}_{\text{from assumption 2}} + (4k + 3)$$

$$= 2k^2 + 5k + 3 = (k + 1)(2k + 3) = \text{RHS}$$

This shows that the statement is true for $n = k + 1$. Accordingly, the statement under discussion is true for all positive integers n, $n \geq 1$.

Example 3. Use mathematical induction to prove that the statement

$$1 + 2 + 2^2 + 2^3 + 2^4 + \cdots + 2^{(n-1)} = 2^n - 1$$

is true for all positive integers n.

Solution.

Step 1. For $n = 1$
Left Hand Side (LHS) $= 1$
Right Hand Side (RHS) $= 2^1 - 1 = 1$
Hence, the statement is true for $n = 1$.
Step 2. Assume that the statement is true for $n = k$, i.e assume that

$$1 + 2 + 2^2 + 2^3 + 2^4 + \cdots + 2^{(k-1)} = 2^k - 1$$

Step 3. We use the equality assumption in Step 2 to show that S_n is true for $n = k + 1$. This means that we should show that

$$1 + 2 + 2^2 + 2^3 + 2^4 + \cdots + 2^{(k-1)} + 2^k = 2^{(k+1)} - 1$$

This can be proved by setting

$$\text{LHS} = 1 + 2 + 2^2 + 2^3 + 2^4 + \cdots + 2^{(k-1)} + 2^k$$

$$= \underbrace{1 + 2 + 2^2 + 2^3 + 2^4 + \cdots + 2^{(k-1)}}_{\text{from assumption 2}} + 2^k$$

$$= \underbrace{2^k - 1}_{\text{from assumption 2}} + 2^k$$

$$= 2^k + 2^k - 1$$

$$= 2 \times 2^k - 1 = 2^{(k+1)} - 1 = \text{RHS}$$

This shows that the statement is true for $n = k + 1$. Accordingly, the statement under discussion is true for all positive integers n, $n \geq 1$.

Example 4. Use mathematical induction to prove that the statement

$$\frac{1}{1 \times 5} + \frac{1}{5 \times 9} + \frac{1}{9 \times 13} + \cdots + \frac{1}{(4n-3)(4n+1)} = \frac{n}{4n+1}$$

is true for all positive integers n.

Solution.

Step 1. For $n = 1$
Left Hand Side (LHS) $= \frac{1}{5}$
Right Hand Side (RHS) $= \frac{1}{4 \times 1 + 1} = \frac{1}{5}$
Hence, the statement is true for $n = 1$.
Step 2. Assume that the statement is true for $n = k$, i.e assume that

$$\frac{1}{1 \times 5} + \frac{1}{5 \times 9} + \frac{1}{9 \times 13} + \cdots + \frac{1}{(4k-3)(4k+1)} = \frac{k}{4k+1}$$

is true for $n = k$.
Step 3. We use the equality assumption in Step 2 to show that S_n is true for $n = k + 1$. This means that we should show that

$$\frac{1}{1 \times 5} + \frac{1}{5 \times 9} + \cdots + \frac{1}{(4k-3)(4k+1)}$$

$$+ \frac{1}{(4(k+1)-3)\,(4(k+1)+1)} = \frac{k+1}{4(k+1)+1}$$

or, equivalently

$$\frac{1}{1 \times 5} + \frac{1}{5 \times 9} + \cdots + \frac{1}{(4k-3)(4k+1)} + \frac{1}{(4k+1)(4k+5)} = \frac{k+1}{4k+5}$$

This can be proved by setting

$$\text{LHS} = \frac{1}{1 \times 5} + \frac{1}{5 \times 9} + \cdots + \frac{1}{(4k-3) \times (4k+1)} + \frac{1}{(4k+1) \times (4k+5)}$$

$$= \underbrace{\frac{1}{1 \times 5} + \frac{1}{5 \times 9} + \cdots + \frac{1}{(4k-3) \times (4k+1)}}_{\text{from assumption 2}} + \frac{1}{(4k+1) \times (4k+5)}$$

$$= \underbrace{\frac{k}{4k+1}}_{} + \frac{1}{(4k+1) \times (4k+5)}$$

from assumption 2

$$= \frac{k(4k+5)+1}{(4k+1)(4k+5)}$$

$$= \frac{4k^2+5k+1}{(4k+1)(4k+5)}$$

$$= \frac{(4k+1)(k+1)}{(4k+1)(4k+5)} = \frac{k+1}{4k+5} = \text{RHS}$$

This shows that the statement is true for $n = k+1$. Accordingly, the statement under discussion is true for all positive integers n, $n \geq 1$.

Example 5. Use mathematical induction to prove that the statement

$$\frac{1}{1 \times 6} + \frac{1}{6 \times 11} + \frac{1}{11 \times 16} + \cdots + \frac{1}{(5n-4)(5n+1)} = \frac{n}{5n+1}$$

is true for all positive integers n.

Solution.

Step 1. For $n = 1$
Left Hand Side (LHS) $= \frac{1}{6}$
Right Hand Side (RHS) $= \frac{1}{5 \times 1 + 1} = \frac{1}{6}$
Hence, the statement is true for $n = 1$.
Step 2. Assume that the statement is true for $n = k$, i.e assume that

$$\frac{1}{1 \times 6} + \frac{1}{6 \times 11} + \frac{1}{11 \times 16} + \cdots + \frac{1}{(5k-4)(5k+1)} = \frac{k}{5k+1}$$

is true for $n = k$.
Step 3. We use the equality assumption in Step 2 to show that S_n is true for $n = k+1$. This means that we should show that

$$\frac{1}{1 \times 6} + \frac{1}{6 \times 11} + \cdots + \frac{1}{(5k-4)(5k+1)}$$
$$+ \frac{1}{(5(k+1)-4)(5(k+1)+1)} = \frac{k+1}{5(k+1)+1}$$

or, equivalently

$$\frac{1}{1 \times 6} + \frac{1}{6 \times 11} + \cdots + \frac{1}{(5k-4)(5k+1)} + \frac{1}{(5k+1)(5k+6)} = \frac{k+1}{5k+6}$$

This can be proved by setting

$$\begin{aligned}
\text{LHS} &= \frac{1}{1 \times 6} + \frac{1}{6 \times 11} + \cdots + \frac{1}{(5k-4)(5k+1)} + \frac{1}{(5k+1)(5k+6)} \\
&= \underbrace{\frac{1}{1 \times 5} + \frac{1}{5 \times 9} + \cdots + \frac{1}{(5k-4)(5k+1)}}_{\text{from assumption 2}} + \frac{1}{(5k+1)(5k+6)} \\
&= \underbrace{\frac{k}{5k+1}}_{\text{from assumption 2}} + \frac{1}{(5k+1)(5k+6)} \\
&= \frac{k(5k+6)+1}{(5k+1)(5k+6)} \\
&= \frac{5k^2 + 6k + 1}{(5k+1)(5k+6)} \\
&= \frac{(5k+1)(k+1)}{(5k+1)(5k+6)} = \frac{k+1}{5k+6} = \text{RHS}
\end{aligned}$$

This shows that the statement is true for $n = k + 1$. Accordingly, the statement under discussion is true for all positive integers n, $n \geq 1$.

Example 6. Use mathematical induction to prove that

$$5^n - 1 \text{ is divisible by } 4$$

for all positive integers $n, n \geq 1$.

Solution.

Step 1. For $n = 1$
$5^1 - 1 = 4$ which is divisible by 4. Therefore, the statement is true for $n = 1$.
Step 2. Assume that the statement is true for $n = k$, i.e assume that

$$5^k - 1 = 4b$$

or equivalently
$$5^k = 4b + 1, \text{ where } b \text{ is an integer.}$$

Step 3. We use the equality assumption in Step 2 to show that the statement is true for $n = k + 1$. This means that we should show that

$$5^{(k+1)} - 1 \text{ is divisible by } 4$$

This can be proved by setting

$$5^{(k+1)} - 1 = \underbrace{5^k}_{\text{from step 2}} \times 5 - 1$$

$$= \underbrace{(4b+1)}_{\text{from step 2}} \times 5 - 1$$

$$= 20b + 4 = 4(5b+1) = 4c$$

where c is an integer given by

$$c = 5b + 1$$

This shows that the statement is true for $n = k + 1$. Accordingly, the statement under discussion is true for all positive integers n, $n \geq 1$.

We point out here that this type of statements can also be proved by using the congruence method. To show that $5^n - 1$ is divisible by 4, it means that we should show that

$$5^n \equiv 1 \bmod 4$$

for all positive integers n. We first set

$$5^1 \equiv 5 \bmod 4$$

which gives

$$5^1 \equiv 1 \bmod 4$$

Raising both sides to power n we find

$$5^n \equiv 1 \bmod 4$$

Example 7. Use mathematical induction to prove that

$$3^{2n+1} + 2^{n+2} \text{ is divisible by } 7$$

for all positive integers $n \geq 1$.

Solution.

Step 1. For $n = 1$
$3^3 + 2^3 = 35$ which is divisible by 7.

This means that the statement is true for $n = 1$.
Step 2. Assume that the statement is true for $n = k$, i.e assume that

$$3^{2k+1} + 2^{k+2} = 7b \text{ where } b \text{ is an integer}$$

Step 3. We use the equality assumption in Step 2 to show that the statement is true for $n = k + 1$. This means that we should show that

$$3^{(2k+3)} + 2^{k+3} \text{ is divisible by 7}$$

This can be proved by setting

$$
\begin{aligned}
3^{(2k+3)} + 2^{k+3} &= 9 \times 3^{2k+1} + 2 \times 2^{k+2} \\
&= 9(3^{2k+1} + 2^{k+2}) - 7 \times 2^{k+2} \\
&= \underbrace{9 \times 7b}_{\text{from step 2}} \quad -7 \times 2^{k+2} \\
&= 7(9b - 2^{k+2}) = 7c
\end{aligned}
$$

where c is an integer given by

$$c = 9b - 2^{k+2}$$

This shows that the statement is true for $n = k + 1$. Accordingly, the statement under discussion is true for all positive integers n, $n \geq 1$.

Example 8. Use mathematical induction to prove that the inequality

$$2^n > n$$

is true for all positive integers $n, n \geq 1$.

Solution. Step 1. For $n = 1$

Left Hand Side (LHS) $= 2^1$
Right Hand Side (RHS) $= 1$. Therefore, the inequality is true for $n = 1$.
Step 2. Assume that the statement is true for $n = k$, i.e assume that

$$2^k > k$$

Step 3. We use the assumption in Step 2 to show that the inequality is true for $n = k + 1$. This means that we should show that

$$2^{k+1} > k + 1$$

This can be proved by setting

$$
\begin{aligned}
\text{LHS} &= 2^{k+1} \\
&= \underbrace{2^k}_{\text{from assumption 2}} \times 2 \\
&> \underbrace{k}_{\text{from assumption 2}} \times 2 \\
&= 2k \\
&= k + k \\
&> k + 1 \\
&= \text{RHS}
\end{aligned}
$$

This shows that the statement is true for $n = k + 1$. Accordingly, the statement under discussion is true for all positive integers n, $n \geq 1$.

Exercises 4.2

Use mathematical induction method to show that each statement of Exercises 1 – 20 is true for every positive integer n, $n \geq 1$, unless otherwise specified:

1. $1 + 3 + 5 + 7 + \cdots + (2n - 1) = n^2$

2. $1 + 5 + 9 + 13 + \cdots + (4n - 3) = n(2n - 1)$

3. $1 + 7 + 13 + 19 + \cdots + (6n - 5) = n(3n - 2)$

4. $3 + 8 + 13 + 18 + \cdots + (5n - 2) = \dfrac{n(5n + 1)}{2}$

5. $2 + 8 + 14 + 20 + \cdots + (6n - 4) = n(3n - 1)$

6. $-2 + 1 + 4 + 7 + \cdots + (3n - 5) = \dfrac{n(3n - 7)}{2}$

7. $1^2 + 2^2 + 3^2 + 4^2 + \cdots + n^2 = \dfrac{n(n + 1)(2n + 1)}{6}$

8. $1^3 + 2^3 + 3^3 + 4^3 + \cdots + n^3 = \left[\dfrac{n(n + 1)}{2}\right]^2$

9. $3 + 3^2 + 3^3 + 3^4 + \cdots + 3^n = \dfrac{3(3^n - 1)}{2}$

10. $1 + 3^2 + 3^4 + 3^6 + \cdots + 3^{2(n-1)} = \dfrac{(3^{2n} - 1)}{8}$

11. $\dfrac{1}{1 \times 2} + \dfrac{1}{2 \times 3} + \dfrac{1}{3 \times 4} + \cdots + \dfrac{1}{n(n+1)} = \dfrac{n}{n+1}$

12. $\dfrac{1}{1 \times 3} + \dfrac{1}{3 \times 5} + \dfrac{1}{5 \times 7} + \cdots + \dfrac{1}{(2n-1)(2n+1)} = \dfrac{n}{2n+1}$

13. $\dfrac{1}{1 \times 7} + \dfrac{1}{7 \times 13} + \dfrac{1}{13 \times 19} + \cdots + \dfrac{1}{(6n-5)(6n+1)} = \dfrac{n}{6n+1}$

14. $\dfrac{1}{2 \times 5} + \dfrac{1}{5 \times 8} + \dfrac{1}{8 \times 11} + \cdots + \dfrac{1}{(3n-1)(3n+2)} = \dfrac{n}{2(3n+2)}$

15. $\dfrac{1}{3 \times 7} + \dfrac{1}{7 \times 11} + \dfrac{1}{11 \times 15} + \cdots + \dfrac{1}{(4n-1)(4n+3)} = \dfrac{n}{3(4n+3)}$

16. $\dfrac{1}{4 \times 9} + \dfrac{1}{9 \times 14} + \dfrac{1}{14 \times 19} + \cdots + \dfrac{1}{(5n-1)(5n+4)} = \dfrac{n}{4(5n+4)}$

17. $3^n > n$

18. $n^2 > 2n + 1, n \geq 4$

19. $2^n > n^2, n \geq 5$

20. Show that $7^n - 1$ is divisible by 6, $n \geq 1$

21. Show that $9^n - 1$ is divisible by 8, $n \geq 1$

22. Show that $11^n - 1$ is divisible by 10, $n \geq 1$

23. Show that $13^n - 1$ is divisible by 6, $n \geq 1$

24. $\dfrac{1^2}{1 \times 3} + \dfrac{2^2}{3 \times 5} + \dfrac{3^2}{5 \times 7} + \cdots + \dfrac{n^2}{(2n-1)(2n+1)} = \dfrac{n(n+1)}{2(2n+1)}$

25. $1 \times 2 + 2 \times 3 + 3 \times 4 + 4 \times 5 + 5 \times 6 + \cdots + n(n+1) = \dfrac{n(n+1)(n+2)}{3}$

26. $1 \times 3 + 2 \times 4 + 3 \times 5 + 4 \times 6 + 5 \times 7 + \cdots + n(n+2) = \dfrac{n(n+1)(2n+7)}{6}$

27. $1 \times 4 + 2 \times 5 + 3 \times 6 + 4 \times 7 + 5 \times 8 + \cdots + n(n+3) = \dfrac{n(n+1)(n+5)}{3}$

28. $1 \times 5 + 2 \times 6 + 3 \times 7 + 4 \times 8 + 5 \times 9 + \cdots + n(n+4) = \dfrac{n(n+1)(2n+13)}{6}$

29. $1 \times 6 + 2 \times 7 + 3 \times 8 + 4 \times 9 + \cdots + n(n+5) = \dfrac{n(n+1)(n+8)}{3}$

30. $2 \times 3 + 3 \times 4 + 4 \times 5 + \cdots + (n+1)(n+2) = \dfrac{n(n^2 + 6n + 11)}{3}$

31. $1 \times 2 \times 3 + 2 \times 3 \times 4 + \cdots + n(n+1)(n+2) = \dfrac{n(n+1)(n+2)(n+3)}{4}$

32. $2 \times 2^0 + 3 \times 2^1 + 4 \times 2^2 + \cdots + (n+1) \times 2^{n-1} = n \times 2^n$

33. $1 \times 2^1 + 2 \times 2^2 + 3 \times 2^3 + \cdots + n \times 2^n = 2 + (n-1)2^{n+1}$

34. $1^2 + 3^2 + 5^2 + \cdots + (2n-1)^2 = \dfrac{n(2n-1)(2n+1)}{3}$

35. Show that $18^n - 5^n$ is divisible by 13, $n \geq 1$

36. Show that $2^{2n-1} + 3^{2n-1}$ is divisible by 5, $n \geq 1$

37. Show that $2^{n+1} < 3^n$, $n \geq 2$

38. Show that $n^3 - n$ is divisible by 3, $n \geq 1$

39. Show that $3^{3n} + 2^{n+2}$ is divisible by 5, $n \geq 1$

40. Show that $2^{5n+1} + 5^{n+2}$ is divisible by 27, $n \geq 1$

41. Show that $3^{n+2} + 4^{2n+1}$ is divisible by 13, $n \geq 1$

42. Show that $5^{2n-1} + 4^{n+1}$ is divisible by 21, $n \geq 1$

43. Show that $2(7^n) + 3(5^n) - 5$ is divisible by 24, $n \geq 1$

44. Show that $5^{2n} + 3(2^{5n-2})$ is divisible by 7, $n \geq 1$

45. Show that $2^{5n} + 30$ is divisible by 31, $n \geq 1$

46. The *Triangle Inequality* is given by

$$|x + y| \leq |x| + |y|$$

Use mathematical induction to show that the generalized triangle inequality

$$|x_1 + x_2 + \cdots + x_n| \leq |x_1| + |x_2| + \cdots + |x_n|$$

is true for any positive integer n

47. Show by MI that $\dfrac{1}{4 \times 1^2 - 1} + \dfrac{1}{4 \times 2^2 - 1} + \dfrac{1}{4 \times 3^2 - 1} + \cdots + \dfrac{1}{4 \times n^2 - 1} = \dfrac{n}{2n+1}$

48. Show by MI that $1 + \frac{1}{1+2} + \frac{1}{1+2+3} + \cdots + \frac{1}{1+2+3+\cdots+n} = \frac{2n}{n+1}$

49. Show by MI that $(1 + \frac{3}{1}) \times (1 + \frac{5}{4}) \times (1 + \frac{7}{9}) \times \cdots \times (1 + \frac{2n+1}{n^2}) = (n+1)^2$

50. Show by MI that $(1 + \frac{1}{1}) \times (1\frac{1}{2}) \times (1 + \frac{1}{3}) \times \cdots \times (1 + \frac{1}{n}) = (n+1)$

51. Show by MI that $\frac{d^n}{dx^n}(xe^{ax}) = a^{n-1}(ax+n)e^{ax}, n \geq 1$

4.3   Finite Differences Method

In mathematical induction, we used the given statement $S(n)$ to conclude that it is true for all positive integers n, where $n \geq n_0$. The finite differences method that we will introduce in this section is practical, reliable and easy to use. Unlike the mathematical induction, this method works only for statements with two equal sides where the right hand side is a polynomial or a rational expression.

In a manner parallel to the mathematical induction, we can assume that the left hand side is equivalent to a polynomial, or to the ratio of two polynomials, where the coefficients of these polynomials are unknown. The coefficients of each polynomial can be completely determined first by forming the following two steps:

Step 1. In this step, using the left hand side, we list the first few partial summations S_1, S_2, S_3, S_4 in the first row. In the second row we list the difference between every pair of S_1, S_2, S_3, S_4. In the third row, we list the difference between any two consecutive terms of the second row. We continue in this manner, and stop at the row where all outcomes are equal. We then conclude that the summation of the sequence of terms is equivalent to a polynomial with degree equal to $(r-1)$, where r is the number of rows resulted in step 1, or the degree of the polynomial is equal to the number of rows of the finite differences.

Step 2. Having established the degree of the polynomial of the right hand side, 1, 2 or 3 for example, we then assume that the polynomial in the right hand side has one of the following standard forms

$$an + b$$

$$an^2 + bn + c$$

or

$$an^3 + bn^2 + cn + d$$

and so on for polynomials of higher degrees, where a, b, c and d are constants to be determined. In a manner parallel to that used in step 1, we construct a new step by evaluating the assumed polynomial for $n = 1, 2, 3, 4$, and 5 and list the results of evaluations in the first row. The successive rows can be established by finding the difference between each two consecutive terms of the preceding row. It is normal that we stop evaluating these differences when we obtain a row of identical results. We point out here that the number of rows constructed in both steps should be the same. The coefficients $a, b, c, \ldots$ are then determined by equating the corresponding entries at the beginning of each row of the two steps, normally starting at the last row.

In what follows we present the difference tables with the general elements in the first row, followed by the subsequent finite differences:

1. For a first degree polynomial: the standard form of a polynomial of the first degree is expressed by

$$p(n) = an + b$$

Evaluating $p(n)$ for $n = 1, 2, 3$ where we find

$$
\begin{aligned}
p(1) &= a + b \\
p(2) &= 2a + b \\
p(3) &= 3a + b
\end{aligned}
$$

Consequently, **step 2** is constructed in the form

$$a + b \qquad 2a + b \qquad 3a + b \qquad 4a + b$$
$$a \qquad\qquad a \qquad\qquad a$$

2. For a second degree polynomial: the standard form of a quadratic polynomial is expressed by

$$p(n) = an^2 + bn + c$$

Evaluating $p(n)$ for $n = 1, 2, 3$ and 4 where we find

$$
\begin{aligned}
p(1) &= a + b + c \\
p(2) &= 4a + 2b + c \\
p(3) &= 9a + 3b + c \\
p(4) &= 16a + 4b + c
\end{aligned}
$$

Consequently, **step 2** is constructed in the form

$$a+b+c \qquad 4a+2b+c \qquad 9a+3b+c \qquad 16a+4b+c$$
$$3a+b \qquad\qquad 5a+b \qquad\qquad 7a+b$$
$$2a \qquad\qquad 2a$$

3. For a third degree polynomial: the standard form of a polynomial of the third degree is expressed by

$$p(n) = an^3 + bn^2 + cn + d$$

Evaluating $p(n)$ for $n = 1, 2, 3$ and 4 where we find

$$
\begin{aligned}
p(1) &= a+b+c+d \\
p(2) &= 8a+4b+2c+d \\
p(3) &= 27a+9b+3c+d \\
p(4) &= 64a+16b+4c+d
\end{aligned}
$$

We point out that we did not evaluate $p(5)$ because we know that equal results will be obtained at the fourth row, thus we reduce the size of calculations. Consequently, step 2 is constructed in the form

$$a+b+c+d \quad 8a+4b+2c+d \quad 27a+9b+3c+d \quad 64a+16b+4c+d$$
$$7a+3b+c \qquad\qquad 19a+5b+c \qquad\qquad 37a+7b+c$$
$$12a+2b \qquad\qquad 18a+2b$$
$$6a$$

For other polynomials of higher degree, we can construct the finite differences pattern in a similar way. It is interesting to point out that mathematicians consider the finite differences method is a technique to determine the formula for the summation of the sequence of terms and not a proof for the validity of this derived formula.

The finite differences method will be explained by discussing the following illustrative examples.

Example 1. Show by using the finite differences method that

$$1 + 5 + 9 + 13 + 17 + \cdots + (4n - 3) = n(2n - 1)$$

Solution. We first start by evaluating the first few partial sums S_1, S_2, S_3, S_4, ... by using the left hand side, where we obtain

$$
\begin{aligned}
S_1 &= 1 \\
S_2 &= 1 + 5 = 6 \\
S_3 &= 1 + 5 + 9 = 15 \\
S_4 &= 1 + 5 + 9 + 13 = 28
\end{aligned}
$$

Consequently, **step 1** is constructed in the form

$$
\begin{array}{ccccccc}
1 & & 6 & & 15 & & 28 \\
& 5 & & 9 & & 13 & \\
& & 4 & & 4 & &
\end{array}
$$

It is clear that the third row of the this step consists of equal entries 4 and 4. This means that the polynomial that best represents the summation of the left hand side is of the second degree. It is well-known that the standard form of the quadratic polynomial is given by

$$
p(n) = an^2 + bn + c
$$

We next form **step 2** by evaluating $p(n)$ for $n = 1, 2, 3$, and 4 where we find

$$
\begin{aligned}
p(1) &= a + b + c \\
p(2) &= 4a + 2b + c \\
p(3) &= 9a + 3b + c \\
p(4) &= 16a + 4b + c
\end{aligned}
$$

Consequently, **step 2** is constructed in the form

$$
\begin{array}{ccccccc}
a+b+c & & 4a+2b+c & & 9a+3b+c & & 16a+4b+c \\
& 3a+b & & 5a+b & & 7a+b & \\
& & 2a & & 2a & &
\end{array}
$$

As stated before, we equate the corresponding entries at the beginning of each row of both steps, starting with the last row. The coefficients a, b and c are determined by setting

$$
\begin{aligned}
2a &= 4 \\
3a + b &= 5 \\
a + b + c &= 1
\end{aligned}
$$

Accordingly, we find

$$a = 2, \, b = -1, \, c = 0$$

Substituting in the polynomial $p(n)$ given above we obtain

$$p(n) = 2n^2 - n = n(2n - 1) = \text{RHS}$$

Example 2. Show by using the finite differences method that

$$3 + 9 + 15 + 21 + \cdots + (6n - 3) = 3n^2$$

Solution. We first start by evaluating the first few partial sums S_1, S_2, S_3, S_4,... by using the left hand side, where we obtain

$$
\begin{aligned}
S_1 &= 3 \\
S_2 &= 3 + 9 = 12 \\
S_3 &= 3 + 9 + 15 = 27 \\
S_4 &= 3 + 9 + 15 + 21 = 48
\end{aligned}
$$

Consequently, step 1 is constructed in the form

$$
\begin{array}{ccccccc}
3 & & 12 & & 27 & & 48 \\
& 9 & & 15 & & 21 & \\
& & 6 & & 6 & &
\end{array}
$$

It is clear that the third row consists of equal entries 6 and 6. This means that the polynomial that best represents the summation of the left hand side is of the second degree. It is well-known that the standard form of the quadratic polynomial is of the form

$$p(n) = an^2 + bn + c$$

We next form step 2 by evaluating $p(n)$ for $n = 1, 2, 3$, and 4 where we find

$$
\begin{aligned}
p(1) &= a + b + c \\
p(2) &= 4a + 2b + c \\
p(3) &= 9a + 3b + c \\
p(4) &= 16a + 4b + c
\end{aligned}
$$

Consequently, step 2 is constructed in the form

$$a + b + c \qquad 4a + 2b + c \qquad 9a + 3b + c \qquad 16a + 4b + c$$
$$3a + b \qquad 5a + b \qquad 7a + b$$
$$2a \qquad 2a$$

As stated before, we equate the corresponding entries at the beginning of each row of both steps, starting with the last row. The coefficients a, b and c are determined by setting

$$2a = 6$$
$$3a + b = 9$$
$$a + b + c = 3$$

Accordingly, we find

$$a = 3, \, b = 0, \, c = 0$$

Substituting in the polynomial $p(n)$ we obtain

$$p(n) = 3n^2 = \text{RHS}$$

Example 3. Show by using the finite differences method that

$$1 \times 2 + 2 \times 3 + 3 \times 4 + 4 \times 5 + \cdots + n(n+1) = \frac{n(n+1)(n+2)}{3}$$

Solution. We first start by evaluating the first few partial sums S_1, S_2, S_3, S_4, ... by using the left hand side, where we obtain

$$S_1 = 2$$
$$S_2 = 2 + 6 = 8$$
$$S_3 = 2 + 6 + 12 = 20$$
$$S_4 = 2 + 6 + 12 + 20 = 40$$
$$S_5 = 2 + 6 + 12 + 20 + 30 = 70$$

Consequently, step 1 is constructed in the form

$$2 \qquad 8 \qquad 20 \qquad 40 \qquad 70$$
$$6 \qquad 12 \qquad 20 \qquad 30$$
$$6 \qquad 8 \qquad 10$$
$$2 \qquad 2$$

It is clear that the fourth row of the configuration consists of equal results. This means that the polynomial that best represents the summation of the left hand side is of the third degree. It is well-known that the standard form of this polynomial is given by

$$p(n) = an^3 + bn^2 + cn + d$$

We next form step 2 by evaluating $p(n)$ for $n = 1, 2, 3, 4$ where we find

$$
\begin{aligned}
p(1) &= a + b + c + d \\
p(2) &= 8a + 4b + 2c + d \\
p(3) &= 27a + 9b + 3c + d \\
p(4) &= 64a + 16b + 4c + d
\end{aligned}
$$

We point out that we did not evaluate $p(5)$ because we know that equal results will be obtained at the fourth row, thus we reduce the size of calculations. Consequently, step 2 is constructed in the form

$$
\begin{array}{cccc}
a + b + c + d & 8a + 4b + 2c + d & 27a + 9b + 3c + d & 64a + 16b + 4c + d \\
& 7a + 3b + c & 19a + 5b + c & 37a + 7b + c \\
& & 12a + 2b & 18a + 2b \\
& & 6a &
\end{array}
$$

As stated before, we equate the corresponding entries at the beginning of each row of both configurations, starting with the last row. The coefficients a, b, c and d are determined by setting

$$
\begin{aligned}
6a &= 2 \\
12a + 2b &= 6 \\
7a + 3b + c &= 6 \\
a + b + c + d &= 2
\end{aligned}
$$

Accordingly, we find

$$a = \frac{1}{3}, \; b = 1, \; c = \frac{2}{3}, \; d = 0$$

Substituting in the polynomial $p(n)$ we obtain

$$
\begin{aligned}
p(n) &= \frac{1}{3}n^3 + n^2 + \frac{2}{3}n \\
&= \frac{1}{3}n(n^2 + 3n + 2) = \frac{1}{3}n(n+1)(n+2) = \text{RHS}
\end{aligned}
$$

It should be noted that the finite differences method can easily provide the polynomial that represents any given summation. This can be clearly observed from step 1, where we can determine the degree of the polynomial by noting the number of rows r of step 1, hence the degree is $(r-1)$ as mentioned before. The polynomial then can be determined by using step 2. This will be illustrated by the following example.

Example 4. Use the idea discussed above to discover a formula for the following summation:

$$1 \times 7 + 2 \times 8 + 3 \times 9 + 4 \times 10 + \cdots + n(n+6)$$

Solution. We first start by evaluating the first few partial sums S_1, S_2, S_3, S_4,... by using the summation given above, where we obtain

$$
\begin{aligned}
S_1 &= 7 \\
S_2 &= 7 + 16 = 23 \\
S_3 &= 7 + 16 + 27 = 50 \\
S_4 &= 7 + 16 + 27 + 40 = 90 \\
S_5 &= 7 + 16 + 27 + 40 + 55 = 145
\end{aligned}
$$

Consequently, step 1 is constructed in the form

$$
\begin{array}{ccccccccc}
7 & & 23 & & 50 & & 90 & & 145 \\
& 16 & & 27 & & 40 & & 55 & \\
& & 11 & & 13 & & 15 & & \\
& & & 2 & & 2 & & &
\end{array}
$$

It is clear that the fourth row consists of equal entries. This means that the polynomial that best represents the summation is of the third degree. It is well-known that the standard form of this polynomial is given by

$$p(n) = an^3 + bn^2 + cn + d$$

We next form step 2 by evaluating $p(n)$ for $n = 1, 2, 3$, and 4 where we find

$$
\begin{aligned}
p(1) &= a + b + c + d \\
p(2) &= 8a + 4b + 2c + d \\
p(3) &= 27a + 9b + 3c + d \\
p(4) &= 64a + 16b + 4c + d
\end{aligned}
$$

Again, we did not evaluate $p(5)$ because we know that equal results will be obtained at the fourth row, thus we reduce the size of calculations. Consequently, step 2 is constructed in the form

$$a+b+c+d \quad 8a+4b+2c+d \quad 27a+9b+3c+d \quad 64a+16b+4c+d$$
$$7a+3b+c \qquad\qquad 19a+5b+c \qquad\qquad 37a+7b+c$$
$$12a+2b \qquad\qquad\qquad 18a+2b$$
$$6a$$

As stated above, we equate the corresponding entries at the beginning of each row of both steps, starting with the last row. The coefficients a, b, c and d are determined by setting

$$
\begin{aligned}
6a &= 2 \\
12a + 2b &= 11 \\
7a + 3b + c &= 16 \\
a + b + c + d &= 7
\end{aligned}
$$

This gives

$$a = \frac{1}{3},\ b = \frac{7}{2},\ c = \frac{19}{6},\ d = 0$$

Substituting in the polynomial $p(n)$ we obtain

$$p(n) = \frac{1}{3}n^3 + \frac{7}{2}n^2 + \frac{19}{6}n = \frac{1}{6}n(n+1)(2n+19)$$

Rational Expressions

For the type of problems that involve rational expressions, we compute the partial sums S_1, S_2, S_3, S_4 in the same manner used before. We then construct two pairs of steps, the first for the resulting numerators of the partial sums, and the second for the resulting denominators. The procedure is identical to that used before, except we have to handle the resulting numerators and the denominators of the partial sums separately. In this case we should determine two polynomials $p(n)$ and $q(n)$ that represent the numerator and the denominator respectively. Since the procedure is the same as discussed before, we therefore skip details.

The technique will be explained by discussing the following example.

Example 5. Show by using the finite differences method that

$$\frac{1}{1 \times 2} + \frac{1}{2 \times 3} + \frac{1}{3 \times 4} + \cdots + \frac{1}{n(n+1)} = \frac{n}{n+1}$$

Solution. We first start by evaluating the first few partial sums $S_1, S_2, S_3, S_4,...$ by using the left hand side, where we obtain

$$S_1 = \frac{1}{2}$$

$$S_2 = \frac{1}{2} + \frac{1}{6} = \frac{2}{3}$$

$$S_3 = \frac{1}{2} + \frac{1}{6} + \frac{1}{12} = \frac{3}{4}$$

$$S_4 = \frac{1}{2} + \frac{1}{6} + \frac{1}{12} + \frac{1}{20} = \frac{4}{5}$$

Consequently, the partial sums S_1, S_2, S_3, S_4 are $\frac{1}{2}, \frac{2}{3}, \frac{3}{4}, \frac{4}{5}$. As stated above, we handle the numerators 1,2,3 and 4 and the denominators 2, 3, 4 and 5 separately to find a polynomial that represents each.

Concerning the partial sums of the numerators, we set

$$
\begin{array}{cccccccc}
1 & & 2 & & 3 & & 4 \\
 & 1 & & 1 & & 1 &
\end{array}
$$

It is clear that we obtained fixed differences after one row. This means that the polynomial that represents the numerator is of first degree. Using the pattern presented before

$$
\begin{array}{cccccccc}
a+b & & 2a+b & & 3a+b & & 4a+b \\
 & a & & a & & a &
\end{array}
$$

We then solve

$$a = 1, a + b = 1$$

This gives $a = 1, b = 0$. Therefore, the polynomial that represents the numerators is

$$p(n) = n$$

Concerning the partial sums of the denominators, we set

$$
\begin{array}{cccccccc}
2 & & 3 & & 4 & & 5 \\
 & 1 & & 1 & & 1 &
\end{array}
$$

It is clear that we obtained fixed differences after one row. This means that the polynomial that represents the numerator is of first degree. Using the pattern presented before

$$a + b \qquad 2a + b \qquad 3a + b \qquad 4a + b$$
$$\qquad a \qquad\qquad a \qquad\qquad a$$

We then solve

$$a = 1, a + b = 2$$

This gives $a = 1, b = 1$. Therefore, the polynomial that represents the numerators is

$$q(n) = n + 1$$

Accordingly, the left hand side is given by

$$\text{LHS} = \frac{p(n)}{q(n)} = \frac{n}{n+1} = \text{RHS}$$

4.3.1   An Alternative Approach

An alternative approach for determining the formula of the summation of any sequence of terms is also used. This approach is not applicable for the statement of the summation of fractions. We first add 0 as the first term of any sequence of terms as presented above. We then create the finite differences pattern exactly in the same way used before, by considering the zero as the first partial sum that corresponds to $n = 0$. Call the first entry of the first row by S_0, the first entry of the second row, which is the first entry of the first row of finite differences, by D_1, the first entry of the third row, which is the first entry of the second row of finite differences, by D_2, the first entry of the fourth row of the finite differences by D_3, and so on. A polynomial formula $f(n)$ can be obtained by using the rule

$$\begin{aligned} f(n) \quad = \quad & S_0 \cdot 1 + D_1 \cdot n + D_2 \cdot \frac{n(n-1)}{2} + D_3 \cdot \frac{n(n-1)(n-2)}{6} \\ & + D_4 \cdot \frac{n(n-1)(n-2)(n-3)}{24} + \cdots \end{aligned}$$

This rule can be formulated by

$$f(n) = S_0 \binom{n}{0} + D_1 \binom{n}{1} + D_2 \binom{n}{2} + D_3 \binom{n}{3} + D_4 \binom{n}{4} + \cdots$$

Notice that the combination

$$\binom{n}{r} = \frac{n!}{r!(n-r)!}$$

will be studied in a coming chapter. The alternative method to determine a formula for the summation will be explained by the following illustrative examples.

Example 6. Use the alternative approach of the finite differences method to find a formula for the summation

$$1 + 3 + 5 + 7 + \cdots + (2n - 1)$$

Solution. We first start by evaluating the first few partial sums S_0, S_1, S_2, S_3, noting that $S_0 = 0$, where we obtain

$$S_0 = 0, \quad S_1 = 1, \quad S_2 = 4, \quad S_3 = 9, \quad S_4 = 16$$

Consequently, step 1 is constructed in the form

$$
\begin{array}{ccccccc}
0 & & 1 & & 4 & & 9 \\
& 1 & & 3 & & 5 & \\
& & 2 & & 2 & &
\end{array}
$$

It is clear $S_0 = 0, D_1 = 1, D_2 = 2$. We substitute these values in the formula

$$f(n) = S_0 \cdot 1 + D_1 \cdot n + D_2 \cdot \frac{n(n-1)}{2}$$

Therefore we obtain

$$f(n) = 1 \cdot n + 2 \cdot \frac{n(n-1)}{2} = n^2$$

Example 7. Use the alternative approach of the finite differences method to find a formula for the summation

$$1 \times 2 + 2 \times 3 + 3 \times 4 + 4 \times 5 + \cdots + n(n+1)$$

Solution. We first start by evaluating the first few partial sums S_0, S_1, S_2, S_3, noting that $S_0 = 0$, where we obtain

$$S_0 = 0, \quad S_1 = 2, \quad S_2 = 8, \quad S_3 = 20, \quad S_4 = 40$$

Consequently, step 1 is constructed in the form

$$
\begin{array}{ccccccccc}
0 & & 2 & & 8 & & 20 & & 40 \\
& 2 & & 6 & & 12 & & 20 & \\
& & 4 & & 6 & & 8 & & \\
& & & 2 & & 2 & & &
\end{array}
$$

It is clear $S_0 = 0, D_1 = 2, D_2 = 4, D_3 = 2$. We substitute these values in the formula

$$
f(n) = S_0 \cdot 1 + D_1 \cdot n + D_2 \cdot \frac{n(n-1)}{2} + D_3 \frac{n(n-1)(n-2)}{6}
$$

Therefore we obtain

$$
f(n) = 0 \times 1 + 2 \times n + 4 \times \frac{n(n-1)}{2} + 2\frac{n(n-1)(n-2)}{6} = \frac{n(n+1)(n+2)}{3}
$$

Exercises 4.3

In exercises 1–14, use the finite differences method to prove the following statements:

1. $1 + 7 + 13 + \cdots + (6n - 5) = n(3n - 2)$

2. $2 + 7 + 12 + 17 + \cdots + (5n - 3) = \frac{1}{2}n(5n - 1)$

3. $-6 + 1 + 8 + 15 + \cdots + (7n - 13) = \frac{1}{2}n(7n - 19)$

4. $1^2 + 2^2 + 3^2 + 4^2 + \cdots + n^2 = \frac{1}{6}n(n + 1)(2n + 1)$

5. $\dfrac{1}{1 \times 6} + \dfrac{1}{6 \times 11} + \dfrac{1}{11 \times 16} + \cdots + \dfrac{1}{(5n - 4)(5n + 1)} = \dfrac{n}{5n + 1}$

6. $\dfrac{1}{1 \times 13} + \dfrac{1}{13 \times 25} + \cdots + \dfrac{1}{(12n - 11)(12n + 1)} = \dfrac{n}{12n + 1}$

7. $\dfrac{1}{2 \times 5} + \dfrac{1}{5 \times 8} + \dfrac{1}{8 \times 11} + \cdots + \dfrac{1}{(3n - 1)(3n + 2)} = \dfrac{n}{2(3n + 2)}$

8. $\dfrac{1}{3 \times 5} + \dfrac{1}{5 \times 7} + \dfrac{1}{7 \times 9} + \cdots + \dfrac{1}{(2n + 1)(2n + 3)} = \dfrac{n}{3(2n + 3)}$

9. $\dfrac{1}{1 \times 4} + \dfrac{1}{4 \times 7} + \dfrac{1}{7 \times 10} + \cdots + \dfrac{1}{(3n - 2)(3n + 1)} = \dfrac{n}{(3n + 1)}$

10. $\dfrac{1}{1 \times 8} + \dfrac{1}{8 \times 15} + \dfrac{1}{15 \times 22} + \cdots + \dfrac{1}{(7n - 6)(7n + 1)} = \dfrac{n}{(7n + 1)}$

11. $\dfrac{1}{1 \times 9} + \dfrac{1}{9 \times 17} + \dfrac{1}{17 \times 25} + \cdots + \dfrac{1}{(8n-7)(8n+1)} = \dfrac{n}{(8n+1)}$

12. $\dfrac{1}{2 \times 9} + \dfrac{1}{9 \times 16} + \dfrac{1}{16 \times 23} + \cdots + \dfrac{1}{(7n-5)(7n+2)} = \dfrac{n}{2(7n+2)}$

13. $\dfrac{1}{4 \times 7} + \dfrac{1}{7 \times 10} + \dfrac{1}{10 \times 13} + \cdots + \dfrac{1}{(3n+1)(3n+4)} = \dfrac{n}{4(3n+4)}$

14. $\dfrac{1}{5 \times 11} + \dfrac{1}{11 \times 17} + \dfrac{1}{17 \times 23} + \cdots + \dfrac{1}{(6n-1)(6n+5)} = \dfrac{n}{5(6n+5)}$

In Exercises 15 – 20, determine a formula that gives the sum of each of the following expressions:

15. $2 + 11 + 20 + 29 + \cdots + (9n - 7)$

16. $3 + 14 + 25 + 36 + \cdots + (11n - 8)$

17. $1 \times 8 + 2 \times 9 + 3 \times 10 + \cdots + n(n+7)$

18. $1 \times 9 + 2 \times 10 + 3 \times 11 + \cdots + n(n+8)$

19. $1 \times 2 \times 3 + 2 \times 3 \times 4 + 3 \times 4 \times 5 + \cdots + n(n+1)(n+2)$

20. $1 \times 3 \times 5 + 2 \times 4 \times 6 + 3 \times 5 \times 7 + \cdots + n(n+2)(n+4)$

21. $1 \times 2 \times 4 + 2 \times 3 \times 6 + 3 \times 4 \times 8 + \cdots + n(n+1)(2n+2)$

22. $1 \times 3 \times 4 + 2 \times 4 \times 5 + 3 \times 5 \times 6 + \cdots + n(n+2)(n+3)$

4.4   The Partial Fractions Method

The partial fractions method works successfully only for statements where rational expressions are involved. This method provides a formula for the summation of a sequences of fractions. This is an easy technique that depends on the so called partial fractions. The idea of the partial fractions depends on expressing any rational expression as the sum of two other partial fractions. This can be explained as follows:

To express the rational expression

$$\frac{1}{n(n+1)}$$

as a sum of two partial fractions, we first write

$$\frac{1}{n(n+1)} = \frac{a}{n} + \frac{b}{n+1}$$

This is true for specific values of the constants a and b. To determine a and b, we note that decomposing the fraction into the sum of two partial fractions is true if we set

$$a(n+1) + bn = 1$$

Setting $n = 0$ gives

$$a = 1$$

and by setting $n = -1$ gives

$$b = -1$$

This means that

$$\frac{1}{n(n+1)} = \frac{1}{n} - \frac{1}{n+1}$$

We point out here that we discussed the simplest case of partial fractions that will cover our need in this course. Other cases will be addressed in calculus courses. Having explained the technique of expressing any fraction into the sum of two partial fractions, we turn to our concern to prove the validity of the statement given in the following example.

Example 1. Show by using the partial fractions method that

$$\frac{1}{1 \times 2} + \frac{1}{2 \times 3} + \frac{1}{3 \times 4} + \cdots + \frac{1}{n(n+1)} = \frac{n}{n+1},$$

is true for all positive integers n, $n \geq 1$.

Solution. To prove that the statement in this example is true, we first express the general term as a sum of two partial fractions. In our example, the general term is given by

$$\frac{1}{n(n+1)}$$

and we have discussed expressing it into the sum of partial fractions where we found

$$\frac{1}{n(n+1)} = \frac{1}{n} - \frac{1}{n+1}$$

Accordingly, each fraction in the statement can be expressed in this way, hence we find

$$\frac{1}{1 \times 2} = \frac{1}{1} - \frac{1}{2}$$

$$\frac{1}{2 \times 3} = \frac{1}{2} - \frac{1}{3}$$

$$\frac{1}{3 \times 4} = \frac{1}{3} - \frac{1}{4}$$

$$\vdots$$

Consequently, the left hand side of the statement is given by

$$\text{LHS} = \frac{1}{1} - \frac{1}{2} + \frac{1}{2} - \frac{1}{3} + \frac{1}{3} - \frac{1}{4} + \cdots + \frac{1}{n} - \frac{1}{n+1}$$

$$= 1 - \frac{1}{n+1} = \frac{n}{n+1} = \text{RHS}$$

It is clear that the first and the last terms of the fractions are kept, whereas the other interior terms are canceled

Example 2. Use the partial fractions method to show that

$$\frac{1}{1 \times 5} + \frac{1}{5 \times 9} + \frac{1}{9 \times 13} + \cdots + \frac{1}{(4n-3)(4n+1)} = \frac{n}{4n+1}$$

Solution. We first express each fraction as the sum of two other partial fractions. Considering the general term at the left hand side, we set

$$\frac{1}{(4n-3)(4n+1)} = \frac{a}{4n-3} + \frac{b}{4n+1}$$

This is true for specific values of the constants a and b. To determine a and b, we note that decomposing the fraction into the sum of two partial fractions is true if we set

$$a(4n+1) + b(4n-3) = 1$$

Setting $n = \frac{3}{4}$ gives

$$a = \frac{1}{4}$$

and by setting $n = -\frac{1}{4}$ gives

$$b = -\frac{1}{4}$$

This means that

$$\frac{1}{(4n-3)(4n+1)} = \frac{1}{4(4n-3)} - \frac{1}{4(4n+1)}$$

This will enable us to decompose each fraction in the left hand side into the sum of two partial fractions, therefore we find

$$\frac{1}{1 \times 5} = \frac{1}{4 \times 1} - \frac{1}{4 \times 5}$$
$$\frac{1}{5 \times 9} = \frac{1}{4 \times 5} - \frac{1}{4 \times 9}$$
$$\vdots$$

Consequently, the left hand side of the statement is given by

$$
\begin{aligned}
\text{LHS} &= \frac{1}{4 \times 1} - \frac{1}{4 \times 5} + \frac{1}{4 \times 5} - \frac{1}{4 \times 9} + \cdots + \frac{1}{4(4n-3)} - \frac{1}{4(4n+1)} \\
&= \frac{1}{4 \times 1} - \frac{1}{4(4n+1)} = \frac{n}{4n+1} = \text{RHS}
\end{aligned}
$$

Example 3. Use the partial fractions method to discover a formula that gives the sum of the expression:

$$S = \frac{1}{2 \times 7} + \frac{1}{7 \times 12} + \frac{1}{12 \times 17} + \cdots + \frac{1}{(5n-3)(5n+2)}$$

Solution. To determine a formula that gives the sum of the given expression, we consider the general term at the left hand side, therefore we set

$$\frac{1}{(5n-3)(5n+2)} = \frac{a}{5n-3} + \frac{b}{5n+2}$$

This is true for specific values of the constants a and b. To determine a and b, we note that decomposing the fraction into the sum of two partial fractions is true if we set

$$a(5n+2) + b(5n-3) = 1$$

Setting $n = \frac{3}{5}$ gives

$$a = \frac{1}{5}$$

and by setting $n = -\frac{2}{5}$ gives

$$b = -\frac{1}{5}$$

This means that

$$\frac{1}{(5n-3)(5n+2)} = \frac{1}{5(5n-3)} - \frac{1}{5(5n+2)}$$

Accordingly, each fraction in the expression can be decomposed into the sum of two partial fractions, therefore we find

$$\frac{1}{2 \times 7} = \frac{1}{5 \times 2} - \frac{1}{5 \times 7}$$

$$\frac{1}{7 \times 12} = \frac{1}{5 \times 7} - \frac{1}{5 \times 12}$$

$$\frac{1}{12 \times 17} = \frac{1}{5 \times 12} - \frac{1}{5 \times 17}$$

$$\vdots$$

Consequently, the sum of the expression S can be written as

$$
\begin{aligned}
S &= \frac{1}{5 \times 2} - \frac{1}{5 \times 7} + \frac{1}{5 \times 7} - \frac{1}{5 \times 12} + \cdots + \frac{1}{5(5n-3)} - \frac{1}{5(5n+2)} \\
&= \frac{1}{5 \times 2} - \frac{1}{5(5n+2)} \\
&= \frac{n}{2(5n+3)} = \text{RHS}
\end{aligned}
$$

Exercises 4.4

In Exercises 1 – 6, use partial fractions method to prove the following statements:

1. $\dfrac{1}{1 \times 6} + \dfrac{1}{6 \times 11} + \dfrac{1}{11 \times 16} + \cdots + \dfrac{1}{(5n-4)(5n+1)} = \dfrac{n}{5n+1}$

2. $\dfrac{1}{1 \times 13} + \dfrac{1}{13 \times 25} + \cdots + \dfrac{1}{(12n-11)(12n+1)} = \dfrac{n}{12n+1}$

3. $\dfrac{1}{2 \times 5} + \dfrac{1}{5 \times 8} + \dfrac{1}{8 \times 11} + \cdots + \dfrac{1}{(3n-1)(3n+2)} = \dfrac{n}{2(3n+2)}$

4. $\dfrac{1}{3 \times 5} + \dfrac{1}{5 \times 7} + \dfrac{1}{7 \times 9} + \cdots + \dfrac{1}{(2n+1)(2n+3)} = \dfrac{n}{3(2n+3)}$

5. $\dfrac{1}{1 \times 4} + \dfrac{1}{4 \times 7} + \dfrac{1}{7 \times 10} + \cdots + \dfrac{1}{(3n-2)(3n+1)} = \dfrac{n}{3n+1}$

6. $\dfrac{1}{1 \times 8} + \dfrac{1}{8 \times 15} + \dfrac{1}{15 \times 22} + \cdots + \dfrac{1}{(7n-6)(7n+1)} = \dfrac{n}{7n+1}$

In Exercises 7 – 12, determine a formula that gives the sum of each of the following expressions:

7. $\dfrac{1}{1 \times 9} + \dfrac{1}{9 \times 17} + \dfrac{1}{17 \times 25} + \cdots + \dfrac{1}{(8n-7)(8n+1)}$

8. $\dfrac{1}{2 \times 9} + \dfrac{1}{9 \times 16} + \dfrac{1}{16 \times 23} + \cdots + \dfrac{1}{(7n-5)(7n+2)}$

9. $\dfrac{1}{4 \times 7} + \dfrac{1}{7 \times 10} + \dfrac{1}{10 \times 13} + \cdots + \dfrac{1}{(3n+1)(3n+4)}$

10. $\dfrac{1}{5 \times 11} + \dfrac{1}{11 \times 17} + \dfrac{1}{17 \times 23} + \cdots + \dfrac{1}{(6n-1)(6n+5)}$

11. $\dfrac{1}{5 \times 9} + \dfrac{1}{9 \times 13} + \dfrac{1}{13 \times 17} + \cdots + \dfrac{1}{(4n+1)(4n+5)}$

12. $\dfrac{1}{5 \times 8} + \dfrac{1}{8 \times 11} + \dfrac{1}{11 \times 14} + \cdots + \dfrac{1}{(3n+2)(3n+5)}$

4.5   Direct Proof Method

In some special cases, a mathematical fact is given in the form of a true statement P. It is required to show that another statement Q is true. This is usually done by using the fact that statement P is true and by using mathematical axioms, identities, theorems and other related relationships that we studied before in other mathematics courses.

It seems reasonable before illustrating the idea of **direct proof** to review some of the well known facts from number theory that will be used frequently in this section. An **even** number n is defined as a number that can be represented by

$$n = 2k, \; k = 0, 1, 2, 3, 4, \cdots$$

On the other hand, an **odd** number n is defined as a number that can be represented by

$$n = 2r + 1, \; r = 0, 1, 2, 3, 4, \cdots$$

In the following examples we will show how we can use the given true statement to prove another one.

Example 1. If n is odd, show that $n^2 + 2$ is an odd number.

Solution. Using the given assumption that n is odd, we therefore set

$$n = 2r + 1, \ r = 0, 1, 2, 3, \cdots$$

Accordingly, we substitute $n = 2r + 1$ in the other statement to obtain

$$
\begin{aligned}
n^2 + 2 &= (2r + 1)^2 + 2 \\
&= 4r^2 + 4r + 1 + 2 \\
&= (4r^2 + 4r + 2) + 1, \text{ by separating the 1,} \\
&= 2(2r^2 + 2r + 1) + 1 \\
&= 2m + 1
\end{aligned}
$$

and this is a representation of an odd number noting that

$$m = 2r^2 + 2r + 1$$

Example 2. If n is odd, show that $n^2 + n$ is an even number.

Solution. Using the given assumption that n is odd, we therefore set

$$n = 2r + 1, \ r = 0, 1, 2, 3, \cdots$$

Accordingly, we substitute $n = 2r + 1$ in the other statement to obtain

$$
\begin{aligned}
n^2 + n &= (2r + 1)^2 + (2r + 1) \\
&= 4r^2 + 4r + 1 + 2r + 1 \\
&= 4r^2 + 6r + 2 \\
&= 2(2r^2 + 3r + 1) \\
&= 2s
\end{aligned}
$$

and this is a representation of an even number noting that

$$s = 2r^2 + 3r + 1$$

Example 3. If $n, n \geq 1$ is odd, show that $3n^3 + 4n^2 + 8n - 18$ is an odd number.

Solution. Using the given assumption that n is odd, we therefore set

$$n = 2r + 1, \; r = 0, 1, 2, 3, \cdots$$

Accordingly, we substitute $n = 2r + 1$ in the other statement to obtain

$$
\begin{aligned}
3n^3 + 4n^2 + 8n - 18 &= 3(2r+1)^3 + 4(2r+1)^2 + 8(2r+1) - 18 \\
&= 24r^3 + 52r^2 + 50r - 3 \\
&= (24r^3 + 52r^2 + 50r - 4) + 1 \\
&= 2(12r^3 + 26r^2 + 25r - 2) + 1 \\
&= 2s + 1
\end{aligned}
$$

and this is a representation of an odd number noting that

$$s = 12r^3 + 26r^2 + 25r - 2$$

Example 4. If n is even, $n \geq 2$, show that $n^2 + 2n - 3$ is an odd number .

Solution. Using the given assumption that n is even, we therefore set

$$n = 2k, \; k = 1, 2, 3, \cdots$$

Accordingly, we substitute $n = 2k$ in the other statement to obtain

$$
\begin{aligned}
n^2 + 2n - 3 &= (2k)^2 + 2(2k) - 3 \\
&= 4k^2 + 4k - 3, \\
&= (4k^2 + 4k - 4) + 1, \; \text{by using} \; -3 = -4 + 1 \\
&= 2(2k^2 + 2k - 2) + 1 \\
&= 2s + 1
\end{aligned}
$$

and this is a representation of an odd number noting that

$$s = 2k^2 + 2k - 2$$

Example 5. Given the congruence

$$4x \equiv 2 \bmod 5$$

show that

$$3x \equiv 4 \bmod 5$$

Solution. Solving the given inequality, we obtain

$$x \equiv 3 \bmod 5$$

and by multiplying both sides of the last result by 3 we find

$$3x \equiv 9 \bmod 5$$

so that

$$3x \equiv 4 \bmod 5$$

Exercises 4.5

Use the direct proof method in the following exercises:

1. If $x > 6$ and $y > 8$ show that $x + y > 10$

2. If $0 < x < 2$ and $0 < y < 8$ show that $x^2 + y^2 < 74$

3. If n is odd, show that n^2 is odd

4. If n is even, show that $n^2 + 4n + 3$ is odd

5. If n is odd, show that $n^2 + n + 2$ is even

6. If n is divisible by 5, show that $n^2 + 5$ is divisible by 5

7. If n is odd, show that $n^3 + n$ is even

8. If n is even, show that $n^3 + n + 1$ is odd

9. If n is even, show that $n^3 - n$ is even

10. If n is odd and m is even, show that $n^2 + m^2$ is odd

11. If n is odd , m is odd, $n > m$ show that $n^2 - m^2$ is even

12. If n is odd , m is odd, $n > m > 1$ show that $n^3 - m^3$ is even

13. If $3x \equiv 1 \bmod 5$ show that $4x \equiv 3 \bmod 5$

14. If n is any positive integer, show that $n^2 + n$ is an even integer

15. Given the equation $x^3 - x^2 + x - 1 = 0$. Show that $x = 1$ is the only real root of this equation

16. Given that r and s are rational numbers. Show that $r + s$ is a rational number

17. If n is even, show that $5n^3 + 6n - 11$ is odd

18. If n is odd, show that $5n^2 + n - 12, n > 2$ is even

19. Use congruences to show that $n^5 - n$ is divisible by 5

20. Use congruences to show that $3^{2n} - 1$ is divisible by 8

21. For all integers $a, b,$ and c, if $a|b$, and $a|c$ show that $a|(b + c)$

22. For all integers $a, b,$ and c, if $a|b$, and $b|c$ show that $a|c$

23. For all integers $a, b,$ and c, if $a + b$ is even, and $b + c$ is even, show that $a + c$ is even

24. For all integers $a, b,$ and c, if $a + b$ is odd, and $b + c$ is odd, show that $a + c$ is even

4.6   Proof by Contradiction

In this section we will discuss another method of proof called **proof by contradiction**. In the preceding section, a true statement P is given where this fact was used to prove another statement Q and hence a conclusion is to be shown. This was done by approaching the problem in a direct way depending on the given true statements and by employing other related mathematical concepts.

In this section, we approach the problem in an indirect way called proof by contradiction. This means that we assume first that the conclusion Q is not true. This can be done by contradicting the given statement Q. Proceeding from this contradictory assumption, and by using other related mathematical concepts we derive that the given statement is not true. This deduction that P is not true contradicts the given fact that P is true. This will immediately indicate that our contradictory assumption is false, hence Q should be true. The concept of the proof by contradiction will be illustrated by the following examples.

Example 1. If $x < 6$, show by contradiction that

$$x + 3 < 9$$

Solution. Assume that

$$x + 3 \nless 9$$

This means that
$$x + 3 \geq 9$$

Solving the last inequality we find
$$x \geq 6$$

But this contradicts the given statement that $x < 6$. Accordingly, the contradictory assumption we made is false. Therefore,
$$x + 3 < 9$$

Example 2. If x is a positive real number, show by contradiction that
$$\frac{x}{x+2} < \frac{x+2}{x+4}$$

Solution. Assume that
$$\frac{x}{x+2} \not< \frac{x+2}{x+4}$$

This means that
$$\frac{x}{x+2} \geq \frac{x+2}{x+4}$$

Because x is positive, then $(x+2)$ and $(x+4)$ are positive. Therefore multiplying both sides of the above inequality by $(x+2)(x+4)$ we get
$$x(x+4) \geq (x+2)(x+2)$$

This gives
$$x^2 + 4x \geq x^2 + 4x + 4$$

This leads to the result
$$0 \geq 4$$

But this contradicts the fact that $0 < 4$. Accordingly, the contradictory assumption we made is false. Therefore,
$$\frac{x}{x+2} < \frac{x+2}{x+4}$$

Example 3. If r is a rational number and s is irrational number, prove by contradiction that the difference $r - s$ is irrational.

Solution. Assume that $r - s$ is rational. This means that if we set

$$r = \frac{a}{b}$$

so that

$$\frac{a}{b} - s = \frac{m}{n}$$

where a, b, m, and n are integers, $b \neq 0$ and $n \neq 0$. This means that

$$s = \frac{a}{b} - \frac{m}{n}$$

or equivalently

$$s = \frac{an - mb}{bn}$$

Note that $an - mb$ and bn are integers and $bn \neq 0$. The last expression indicates that s is a rational number. This contradicts the given statement that s is irrational. This implies that the contradictory assumption we made is false. Therefore, the difference $r - s$ is irrational.

Exercises 4.6

Use the method of proof by contradiction to prove each of the following exercises:

1. If $x < 4$ show that $x + 6 < 10$

2. If $x \geq 6$ show that $x + 4 \geq 10$

3. If $|x| = 4$ show that $|x + 3| \leq 7$

4. If $|x| = 4$ show that $|x - 3| \leq 7$

5. If $x \equiv 1 \bmod 2$ show that x is an odd number

6. If $x \equiv 0 \bmod 2$ show that x is an even number

7. If r is a nonzero rational number and s is any irrational number. Show that the sum $r + s$ is irrational

8. If r is a nonzero rational number and s is any irrational number. Show that the product rs is irrational

9. If r and s are nonzero rational numbers. Show that $\frac{r}{s}$ is irrational

10.   Show that $\frac{1}{5}\sqrt{2} + 7$ is an irrational number

11. If x is a positive real number, show that $\frac{x}{x+1} < \frac{x+1}{x+2}$

Chapter 5

Set Theory

5.1   Sets

A **set** is a well-defined collection of objects called the elements of the set. A set is commonly denoted by capital letters such as A, B, and C. The elements of a set are usually listed inside the curled braces { }. The order in which the elements of a set might be listed is not important. For example, the set

$$A = \{1, 3, 5, 7\}$$

may also be expressed by

$$A = \{3, 7, 1, 5\}$$

There are commonly two different ways for specifying sets, namely:

1. The Roster Form (the list form): The set in this case is described by listing the distinct elements of the set inside the curled braces. This is mostly used if the number of elements of a set is finite and not large. The following examples illustrate the **roster form** of a set.

Example 1. Write the sets that characterize the following descriptions:

(a) Colors of the American flag.

(b) Odd positive integers ≤ 9.

(c) Prime numbers ≤ 11.

(d) Roots of the equation $x^2 - 7x + 6 = 0$.

(e) Outcomes of rolling one die.

Solution.

(a) It is well-known that red, blue and white are the colors of the American flag, therefore we write the set by

$$A = \{\text{red, blue, white}\}$$

(b) It is required to list the first five odd integers, hence we find the set

$$B = \{1, 3, 5, 7, 9\}$$

(c) As discussed in Chapter 2, the first five primes are given by the set

$$C = \{2, 3, 5, 7, 11\}$$

(d) The equation
$$x^2 - 7x + 6 = 0$$

can be written as
$$(x-1)(x-6) = 0$$

so that the set of solutions of the given equation is given by

$$D = \{1, 6\}$$

(e) A standard die has six faces marked $1, 2, 3, 4, 5$ and 6. Accordingly, the set of outcomes when rolling a die one time is

$$E = \{1, 2, 3, 4, 5, 6\}$$

It is important to note that sometimes the given description of the elements of a set does not provide any element that characterizes the described property. In this case, we call the resulting set as the **empty set**, the **null set**, or the **void set**. The empty set F is expressed by

$$F = \{\ \ \}$$

or by

$$F = \phi$$

where ϕ is a Greek letter. Furthermore, in other cases, the given description of the elements are justified by an infinite number of elements. In this case, we write the first few terms of the pattern followed by the ellipsis (...) to indicate that the pattern is continuing. The following example illustrates the last specific types of sets where the number of elements is zero or infinite.

Example 2. Write the sets that characterize the following descriptions:

(a) The Names of students in your class with height 20 ft.

(b) The non-zero positive integers.

(c) The even primes that are > 3.

(d) The square of non-zero positive even integers.

(e) The real roots of the equation $x^2 + 4 = 0$.

Solution.

(a) Since we cannot find a student with height 20 ft, hence the set with elements that characterize this condition is given by

$$F = \phi, \text{ or } F = \{ \ \ \}$$

As mentioned above, there is no element listed inside the braces. For this reason it is called empty set.

(b) There is an infinite number of non-zero positive integers. The set is called the set of natural numbers and is given by

$$N = \{1, 2, 3, 4, 5, \cdots\}$$

(c) As mentioned before in Chapter 2, the only even prime number is 2. Based on this, we cannot find an even prime that is > 3. Hence, we write

$$G = \phi, \text{ or } G = \{ \ \ \}$$

(d) The non-zero even integers are 2, 4, 6, etc., hence the set of squares of these integers is given by

$$H = \{4, 16, 36, 64, 100, \cdots\}$$

(e) The equation $x^2 + 4 = 0$ has no real roots. Therefore, the set of solutions is defined by

$$K = \phi, \text{ or } K = \{ \ \ \}$$

It should be noted that most of the sets defined above have finite number of elements. If X is a finite set, we define

$$n(X) = \text{number of elements of } X$$

Considering Example 1 and Example 2, we can write

$$n(A) = 3,\ n(B) = n(C) = 5,\ n(D) = 2,\ n(E) = 6,\ n(F) = n(G) = n(K) = 0$$

On the other hand, there is an infinite number of elements in the sets N and H described above.

2. The Set-Builder Notation (the description form): In this case, the set is defined by describing the elements of the set by a property that distinguishes these elements. This notation is especially used when describing infinite sets. For example, the set

$$A = \{x | x \text{ is an integer, } 2 \le x \le 6\}$$

defines the set A that contains all integers from 2 to 6. The vertical bar ($|$) is read "such that". In other words, the set A can be read as: "A is the set of all elements x such that x is an integer greater than or equal to 2 but less than or equal to 6". The set A can be rewritten as

$$A = \{2, 3, 4, 5, 6\}$$

where all elements are listed.

The type of sets that are characterized by the **description format** will be illustrated by the following examples.

Example 3. Write each statement in a **set-builder notation**:

(a) The even integers between 3 and 47.

(b) The integers 2, 3, 5, 7, 11, 13.

(c) The integers 1, 4, 9, 16, 25.

(d) The integers 1, 11.

(e) The arithmetic sequence 1,5,9,13,... .

Solution.

(a) $A = \{x \,|\, x \text{ is even, } 4 \le x \le 46\}$

(b) $B = \{y \mid y \text{ is a prime number, } y \le 13\}$

(c) $C = \{z \mid z \text{ is a perfect square, } 1 \le z \le 25\}$

(d) $D = \{x \mid x \text{ is a solution of } (x-1)(x-11) = 0\}$

(e) $E = \{w \mid w \text{ satisfies the congruence } w \equiv 1 \bmod 4\}$.

Example 4. Write each of the following set-builder sets (the description format sets) by an equivalent roster set (list format set). Write the number of elements of each set.

(a) $G = \{a \mid a \text{ is a prime number, and } a \text{ is even}\}$.

(b) $H = \{b \mid b \text{ is an odd number, } 1 \le b \le 7\}$.

(c) $K = \{x \mid x \text{ is a real solution of } x^2 + 1 = 0\}$.

(d) $M = \{x \mid x \text{ is a real solution of } x^2 - x = 0\}$.

(e) $N = \{x \mid x \equiv 2 \bmod 5, \ 2 \le x \le 27\}$.

Solution.

(a) $G = \{2\}$, $n(G) = 1$

(b) $H = \{1, 3, 5, 7\}$, $n(H) = 4$

(c) $K = \phi$, $n(K) = 0$

(d) $M = \{0, 1\}$, $n(M) = 2$

(e) $N = \{2, 7, 12, 17, 22, 27\}$, $n(N) = 6$

Exercises 5.1

1. Write the sets that characterize the following descriptions:

(a) Colors of deck of cards.

(b) The non-zero even integers less than or equal to 12.

(c) The roots of the equation $x^2 - 4x = 0$.

(d) The positive integers ≤ 19 that satisfy $x \equiv 3 \bmod 7$.

(e) The outcomes of even numbers when rolling a die.

2. Write the sets that characterize the following descriptions:

(a) Names of students with weight 1000 pounds.

(b) Perfect squares of non-zero positive integers.

(c) Real solutions of the equation $x^2 + x + 1 = 0$.

(d) Solutions of the congruence $3x \equiv 1 \bmod 5$.

(e) The outcomes of tossing a coin.

3. Write each of the following set-builder sets (description format sets) by an equivalent roster set (list format set). Write the number of elements of each set.

(a) $A = \{x \mid x$ is a vowel letter$\}$.

(b) $B = \{x \mid x$ is a solution of $x^2 - 13x + 12 = 0\}$.

(c) $C = \{x \mid x$ is a decimal number equivalent to $11_{(2)}\}$.

(d) $D = \{y \mid y$ is an integer solution of $3y - 1 = 4\}$.

(e) $E = \{z \mid 5z \equiv 1 \bmod 13\}$.

4. Write the sets that characterize the following descriptions:

(a) The set of all months in a year that starts with the letter J.

(b) The set of all days in a week that starts with the letter T.

(c) The set of all multiples of 4 between 10 1nd 25.

(d) The set of all positive factors of 6.

(e) The set of all months in a year that starts with the letter R.

5. Show that $\phi \neq \{\phi\}$.

5.2   Subsets

If a is an element of a set A, we usually use the notation

$$a \in A$$

to indicate that the element a belongs to the set A. On the other hand, the expression

$$a \notin A$$

indicates that a is not an element of A. For example, given the set

$$A = \{1, 2, 3, 4, 5\}$$

then

$$1 \in A, \text{ but } 6 \notin A$$

If the sets A and B contain the same elements, we say that A and B are equal, and this is expressed by

$$A = B$$

For example, the set

$$A = \{x \mid x \text{ is a solution of } x^2 - 5x + 4 = 0\}$$

and the set

$$B = \{1, 4\}$$

are equal.

We turn now to another way of defining another possible relation between two sets. Given two sets A and B. If every element of the set B is an element of the set A, we say that B is a subset of A. This is denoted by the following notation:

$$B \subset A$$

If there is at least one element of B that is not an element of A, then we say that B is not a subset of A, and this is expressed by:

$$B \not\subset A$$

Furthermore, if B is a subset of A, and $A \neq B$, then B is called a proper subset of A.

The concepts presented above will be illustrated by discussing the following examples.

Example 1. Which of the following are subsets, and which are not, of the other sets:

(a) $A = \{1, 2, 3, 4, 5, 6\}$

(b) $B = \{1, 3, 5\}$

(c) $C = \{3, 5\}$

(d) $D = \{5, 6, 7\}$

Solution.

(a) Every element of the set B is an element of the set A, hence we have

$$B \subset A$$

(b) Every element of C is an element of B, and therefore an element of A. Hence, we have

$$C \subset B, \text{ and } C \subset A$$

(c) $7 \in D$, but $7 \notin A$, hence

$$D \not\subset A$$

(d) $3 \in C$, but $3 \notin D$, hence

$$C \not\subset D$$

Example 2. Given the following sets:

$A = \{2, 4, 6, 8, 10\}$

$B = \{x | x \text{ is an even integer}, 2 < x < 9\}$

$C = \{4, 6, 8\}$

Classify the following statements as True or False:

(a) $2 \subset A$

(b) $C \in A$

(c) $C \subset A$

(d) $C = B$

(e) $\{1, 2, 3, 4\} \not\subset A$

Solution.

(a) False. The number 2 is an element of A. The proper expression is $2 \in A$. The sign $\subset$ governs the relation between two sets only.

(b) False. The sign $\in$ governs the relation between an element and the set that contains this element. C is a set and not an element.

(c) True. Every element of C is an element of A.

(d) True. The set B is given by $B = \{4, 6, 8\}$, hence $B = C$.

(e) True. The element $1 \in \{1, 2, 3, 4\}$, but $1 \notin A$.

5.2.1 Number of Subsets

In the following we will introduce some identities related to the concept of subsets. The proofs of these facts can be easily verified, and left as exercises to the reader.

1. The empty set ϕ is a subset of every set. In other words, we write

$$\phi \subset A, \text{ for every set } A$$

2. Any set A is a subset of itself. This can be expressed by $A \subset A$.

3. If a set A contains n elements, then we can establish 2^n subsets of A. Notice that the subsets must include the empty set ϕ and the set A itself. The last property can be illustrated by using the following examples.

Example 3. Form all possible subsets of the set

$$A = \{1, 2\}$$

Solution. The subsets are given by

$$\phi, \{1\}, \{2\}, A$$

Notice that the number of subsets is $2^2 = 4$ subsets.

Example 4. Form all possible subsets of the set

$$B = \{2, 4, 6\}$$

Solution. The subsets are:

(a) ϕ, no elements in this subset

(b) $\{2\}, \{4\}, \{6\}$, each subset has one element

(c) $\{2, 4\}, \{2, 6\}, \{4, 6\}$, each subset has two elements

(d) B= $\{2, 4, 6\}$. The number of subsets is $2^3 = 8$.

Example 5. How many subsets can be formed from the following sets:

(a) $A = \{1, 2, 3, 4\}$

(b) ϕ

(c) $D = \{1, \phi\}$

(d) $C = \{1, 3, 5, 7, \cdots\}$

Solution. The number of subsets is given by:

(a) $2^4 = 16$

(b) $2^0 = 1$

(c) $2^2 = 4$

(d) An infinite number of subsets can be formed

Exercises 5.2

1. Given the following sets

$U = \{1, 2, 3, 4, \cdots, 9, 10\}$

$A = \{1, 3, 5, 7, 9\}$

$B = \{x | x \in U, x \text{ is odd}\}$

Classify the following statements as True or False. Explain your answer.

(a) $\{1, 3\} \in U$

(b) $\{1, 3\} \subset A$

(c) $B \subset U$

(d) $B \neq A$

(e) $\{1, 3, 5\} \subset B$

2. Given the following sets

$X = \{2, 4, 6, 8, 10\}$

$Y = \{2, 6, 10\}$

$Z = \{x | x \equiv 2 \bmod 4\}$

$F = \phi$

Classify the following statements as True or False. Explain your answer.

(a) $Y \subset X$

(b) $Z \neq Y$

(c) $Z \not\subset X$

(d) $F \not\subset Y$

(e) $F \in X$

3. Form all possible subsets of the set

$$K = \{a, b, c\}$$

4. How many subsets we can form from the following sets:

(a) $F = \{x \mid x \text{ is an integer solution of } 3x - 1 = 0\}$

(b) $K = \{x \mid x^2 - 3x + 2 = 0\}$

(c) $L = \{y \mid y \text{ is an integer}, 1 \le y \le 5\}$

(d) $M = \{z \mid z \text{ is an odd outcome when rolling a die}\}$

(e) $N = \{n \mid n \equiv 1 \bmod 3, \, 2 \le n \le 6\}$

5. Which of the following is an empty set:

(a) $A = \{x \mid x^2 + 4 = 0, \, x \in R\}$

(b) $B = \{x \mid x \text{ is an integer solution of } 2x - 1 = 0\}$

(c) $C = \{x \mid x > 0 \text{ and } x < 0\}$

(d) $D = \{x \mid x \text{ is a solution of } 3x \equiv 1 \bmod 6\}$

6. The set of all possible outcomes of rolling a die is given by

$$U = \{1, 2, 3, 4, 5, 6\}$$

Write the following subsets of U

(a) The subset that the die shows an even prime

(b) The subset that the die shows a prime number

(c) The subset that the die shows an even number

(d) The subset that the die shows a number ≤ 4

(e) The subset that the die shows a number ≥ 8

5.3 Operations on Sets

In this section, we will discuss three important operations that will be used to establish new sets. The operations that will relate two or more sets are:

1. Union of two sets.

2. Intersection of two sets.

3. Complement of a set.

To use these operations, it seems reasonable to consider the following two sets

$A = \{1, 2, 3, 4, 5\}$

$B = \{4, 5, 6, 7\}$

In the following, we will introduce the definition of each operation and will apply the operation on the sets A and B given above:

1. The Union of Two Sets: The union of two sets A and B is denoted by $A \cup B$ and is defined by

$$A \cup B = \{x | x \in A, \textbf{ or}, x \in B \text{ (or } x \text{ in both)}\}$$

In other words, the union of two sets is a new set that consists of all elements that belong to either the set A or to the set B (or to both). Using the sets A and B given above we find

$$A \cup B = \{1, 2, 3, 4, 5, 6, 7\}$$

Notice that the new set $A \cup B$ contains all elements of the sets A and B where the elements 4 and 5 that belong to both sets are listed once in the resulting set. The union operation unite the elements of the sets A and B where identical elements are listed once.

At this point, it is worth noting the following relations:

1. $A \cup B = B \cup A$

2. $A \cup \phi = A$

3. $A \cup A = A$

2. The Intersection of Two Sets: The intersection of two sets A and B is denoted by $A \cap B$ and is defined by

$$A \cap B = \{x | x \in A, \textbf{ and}, x \in B\}$$

In other words, the intersection of two sets is a new set that consists of the common elements that belong to the set A and to the set B. Using the sets A and B given above we find

$$A \cap B = \{4, 5\}$$

At this point, it is worth noting the following relations:

1. $A \cap B = B \cap A$

2. $A \cap \phi = \phi$

3. $A \cap A = A$

It should be noted that the intersection of two sets may sometimes result in an empty set. This occurs if the two sets under discussion have no common elements. For example, consider the sets

$$C = \{5, 6, 7\}$$

and

$$D = \{10, 11, 12\}$$

It is easily observed that C and D have no common elements, hence we write

$$C \cap D = \phi$$

In this case, the two sets are therefore called **disjoint** sets. On the other hand, the union of these two sets exists and is given by

$$C \cup D = \{5, 6, 7, 10, 11, 12\}$$

3. The Complement of a Set: The complement of a set A is denoted by A^c. To define the complement set A^c, it is necessary that we define first the notion of a **Universal set** U. A universal set U is the set that contains all elements to which we restrict our study and concern. This means that all sets under discussion are subsets of U. For example, if our consideration is focused on non-zero integers from 1 to 10, the universal set U is therefore

$$U = \{1, 2, 3, 4, 5, 6, 7, 8, 9, 10\}$$

As a second example, rolling a die will result in the following universal set given by

$$U = \{1, 2, 3, 4, 5, 6\}$$

Notice that the universal set is not the same for all cases, but it depends only on the problem under discussion.

Having defined a universal set U, the complement set A^c is therefore defined by

$$A^c = \{x | x \in U, \text{ and } x \notin A\}$$

In other words, A^c is the set that contains all elements of U that are not in A. The complement set A^c is simply obtained by eliminating the elements of A from the elements of U.

At this point, we note the following relations:

1. $A \cup A^c = U$
2. $A \cap A^c = \phi$
3. $(A^c)^c = A$

5.3.1 Properties of Sets

We now present the properties of the sets operations that were discussed above:

1. **Idempotence Laws:** For any set A

   (i) $A \cup A = A$,   (ii) $A \cap A = A$

2. **Commutative laws:** For any two sets A and B

   (i) $A \cup B = B \cup A$ (ii) $A \cap B = B \cap A$

3. **Associative laws:** For any three sets A, B and C

   (i) $A \cup (B \cup C) = (A \cup B) \cup C$ (ii) $A \cap (B \cap C) = (A \cap B) \cap C$

4. **Distributive laws:**

   (i) $A \cup (B \cap C) = (A \cup B) \cap (A \cup C)$ (ii) $A \cap (B \cup C) = (A \cap B) \cup (A \cap C)$

5. **DeMorgan's laws:** For any two sets A and B

   (i) $(A \cup B)^c = A^c \cap B^c$ (ii) $(A \cap B)^c = A^c \cup B^c$

6. **Complement laws:** For any set A

   (i) $A \cup A^c = U$ (ii) $A \cap A^c = \phi$

7. **Identity Laws:** For any set A

   (i) $A \cup \phi = A$ (ii) $A \cap U = A$

The operations discussed above will be illustrated by the following examples.

Example 1. Given a universal set U defined by

$$U = \{1, 2, 3, 4, \cdots, 11, 12\}$$

and the subsets of U given by

$$A = \{1, 3, 5, 7, 9, 11\}$$

$$B = \{8, 9, 10, 11, 12\}$$

Answer the following:

(a) $A \cap B$

(b) $A \cup B$

(c) A^c

(d) $(A \cup B)^c$

Solution.

(a) $A \cap B = \{9, 11\}$

(b) $A \cup B = \{1, 3, 5, 7, 8, 9, 10, 11, 12\}$

(c) $A^c = \{2, 4, 6, 8, 10, 12\}$
    A^c is obtained by eliminating the elements of A from U

(d) $(A \cup B)^c = \{2, 4, 6\}$

Example 2. The universal set S that results from rolling one die is

$$S = \{1, 2, 3, 4, 5, 6\}$$

and the subsets of S defined by

$$C = \{x | x \text{ is a prime number}\}$$

$$D = \{y | y \text{ is a multiple of 2}\}$$

Answer the following:

(a) $C \cap D$

(b) $C \cup D$

(c) C^c and D^c

(d) $(C \cap D)^c$

(e) $C^c \cap D^c$

Solution. The subsets C and D of the universal set S may be written in a list format by

$$C = \{2, 3, 5\}$$
$$D = \{2, 4, 6\}$$

(a) $C \cap D = \{2\}$

(b) $C \cup D = \{2, 3, 4, 5, 6\}$

(c) $C^c = \{1, 4, 6\}, \quad D^c = \{1, 3, 5\}$

(d) $(C \cap D)^c = \{1, 3, 4, 5, 6\}$

(e) $C^c \cap D^c = \{1\}$, obtained by using the answer in (c).

Example 3. The universal set S that results from tossing a coin twice is given by

$$S = \{HH, HT, TH, TT\}$$

where H and T indicate that the outcomes are head and tail respectively. The element HT for example means that the outcome of the first die is a H and the second die shows a T. The following are subsets of S defined by

$$
\begin{aligned}
C &= \{x \mid x \text{ is at least one tail occurs}\} \\
D &= \{y \mid y \text{ is at least one head occurs}\}
\end{aligned}
$$

Find

(a) $C \cap D$

(b) $C \cup D$

(c) C^c

(d) D^c

(e) $C^c \cap D^c$

Solution. Notice that the given description of the set C that at least 1 tail occurs means that we should list the outcomes when 1 or 2 tails occur. Similarly, the elements of D contain 1 or 2 heads. Therefore, the subsets C and D of the universal set S may be written in a list format by

$$C = \{HT, TH, TT\}$$

$$D = \{HH, HT, TH\}$$

(a) $C \cap D = \{HT, TH\}$

(b) $C \cup D = S$

(c) $C^c = \{HH\}$

(d) $D^c = \{TT\}$

(e) $C^c \cap D^c = \phi$, obtained by using (c) and (d).

Example 4. A student committee interviewed 500 students about their car color preferences. The results obtained are given in the following table.

	Red R	Silver S	White W
Boys B	90	80	60
Girls G	120	100	50

Find

(a) $n(B)$

(b) $n(B \cap S)$

(c) $n(B \cup R)$

(d) $n(W \cap G)$

(e) $n(R^c)$

Solution. The 500 students interviewed form the universal set. We can easily observe that there are 5 distinct subsets that are derived from the universal set. Two distinct categories were use, namely gender and car colors.

There are 230 boys that form the set B defined by the first row, and 270 girls that form the set G defined by the second row.

In addition, there are 210 students, who prefer the red color, that form the set R given by the first column. In a parallel manner, we observe the sets S and W given by the second and the third column respectively, where $n(S) = 180$, and $n(W) = 110$.

As discussed before, we can construct other subsets by subjecting these sets to the union, intersection and the complement operations. Accordingly, we find

(a) $n(B) = 230$

(b) $n(B \cap S) = 80$

(c) $n(B \cup R) = 350$

(d) $n(W \cap G) = 50$

(e) $n(R^c) = 500 - 210 = 290$

Example 5. Given a universal set by

$$U = \{1, 2, 3, 4, 5, 6, 7, 8, 9, 10\}$$

and the subsets

$$
\begin{aligned}
A &= \{x \mid x \text{ is an integer, } 1 \le x \le 6\} \\
B &= \{y \mid y \text{ is and odd prime, } y < 10\}
\end{aligned}
$$

Prove DeMorgan's laws by using the sets A and B.

Solution. We first write the roster format (list format) of each set by

$$
\begin{aligned}
A &= \{1, 2, 3, 4, 5, 6\} \\
B &= \{3, 5, 7\}
\end{aligned}
$$

Recall that DeMorgan's laws are:

(i) $(A \cup B)^c = A^c \cap B^c$

(ii) $(A \cap B)^c = A^c \cup B^c$

Considering A and B we find

$$
\begin{aligned}
A \cup B &= \{1, 2, 3, 4, 5, 6, 7\} \\
A \cap B &= \{3, 5\} \\
A^c &= \{7, 8, 9, 10\} \\
B^c &= \{1, 2, 4, 6, 8, 9, 10\}
\end{aligned}
$$

For (i) we find

$$(A \cup B)^c = \{8, 9, 10\} = A^c \cap B^c$$

For (ii) we get

$$(A \cap B)^c = \{1, 2, 4, 6, 7, 8, 9, 10\} = A^c \cup B^c$$

Example 6. Given a universal set by

$$U = \{1, 2, 3, 4, 5, 6, 7, 8, 9, 10, 11, 12\}$$

and the following subsets of U

$$A = \{1, 2, 3, 4, 5, 6, 7\}$$
$$B = \{x | x \text{ is a prime number}\}$$
$$C = \{x | x \text{ is an even integer}\}$$

Find

(a) $A \cap B \cap C =$

(b) $A \cup B \cup C =$

(c) $(A \cup C)^c =$

(d) $(A^c \cup B)^c =$

(e) How many subsets we can make from the answer in (c)

Solution. We first write the roster format (list format) of each set by

$$
\begin{aligned}
A &= \{1, 2, 3, 4, 5, 6, 7, 8, 9, 10, 11, 12\} \\
B &= \{2, 3, 5, 7, 11\} \\
C &= \{2, 4, 6, 8, 10, 12\}
\end{aligned}
$$

(a) $A \cap B \cap C = \{2\}$

(b) $A \cup B \cup C = \{1, 2, 3, 4, 5, 6, 7, 8, 10, 11, 12\}$

(c) $(A \cup C)^c = \{1, 2, 3, 4, 5, 6, 7, 8, 10, 12\}^c = \{1, 9\}$

(d) $(A^c \cup B)^c = (\{8, 9, 10, 11, 12\} \cup \{2, 3, 5, 7, 11\})^c = \{1, 4, 6\}$

(e) number of subsets$= 2^2 = 4$

Exercises 5.3

1. Given a universal set U defined by

$$U = \{1, 2, 3, 4, \cdots, 9, 10\}$$

and the subsets of U

$$A = \{x \,|\, x \text{ is even}\}$$

$$B = \{y \,|\, y \equiv 1 \bmod 3\}$$

Find

(a) A^c

(b) $A \cup B$

(c) $A \cap B$

(d) $(A \cup B)^c$

(e) $(A \cap B)^c$

2. Given a universal set U defined by

$$U = \{1, 2, 3, 4, \cdots, 10, 11, 12\}$$

and the subsets of U

$$A = \{x \,|\, x \text{ is odd}\}$$
$$B = \{y \,|\, y \text{ is even}\}$$

Show that

(a) $A^c = B$

(b) $B^c = A$

(c) $A \cup B = U$

(d) $(A \cap B) = \phi$

(e) $(A^c \cup B) = B$

3. If S is the universal set that contains all possible outcomes when rolling a die given by

$$S = \{1, 2, 3, 4, 5, 6\}$$

The following subsets of S are defined by

$$\begin{aligned} A &= \{x \,|\, x \text{ is an even outcome}\} \\ B &= \{y \,|\, y \text{ is a multiple of } 3\} \end{aligned}$$

Find

(a) A^c

(b) $A \cap B$

(c) $(A \cup B)^c$

(d) $(A \cup B)^c \cap B$

(e) $A^c \cup B^c$

4. Tossing a coin twice. The following subsets of the universal set S of all possible outcomes are defined by

$$C = \{x \mid x \text{ is at least one head occurs}\}$$

$$D = \{y \mid y \text{ is exactly one tail occurs}\}$$

Find

(a) C^c

(b) $C \cap D$

(c) $(C \cap D)^c$

(d) $C^c \cap D^c$

(e) D^c

5. A student committee interviewed 1000 students, boys and girls, about the courses they registered for one term. The only courses needed for this survey were Mathematics M, History H, and English E. The results obtained are given in the following table.

	Math M	History H	English E
Boys B	100	120	180
Girls G	240	170	190

Find

(a) $n(B^c)$

(b) $n(B \cap H)$

(c) $n(G \cup M)$

(d) $n(M \cap H)$

(e) $n(G \cap E)$

6. The following table shows the number of voters in a small town:

	Republican R	Democratic D	Independent I
Men M	184	87	29
Women W	230	130	40

Find

(a) $n(W^c)$ (b) $n(M \cap R)$ (c) $n(W \cup D)$

(d) $n(R \cap D)$ (e) $n(W \cup I)$

7. Given a universal set U by

$$U = \{1, 2, 3, 4, 5, 6, 7, 8\}$$

and the subsets of U

$$A = \{1, 2, 3, 4\}$$
$$B = \{3, 4, 5\}$$
$$C = \{6, 7\}$$

Find

(a) $A \cap B \cap C$

(b) $A^c \cap B$

(c) $A \cap B^c$

(d) $(A \cup C)^c \cap B$

(e) $(A^c \cap C) \cup B$

8. Use Exercise 7 to prove DeMorgan's laws for the sets A and B

9. Given the universal set

$$U = \{1, 2, 3, 4, 5, 6, 7, 8, 9, 10, 11\}$$

and the subsets

$$A = \{1, 2, 3, 4, 5\}$$
$$B = \{x \mid x \text{ is an odd prime number}\}$$
$$C = \{2, 4, 6, 8, 10\}$$

Find

(a) $A \cap B \cap C$

(b) $A \cup B \cup C$

(c) $(A \cup C)^c$

(d) $A^c \cap C^c$

(e) How many subsets you can construct from the answer in (c)

10.   Given the universal set

$$U = \{1, 2, 3, 4, 5, 6, 7, 8, 9, 10, 11, 12, 13, 14\}$$

and the subsets

$$A = \{1, 2, 3, 4, 5\}$$
$$B = \{x | x \text{ is a prime number}\}$$
$$C = \{2, 4, 6, 8, 10, 12\}$$

Find

(a) $A \cap B \cap C$

(b) $(A \cup B) \cap C$

(c) $(A \cup C)^c$

(d) $A^c \cap C^c$

(e) How many subsets you can construct from the answer in (a)

11.   Given the universal set

$$U = \{1, 2, 3, 4, 5, 6, 7, 8, 9, 10\}$$

and the following subsets of U

$$A = \{1, 2, 3, 4\}$$
$$B = \{x | \ 3 \le x < 9\}$$
$$C = \{x | \ 6 < x \le 10\}$$

Find

(a) $(A \cap B \cap C)^c$

(b) $(A \cup B \cup C)^c$

(c) $A \cap B$

(d) $A \cup C^c$

(e) How many subsets we can make from the answer in (d)

12.   Given the universal set

$$U = \{1, 2, 3, 4, 5, 6, 7, 8, 9, 10, 11, 12\}$$

and the following subsets of U

$$A = \{1, 2, 3, 10, 11, 12\}$$
$$B = \{x| \ 3 \le x \le 8\}$$
$$C = \{x| \ 7 \le x \le 11\}$$

Find

(a) $A \cup B \cup C$

(b) $(A \cap B \cap C)^c$

(c) $C^c \cap B^c$

(d) $(A^c \cup B^c)^c$

(e) How many subsets we can make from $(A \cup B \cup C)^c$

5.4   The Venn Diagram

We usually make graphs for functions to give an insight about the functions that represent. John Venn established a useful visual representation, called **Venn Diagram**, to represent sets and to picture sets relationships as follows:

1. A universal set U is represented by a rectangle where its interior region contains all elements of the universal set U. This is shown by Figure 5.1.

Figure 5.1

2. Any subset of U is represented by a circular disc drawn inside the rectangle as shown by Figure 5.2. The circle contains all elements of A.

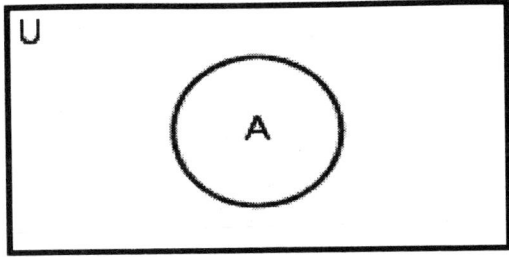

Figure 5.2

3. The complement of a set A, denoted by A^c, is represented by the shaded area as shown in Figure 5.3.

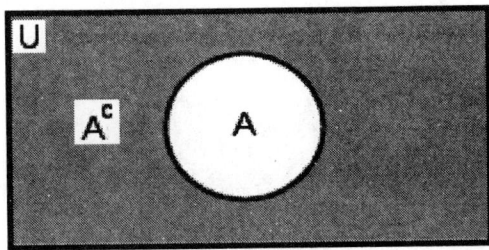

Figure 5.3

4. The operations $\cup$ and $\cap$ between two sets A and B can be easily represented according to the following cases:

(i) Disjoint sets: When sets A and B have no common elements, i.e. $A \cap B = \phi$, the sets A and B are called disjoint sets and represented by disjoint circles as shown by Figure 5.4.

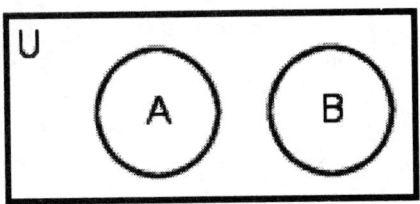

Figure 5.4

(ii) Overlapped sets: When $A \cap B \neq \phi$, the sets are overlapped sets and represented by overlapped circles as shown by Figure 5.5.

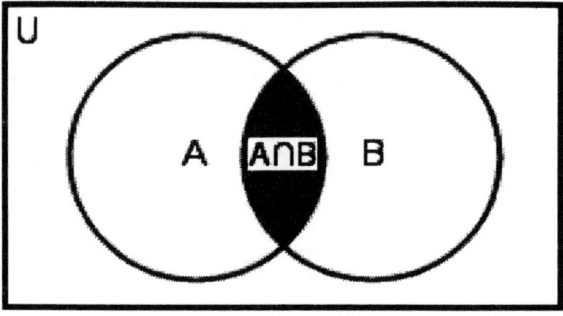

Figure 5.5

Notice that the shaded area in Figure 5.5 represents $A \cap B$. However, the shaded area in Figure 5.6 represents $A \cup B$.

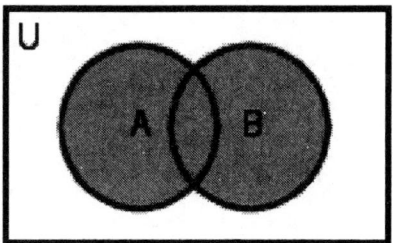

Figure 5.6

(iii) If $B \subset A$ **:** In this case, the set B will be represented by an interior circle inside the circle that represents A as shown by Figure 5.7.

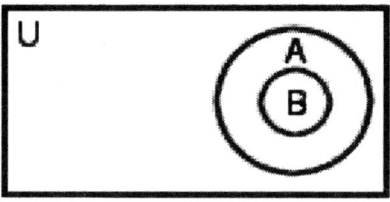

Figure 5.7

The above concepts of the Venn diagram can be illustrated by the following examples.

Example 1. Given the universal set U and the subsets A and B by:

$$U = \{1, 2, 3, 4, \cdots, 9, 10\}$$

$$A = \{1, 2, 3, 4, 5\}$$

$$B = \{4, 5, 6, 7\}$$

Represent U, A and B by a Venn diagram.

Solution. We first start by finding $A \cap B$, where we obtain

$$A \cap B = \{4, 5\}$$

Accordingly, the Venn diagram that represents these sets consists of a rectangle and two overlapped circles. This is shown by Figure 5.8.

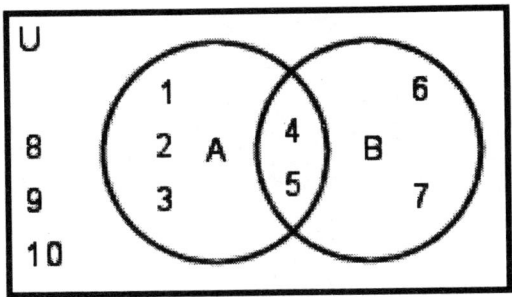

Figure 5.8

Example 2. Given the universal set U and the subsets A and B by:

$$U = \{1, 2, 3, 4, \cdots, 9, 10\}$$

$$A = \{1, 3, 5\}$$

$$B = \{2, 4, 6, 8\}$$

Represent U, A and B by a Venn diagram.

Solution. We first start by finding $A \cap B$, where we obtain

$$A \cap B = \phi$$

Accordingly, the Venn diagram that represents these sets is given by two disjoint circles as shown by Figure 5.9.

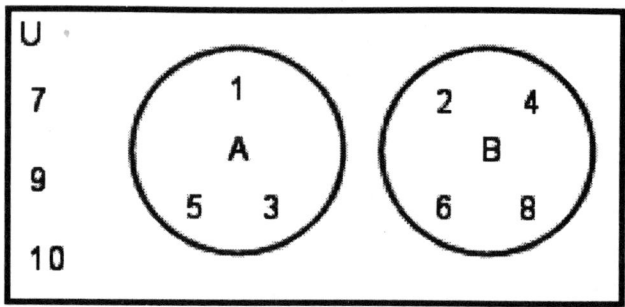

Figure 5.9

Example 3. Given the following Venn diagram

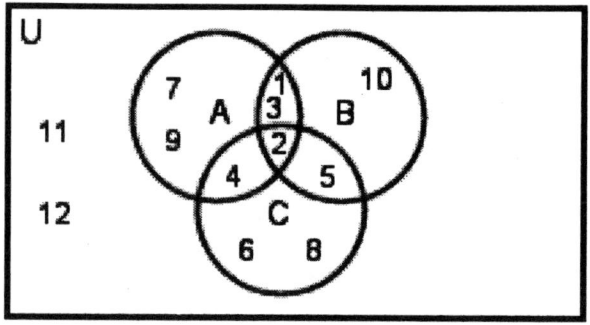

Figure 5.10

Use this diagram to answer the following questions:

(a) Find the universal set U

(b) Find the sets A, B, and C

(c) Find $A \cap B$

(c) Find $A \cap B \cap C$

(e) Find $(A \cup B \cup C)^c$

Solution.

(a) $U = \{1, 2, 3, \cdots, 11, 12\}$

(b) $A = \{1, 2, 3, 4, 7, 9\}$, $B = \{1, 2, 3, 5, 10\}$, $C = \{2, 4, 5, 6, 8\}$

(c) $A \cap B = \{1, 2, 3\}$

(d) $A \cap B \cap C = \{2\}$

(e) $(A \cup B \cup C)^c = \{11, 12\}$

5.4.1 Survey Problems

For survey problems of three distinct objects, such as three distinct courses and three different sport games, it is useful to solve the problem by using Venn diagram. It is normal to draw the three overlapped circles with a common overlapped region that represents the intersection of the three sets as shown by Figure 5.11 below.

The usual way is to fill in the number that represents the common intersection of the three sets $A \cap B \cap C$ in the innermost region VII. We then go outward through the Venn diagram filling the regions IV, V and VI by subtracting the number in region VII from the numbers in the intersections $A \cap C$, $A \cap B$ and $B \cap C$. We complete filling the remaining parts I, II and III of each circle by subtracting the numbers that were filled already in each circle from the numbers given in the sets A, B and C. Notice that $(A \cup B \cup C)^c$ should be filled inside the rectangle and out of all circles.

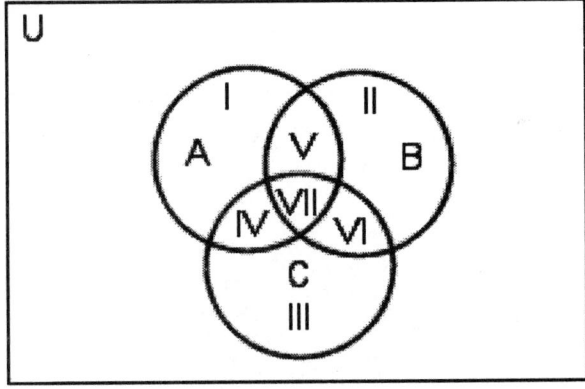

Figure 5.11

In Figure 5.11, notice that:
(i) Regions I, II, and III represents only A, only B and only C respectively.
(ii) Regions IV, V, and VI represents only $A \cap C$, only $A \cap B$, and only $B \cap C$ respectively.

(iii) Regions I,II,IV,V VI and VII represents $A \cup B$. However, regions I, II and V represents only $A \cup B$. Similar conclusions can be made for other regions.

The survey concepts will be explained by the following examples.

Example 4. The following data was obtained by interviewing several students:

60 students study Physics P.

60 students study Chemistry C.

40 students study Biology B.

30 students study Physics and Chemistry.

22 students study Biology and Chemistry.

15 students study Physics and Biology.

10 students study all three courses.

8 students do not study any of these courses.

Answer the following:

(a) Represent this data by using Venn diagram.

(b) How many students were interviewed.

(c) Use the diagram to find how many students study only Physics.

(d) How many students study only Physics and Biology.

(e) How many students study only Chemistry.

Solution. To answer these questions, we usually find the set of common elements of the three sets. The data given shows that 10 students study all three courses, therefore, we have

$$n(P \cap B \cap C) = 10$$

We next note that 30 students study Physics and Chemistry, 22 students study Biology and Chemistry, and 27 students study Physics and Biology. This means that

$$n(P \cap C) = 30, \ n(B \cap C) = 22, \ n(P \cap B) = 15$$

It should be noted here that the common region between P and C contains 30 students, where 10 students belong to $P \cap C \cap B$. This means that 20 students study only Physics and Chemistry. Similar conclusions can be made to overlapped regions between other sets.

Accordingly, the three sets P, B, and C are drawn as three overlapped circles inside the rectangle where the common regions are now determined.

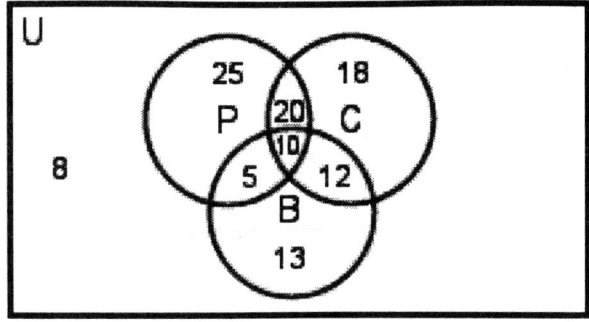

Figure 5.12

(a) The Venn diagram shows the three overlapped circles by Figure 5.12.

(b) $n(U) = 25 + 20 + 10 + 5 + 18 + 12 + 13 + 8 = 111$

(c) 25 students study only Physics.

(d) 5 students study only Physics and Biology.

(e) 18 students study only Chemistry.

Example 5. A survey of 100 selected persons gave the following data:
10 persons watch only basketball games.
12 persons watch only football games.
8 persons watch only volleyball games.
30 persons watch basketball and football games.
26 persons watch football and volleyball games.
20 persons watch basketball and volleyball games.
6 persons watch all three games.

Answer the following:

(a) Represent this data by using Venn diagram.

(b) How many persons watch basketball games.

(c) How many persons watch only basketball and football.

(d) How many persons watch basketball or football games.

(e) How many persons watch only basketball or football games.

(f) How many persons are not watching any of these games.

Solution. To answer these questions, we usually find the set of common elements of the three sets.

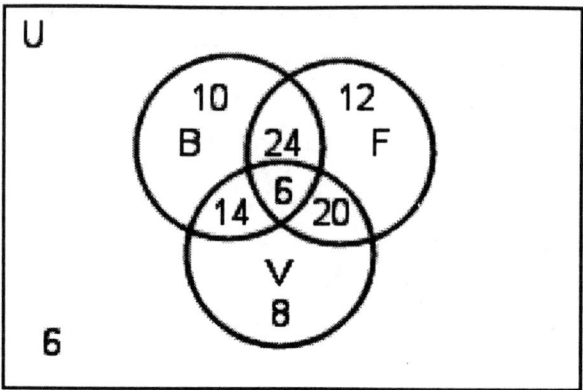

Figure 5.13

The data given shows that 6 persons are watching all three games, therefore, we have

$$n(B \cap F \cap V) = 6$$

This number should be placed in the common region between all three circles. We next note that

$$n(B \cap F) = 30, \; n(F \cap V) = 26, \; n(B \cap V) = 20$$

By excluding 6 from all these three numbers, we fill the regions V, VI and IV respectively in Figure 5.11.

Moreover, it should be noted here that for the sets B, F and V, only $n(B)$, only $n(F)$ and only $n(V)$ are given. This means that the numbers that represent the regions I, II and III in Figure 5.11 are given and not for the numbers of the sets.

(a) The Venn diagram is shown above by Figure 5.13.

(b) $n(B) = 10 + 24 + 14 + 6 = 54$

(c) 24 persons watch only basketball and football.

(d) $n(B \cup F) = 10 + 24 + 12 + 14 + 6 + 20 = 86.$

(e) $10 + 24 + 12 = 46$ persons watch only basketball or football.

(f) 6 persons are not watching any of these games

Exercises 5.4

1. Given the following sets:

$$U = \{1, 2, 3, \cdots, 11, 12\}$$
$$A = \{1, 3, 5, 7, 9\}$$
$$B = \{2, 4, 6, 8\}$$
$$C = \{6, 8, 10\}$$

Represent these sets by a proper Venn diagram.

2. Given the following sets:

$$U = \{1, 2, 3, \cdots, 10\}$$
$$A = \{1, 2, 3, 4, 5\}$$
$$B = \{3, 5, 7, 9\}$$
$$C = \{3, 4, 6, 7, 8\}$$

Represent these sets by a proper Venn diagram.

3. The results of a survey of a number of students were as follows:
80 students are taking a Math course.
70 students are taking a Science course.
90 students are taking an English course.
40 students are taking a Math and a Science courses.
34 students are taking a a Science and English courses.
37 students are taking a Math and English courses.
23 students are taking all three courses.
33 students are not taking any of these courses.

Answer the following:

(a) Represent this data by using Venn diagram.

(b) How many students were interviewed.

(c) Use the diagram to find how many students study only Science.

(d) How many students study only English and Math.

(e) How many students study only Math.

4. A researcher surveyed 80 persons to determine whether they like orange juice, apple juice or soda products. The following results were obtained:
42 persons like Orange juice.

39 persons like Apple juice.

38 persons like Soda products.

20 persons like Orange juice and Apple juice.

17 persons like Apple juice and Soda products.

18 persons like Orange juice and Soda products.

11 persons like all three products.

Answer the following:

(a) Represent this data by using Venn diagram.

(b) How many persons were interviewed.

(c) How many persons like only soda.

(d) How many persons like orange or apple.

(e) How many persons like only soda and apple.

5. Given a universal set

$$U = \{1, 2, 3, \cdots, 11, 12\}$$

and the subsets

$$
\begin{aligned}
A &= \{1, 2, 3, 4, 5, 6\} \\
B &= \{4, 5, 6, 7, 8, 9\} \\
C &= \{8, 9, 10, 11\}
\end{aligned}
$$

Represent these sets by a proper Venn diagram.

6. Given a universal set

$$U = \{a, b, c, d, e, f, g, h, i, j\}$$

and the subsets

$$
\begin{aligned}
A &= \{a, b, c, d, h\} \\
B &= \{c, d, e, f, g\} \\
C &= \{c, g, h, i\}
\end{aligned}
$$

Represent these sets by a proper Venn diagram.

7. Rolling a die one time. Let S be the universal set given by

$$S = \{1, 2, 3, 4, 5, 6\}$$

Consider the following subsets of S:

$$A = \{x \mid x \text{ is a prime number}\}$$
$$B = \{y \mid y \text{ is an even prime}\}$$
$$C = \{z \mid z \text{ is a multiple of 3}\}$$

Represent these sets by a proper Venn diagram.

8. Given the following Venn diagram

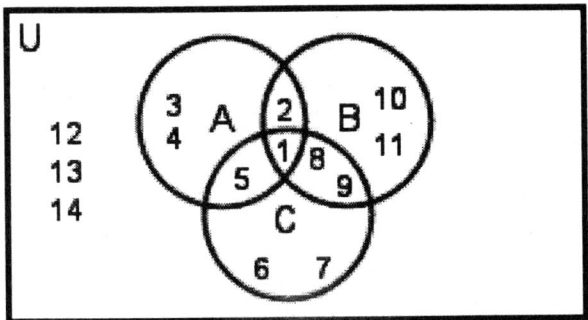

Use this diagram to answer the following questions:

(a) Find the universal set U

(b) Find $A \cap B \cap C$

(c) Find $A \cap B$

(d) Find $(A \cup B)^c$

(e) Find $(A \cup B \cup C)^c$

9. Given the following Venn diagram

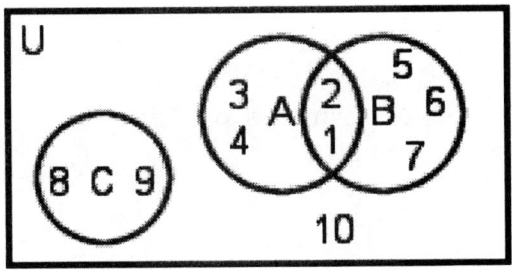

Use this diagram to answer the following questions:

(a) Find the universal set U

(b) Find $A \cap B \cap C$

(c) Find $A \cap B$

(d) Find $(A \cup B)^c$

(e) Find $(A \cap B)^c$

10. Given the following Venn diagram

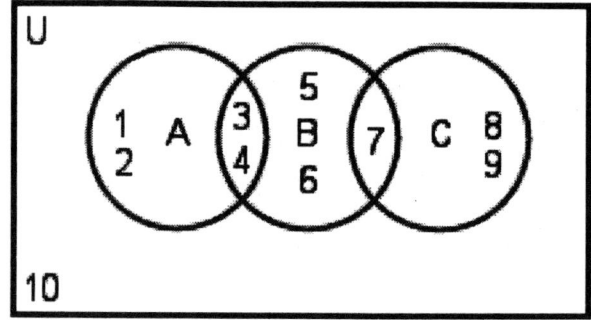

Use this diagram to answer the following questions:

(a) Find $A \cap B \cap C$

(b) Find $A \cap B$

(c) Find $(B \cap C)$

(d) Find $(A \cup B \cup C)^c$

(e) Find $(A \cap B)^c$

11. In a survey of 200 students, the following information was obtained:

78 students study Mathematics M.
82 students study Biology B.
91 students study Chemistry.
25 students study only Mathematics and Biology.
12 students study only Biology and Chemistry.
18 students study only Mathematics and Chemistry.
15 students study all three courses.

Answer the following:

(a) Represent this data by using the Venn diagram.

(b) How many students are not taking these courses.

(c) Find how many students study Mathematics and Chemistry.

(d) How many students study only Mathematics or Biology.

(e) How many students study only Chemistry.

12. A survey was conducted and the following information was obtained:

57 persons buy only cereal A.
40 persons buy only cereal B.
50 persons buy only cereal C.
50 persons buy cereal A and cereal B.
42 persons buy cereal B and cereal C.
35 persons buy cereal A and cereal C.
22 persons buy all three.
20 persons do not buy any.

Answer the following:

(a) Represent this data by using the Venn diagram.

(b) How many persons were interviewed.

(c) How many persons buy cereal A.

(d) How many persons buy only cereal A and cereal C.

(e) How many persons buy only cereal B or cereal C.

13. In a survey of students, it was found that:

170 students buy toothpaste A.
145 students buy toothpaste B.
120 students buy toothpaste C.
28 students buy only toothpaste A and toothpaste B.
15 students only buy toothpaste B and toothpaste C.
10 students only buy toothpaste A and toothpaste C.
20 students buy all three tooth pastes.
158 students buy none of the three.

Answer the following:

(a) Represent this data by using the Venn diagram.

(b) How many students were interviewed.

(c) Find how many students buy toothpaste A and toothpaste B.

(d) How many students buy toothpaste A or toothpaste B.

(e) How many students buy only toothpaste C.

14. A survey of 400 farmers was conducted where we obtained :

98 grew only wheat W.
74 grew only oats O.
86 grew only corn C.
32 grew wheat W and oats O.
26 grew oats O and corn C.
24 grew wheat W and corn C.
10 grew all three.

Answer the following:

(a) Represent this data by using the Venn diagram.

(b) How many farmers did not grow any corn, oats or wheat.

(c) How many farmers grew wheat W.

(d) How many farmers grew only oats O and corn C.

(e) How many farmers grew only wheat W or corn C.

15. In a survey of students about the courses they are taking, it was found that:

35 students take only Math M.
40 students take Biology B.
11 students take only Chemistry C.
10 students take Math and Biology.
10 students take only Biology B and Chemistry C.
9 students take Math M and Chemistry.
4 students take all three courses Math M, Biology B, and Chemistry C. 100 students do not take anyone of these three courses.

Answer the following:

(a) Represent this data by using the Venn diagram.

(b) Find how many students take Math M.

(c) How many students take only Math M and Biology B.

(d) How many students take Math M or Biology B.

(e) How many students were interviewed.

16. In a survey of students about the courses they are taking, it was found that:

60 students take Math M.
25 students take only Biology B.

60 students take Chemistry C.
8 students take only Math and Biology.
17 students take Biology B and Chemistry C.
12 students take only Math M and Chemistry.
10 students take all three courses Math M, Biology B, and Chemistry C. 100 students do not take anyone of these three courses.

Answer the following:

(a) Represent this data by using the Venn diagram.

(b) Find how many students take only Chemistry C.

(c) How many students take only Chemistry C and Biology B.

(d) How many students take only Math M or Biology B.

(e) How many students were interviewed.

17. In a survey of 430 students about the courses they are taking, it was found that:

120 students take Math M.
50 students take only Biology B.
40 students take only Chemistry C.
60 students take only Math M.
30 students take Math M and Biology B.
30 students take Biology B and Chemistry C.
40 students take Math M and Chemistry C.

Answer the following:

(a) Represent this data by using the Venn diagram.

(b) Find how many students take all three courses.

(c) How many students take Chemistry C.

(d) How many students take Math M or Biology B.

(e) How many students do not take anyone of these three courses.

18. In a survey of 200 students about the courses they are taking, it was found that:

38 students take Biology B.
10 students take only Math M.
20 students take only Biology B.

30 students take only Chemistry C.

13 students take Math M and Biology B.

10 students take Biology B and Chemistry C.

12 students take Math M and Chemistry C.

Answer the following:

(a) Represent this data by using the Venn diagram.

(b) Find how many students take all three courses.

(c) How many students take Biology B.

(d) How many students take only Math M or Biology B.

(e) How many students do not take anyone of these three courses.

Chapter 6

Logic and Boolean Algebra

6.1 Statements

In this section, we will study two types of statements usually used in logic. The statements are defined as follows:

1. Simple Statements: A statement is a declarative sentence that can be judged either true or false. The statement is called a **simple statement** because it introduces one idea or proposition. The following sentences are examples of simple statements:

(a) Newton is a famous mathematician.
(b) $4 + 7 = 11$.
(c) $1 + 1 = 3$.
(d) February is always twenty nine days.

The statements (a) and (b) are true statements, and (c) and (d) are false statements. Examples of the sentences that are not simple statements are given by:

(a) How old are you?
(b) What is the capital of Canada?

It should be noted that we commonly represent simple statement with a lowercase letter such as p, q, r or s. As will be seen later, these symbols will be subjected to mathematical operations in analogy with the operations performed on sets.

2. Compound Statements: A statement that is built from two or more simple statements is called a **compound statement**. Examples of compound statements are given by:

(a) Today is Monday and the sky is blue.

(b) Jack is at school or Jack is taking lunch.

It is easily observed that in each example, the simple statements "Today is Monday" and "The sky is blue" are combined by using the connective "and". In the second example, the connective "or" is used to connect the two simple statements.

Having introduced the definitions of the simple and the compound statements, it is useful to discuss all possible operations that work on statements.

6.1.1   Logical Connectives:

In this section, we will discuss three distinct connectives that can be used to join two simple statements to form compound statements. Our concern will be focused on studying whether the resulting compound statement is true when subjected to a specific connective. The three distinct connectives used on simple statements are:

1. **Conjunction**: the conjunction is symbolized by $\wedge$ and read "**and**". The compound statement $p \wedge q$ is read "p and "q".

2. **Disjunction**: the disjunction is symbolized by $\vee$ and read "**or**". The compound statement $p \vee q$ is read "p or "q".

3. **Negation**: the negation is symbolized by $\sim$ and read "**not**". The compound statement $\sim p$ is read "not p".

In the following we will employ a useful technique, called the **truth table** to make our judgment on whether the compound statement is true. To construct a truth table for a compound statement, we first note that the number of distinct cases, represented by rows, in a truth table with n different simple statements is 2^n. We will now discuss how a truth table is constructed if one or more logical connectives $\wedge$, $\vee$ or $\sim$ is used.

1. **The truth table for p $\wedge$ q:** The truth table for the compound statement $p \wedge q$ is constructed as follows:

   1. Number of cases is $2^2 = 4$. This means that there are four different cases represented by four rows in the truth table shown below.

   2. In the first column under p, we write two Ts followed by two Fs. The symbol T stands for true, and the symbol F stands for false.

3. In the second column under q, T alternates with F as shown in the truth table.

4. In the third column, the answer column, under the compound statement $p \wedge q$ we write the judgment by noting that $p \wedge q$, is true only when **both p and q are true.**

The four steps discussed above are represented by Table 6.1.

p	q	$p \wedge q$
T	T	T
T	F	F
F	T	F
F	F	F

Table 6.1

2. The truth table for p ∨ q: The truth table for the statement $p \vee q$ is constructed as follows:

1. Number of cases is $2^2 = 4$. This means that there are four different cases represented by four rows in the truth table.

2. In the first column under p, we write two Ts followed by two Fs. The symbol T stands for true, and the symbol F stands for false.

3. In the second column under q, T alternates with F as shown below in the truth table. This means that the first two columns are the same as discussed before, but of course the result is different in the output column because the connective used $\vee$ is different from the connective $\wedge$.

4. In the third column, the answer column, under the compound statement $p \vee q$, we write the judgment by noting that $p \vee q$ is true when **only p or q is true.**

The four steps discussed above for constructing the truth table for **p∨q** are explained by Table 6.2 below.

p	q	$p \vee q$
T	T	T
T	F	T
F	T	T
F	F	F

Table 6.2

3. The truth table for $\sim$ p: The truth table for the negation connective is constructed as follows:

1. Number of cases is $2^1 = 2$. This means that there are two different cases represented by two rows in the truth table.

2. In the first column under p, T alternates with F.

3. In the second column under $\sim p$, we write the judgment. This can be easily realized by noting that the negation of true is false and vice versa.

The steps discussed above are illustrated by Table 6.3.

p	$\sim p$
T	F
F	T

Table 6.3

It should be noted that the three truth tables discussed above can be used to construct other truth tables for compound statements that consist of more than two simple statements. In addition, notice that the operator $\wedge$ gives T if both statements are true, whereas the operator $\vee$ gives T if at least one statement is true. The operation $\sim$ produces the negation of the statement under discussion.

To construct the truth tables for compound statements, we should consider first the priority rule that controls the operations of the connectives used. The preference of operations for the three connectives discussed is summarized as follows:

1. We first evaluate the parenthesis if there is one.
2. We next apply the negation on the statements involved.
3. The connectives $\wedge$ and $\vee$ are then evaluated from left to right.

The construction of the truth tables for compound statements will be illustrated by discussing the following examples:

Example 1. Construct a truth table for the compound statement:

$$(p \wedge q) \wedge r$$

Solution. The following steps should be observed:

(a) The compound statement consists of three simple statements p, q and r, hence, number of cases is $2^3 = 8$. This implies that eight different cases arise, and will be represented by eight rows as shown below.

(b) In the first column under p, we write four Ts followed by four Fs. Notice that the number of Ts is half the number of rows needed to construct the table.

(c) In the second column under q, two Ts alternate with two Fs.

(d) In the third column under r, one T alternates with one F.

p	q	r	$(p \wedge q)$	$(p \wedge q) \wedge r$
T	T	T	T	T
T	T	F	T	F
T	F	T	F	F
T	F	F	F	F
F	T	T	F	F
F	T	F	F	F
F	F	T	F	F
F	F	F	F	F

(e) In the fourth column, we apply the conjunction $\wedge$ on p and q, thus evaluating first the statement inside the parenthesis. Notice that we obtain

T only if the two statements p and q are true. This can be easily seen in the table.

(f) In the answer column, we employ the conjunction $\wedge$ on the result in the fourth column and the third column that represents all possibilities of the statement r, to obtain the result of this compound statement that was built of three simple statements.

The steps presented before can be easily examined in the following table.

Example 2. Construct a truth table for the compound statement:

$$(p \vee q) \vee r$$

Solution. Following the steps of Example 1, noting that we should evaluate first the compound statement inside the parenthesis, we obtain the following truth table.

p	q	r	$(p \vee q)$	$(p \vee q) \vee r$
T	T	T	T	T
T	T	F	T	T
T	F	T	T	T
T	F	F	T	T
F	T	T	T	T
F	T	F	T	T
F	F	T	F	T
F	F	F	F	F

Example 3. Write the truth table for the compound statement:

$$\sim (p \wedge \sim q)$$

Solution. In this example, two simple statements are involved, hence, number of rows is $2^2 = 4$. Notice that we first list a column for $\sim q$. Next we evaluate the compound statement inside the parenthesis. We then list our judgment in the last column by finding the negation of every case in column four.

p	q	$\sim q$	$(p \wedge \sim q)$	$\sim (p \wedge \sim q)$
T	T	F	F	T
T	F	T	T	F
F	T	F	F	T
F	F	T	F	T

Example 4. Write the truth table for the compound statement:

$$\sim p \wedge (q \vee r)$$

Solution. Notice that we first evaluate the statement of the parenthesis $(q \vee r)$. We next list the negation of p in the next column. We then list our judgment in the last column.

p	q	r	$(q \vee r)$	$\sim p$	$\sim p \wedge (q \vee r)$
T	T	T	T	F	F
T	T	F	T	F	F
T	F	T	T	F	F
T	F	F	F	F	F
F	T	T	T	T	T
F	T	F	T	T	T
F	F	T	T	T	T
F	F	F	F	T	F

Example 5. If p is true, q is false and r is false. Find the **truth value** of the compound statement $\sim r \vee (p \wedge q)$

Solution. Proceeding as before we find

p	q	r	$\sim r$	$(p \wedge q)$	$\sim r \vee (p \wedge q)$
T	F	F	T	F	T

Exercises 6.1.1

In Exercises 1 – 15, construct a **truth table** for each of the following compound statements:

1. $p \vee \sim q$

2. $\sim p \vee (p \wedge \sim q)$

3. $\sim (\sim p \vee \sim q)$

4. $p \vee (q \wedge r)$

5. $(p \vee \sim q) \wedge r$

6. $(r \wedge \sim p) \vee \sim q$

7. $(p \wedge q) \wedge (p \wedge \sim q)$

8. $(q \wedge \sim p) \vee p$

9. $(q \vee \sim p) \vee r$

10. $(p \wedge \sim q) \vee \sim r$

11. $p \vee (\sim q \wedge \sim r)$

12. $p \wedge (\sim q \vee r)$

13. $q \vee (p \wedge \sim r)$

14. $\sim r \wedge (\sim p \vee \sim q)$

15. $r \vee (p \wedge \sim q)$

In exercises 16 – 20, if p and r are false, q is true, find the **truth value** of the following statements:

16. $(\sim p \vee r) \vee q$

17. $(\sim p \wedge r) \wedge q$

18. $(p \wedge q) \vee \sim r$

19. $\sim p \wedge (q \vee r) \vee q$

20. $(\sim p \wedge q) \vee \sim r$

6.1.2 Conditional and Biconditional Statements:

In this section, we will discuss two other logical connectives that will combine simple statements to form a compound statement. The truth tables for these connectives are not as simple as the truth tables discussed before for the connectives $\wedge, \vee$ and $\sim$. However, the construction of these tables will be examined in this section. It seems reasonable at this moment to explain the two logical connectives:

1. **The conditional connective**: the conditional connective is symbolized by $\rightarrow$ and read "if ... then". An example of this conditional statement is: If you study hard, then you will get A. In symbols notation, this statement is expressed as "$p \rightarrow q$", and read as " if p then q " or "p implies q". The truth table for this conditional statement $p \rightarrow q$ is shown in Table 6.4.

p	q	$p \rightarrow q$
T	T	T
T	F	F
F	T	T
F	F	T

Table 6.4

The truth table given above indicates that the conditional statement $p \rightarrow q$ is true for every case except when p is true and q is false.

The ambiguous part of this table is how a false statement can lead to a true statement as given by the third row. This can simply be shown by introducing the false statement

$$1 = 2$$

This statement can be rewritten as

$$2 = 1$$

Adding the two statement gives

$$3 = 3$$

and the result is a true statement. This proves the third part of the table.

The truth table for the implication connective → is illustrated by the following examples.

Example 6. Construct a truth table for the statement

$$p \rightarrow (p \vee q)$$

Solution. Note that the implication connective → is the last operation to be used in constructing the truth table. We first evaluate $(p \vee q)$, then we apply the conditional connective between p and $(p \vee q)$. Accordingly, we find

p	q	$p \vee q$	$p \rightarrow (p \vee q)$
T	T	T	T
T	F	T	T
F	T	T	T
F	F	F	T

The result in the last column is true for every case. We usually call every compound statement that is always true a **tautology**.

Example 7. Construct a truth table for the statement

$$(p \rightarrow q) \wedge (\sim q \wedge p)$$

Solution.

p	q	$(p \rightarrow q)$	$\sim q$	$(\sim q \wedge p)$	$(p \rightarrow q) \wedge (\sim q \wedge p)$
T	T	T	F	F	F
T	F	F	T	T	F
F	T	T	F	F	F
F	F	T	T	F	F

Note that we first evaluate the parenthesis $(p \to q)$ where the statements are subjected to the conditional connective $\to$. Next we evaluate the parenthesis $(\sim q \wedge p)$. We then apply the $\wedge$ connective on the two compound statements inside the parenthesis.

Notice that the result in the last column is false for every case. We usually call every compound statement that is always false a **contradiction**.

2. **The biconditional connective**: the biconditional connective is symbolized by $\leftrightarrow$ and read "if and only if". The statement $p \leftrightarrow q$ means that $p \to q$ and $q \to p$. The truth table for this biconditional statement $p \leftrightarrow q$ is shown by Table 6.5.

p	q	$p \leftrightarrow q$
T	T	T
T	F	F
F	T	F
F	F	T

Table 6.5

This means that the biconditional statement $p \leftrightarrow q$ is true only when both p and q are true or both are false. The truth table can be illustrated by considering the following example.

Example 8. Construct a truth table for the statement

$$\sim (p \wedge q) \leftrightarrow (\sim p \vee \sim q)$$

Solution. Note that we first evaluate the parenthesis $(p \wedge q)$, then $\sim (p \wedge q)$ where the statements are subjected to the connective $\wedge$ then to the negation operator. Next we evaluate the parenthesis $(\sim p \vee \sim q)$. We then apply the biconditional connective $\leftrightarrow$ on the two compound statements inside the parenthesis.

Accordingly, we find

p	q	$(p \wedge q)$	$\sim (p \wedge q)$	$(\sim p \vee \sim q)$	$\sim (p \wedge q) \leftrightarrow (\sim p \vee \sim q)$
T	T	T	F	F	T
T	F	F	T	T	T
F	T	F	T	T	T
F	F	F	T	T	T

The last column of the outcome indicates that the biconditional statement is a tautology.

Exercises 6.1.2

In Exercises 1 – 8, construct a truth table for the statement.

1. $p \rightarrow \sim q$

2. $\sim p \rightarrow \sim q$

3. $\sim (q \rightarrow \sim p)$

4. $\sim q \rightarrow \sim p$

5. $\sim p \leftrightarrow q$

6. $\sim p \leftrightarrow \sim q$

7. $(p \wedge q) \leftrightarrow (p \vee q)$

8. $(p \leftrightarrow q) \vee (q \rightarrow \sim p)$

In Exercises 9 – 14, determine whether the statement is tautology, contradiction or neither.

9. $(p \wedge q) \wedge (\sim p)$

10. $(\sim p \rightarrow q) \vee \sim p$

11. $\sim q \rightarrow p$

12. $p \vee \sim (p \wedge q)$

13. $(p \vee q) \leftrightarrow (p \wedge q)$

14. $p \wedge (\sim p \wedge q)$

In Exercises 15 – 20, construct a truth table for the statement.

15. $p \rightarrow (p \vee q)$

16. $(p \wedge q) \leftrightarrow ((\sim q) \vee \sim p)$

17. $(p \wedge q) \rightarrow (\sim r)$

18. $(p \wedge q) \leftrightarrow (q \vee r)$

19. $(p \wedge q) \leftrightarrow (p \rightarrow r)$

20. $(p \wedge \sim q) \leftrightarrow (\sim p \rightarrow r)$

6.2 Switching Circuits

In this section we will discuss the concept of switching circuits. Switches are defined as electric devices usually contained in the central processing unit of any computer. The switches are either **open** or **closed**. In Figure 6.1, the switch A is either **ON**, thus allowing electric current to flow, and hence switch A is assigned the state value 1, written by $A = 1$.

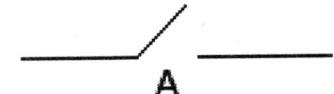

Figure 6.1

On the other hand, if the switch A is **OFF**, thus electric current is not allowed to flow, therefore switch A is assigned the state value 0 written by $A = 0$.

It is important to note that two switches or more can be connected together to make what we usually call **switching circuits**. There are two types of switching circuits: **series circuits** and **parallel circuits**:

1. **Series Circuits:**

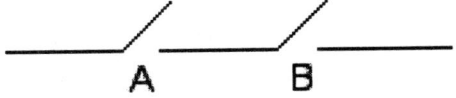

Figure 6.2

Figure 6.2 above shows clearly the structure of the series connection. In a series circuit the electric current will flow in only one path. It is obvious that the electric current will flow across only if these switches are ON. In this case, each switch is assigned the state 1. If either switch, A, B or both are OFF, current will not flow. This is identical to the AND condition that was discussed above in the logic section. Therefore, the series circuit will be characterized by the $\wedge$ notation. The following **switching table** gives the output of the current for all states of the switches A and B if connected in series.

		circuit output
A	B	$A \wedge B$
0	0	0
0	1	0
1	0	0
1	1	1

Notice that the switching table for $A \wedge B$ is constructed in a manner similar to that used in the truth table for $p \wedge q$. Recall that switching circuits are applications of logic. The number of rows used in the switching circuit is $2^2 = 4$, because two switches are used. In the first column under A, we list two 0s and two 1s. In the second column, 0 alternates with 1 as shown above. In the third column, we list the conclusion that the current flows in the circuit only if A and B are ON.

2. **Parallel Circuits:**

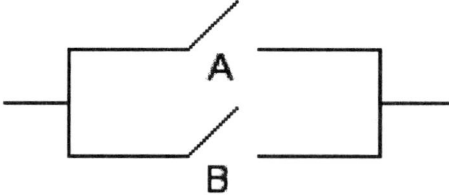

Figure 6.3

Figure 6.3 above shows the structure of the parallel connection. It is easily observed that the current will flow across when either switch A or switch B is ON. The current will not be permitted to flow when both A and B are OFF. This is identical to the OR condition we studied before. For this reason, the parallel circuit is characterized by the $\vee$ notation. The following switching table gives the output of the current for all states of the switches A and B if connected in parallel.

		circuit output
A	B	$A \vee B$
0	0	0
0	1	1
1	0	1
1	1	1

This table is similar to the truth table describing the logical connective OR for $p \vee q$ upon setting F by 0 and T by 1. In addition to the above discussion, we note that A' means the negation of the switch A. If A is ON, then A' is OFF.

As mentioned earlier, the switching circuit consists of two or more switches connected in parallel or in series. The circuit output is determined by constructing the switching table as discussed above. The switching circuits and the construction of the switching tables will be explained by introducing the following examples.

Example 1. Construct the switching table for the switching circuit described by the following figure.

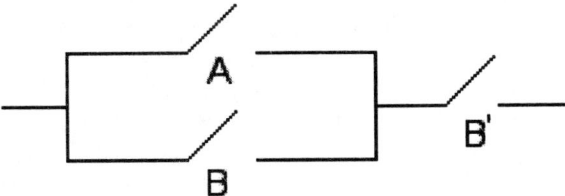

Solution. It is easily observed that A and B are connected in parallel, therefore $(A \vee B)$ represents the left part of the circuit. The left connection

$(A \vee B)$ is connected in series with B'. Accordingly, the switching circuit is represented by the expression $(A \vee B) \wedge B'$. As a result of this compound connection, the switching circuit is thus described by the following switching table.

A	B	$A \vee B$	B'	circuit output $(A \vee B) \wedge B'$
0	0	0	1	0
0	1	1	0	0
1	0	1	1	1
1	1	1	0	0

Concerning the evaluation process, we apply the preference rule used before in the logic section.

Example 2. Construct the switching table for the switching circuit described by the figure

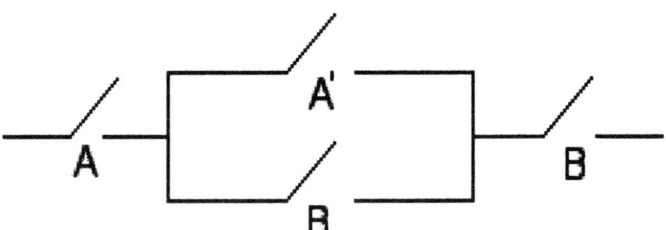

Solution. It is easily observed that the switch A' and the switch B are connected in parallel, hence represented by $(A' \vee B)$. The switches A and B are connected with the middle part in series. Accordingly, the switching circuit is represented by the following compound expression

$$A \wedge (A' \vee B) \wedge B$$

We construct a column for $(A' \vee B)$, then we evaluate $A \wedge (A' \vee B)$ and the circuit output can be listed by evaluating $A \wedge (A' \vee B) \wedge B$. Consequently, the following switching table can be easily constructed.

A	B	A'	$A' \vee B$	$A \wedge (A' \vee B)$	circuit output $A \wedge (A' \vee B) \wedge B$
0	0	1	1	0	0
0	1	1	1	0	0
1	0	0	0	0	0
1	1	0	1	1	1

Example 3. Construct the switching table for the switching circuit described by the figure

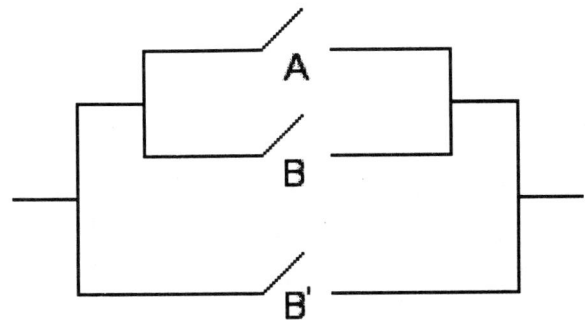

Solution. The switching circuit is represented by the following expression

$$(A \vee B) \vee B'$$

A	B	$A \vee B$	B'	circuit output $(A \vee B) \vee B'$
0	0	0	1	1
0	1	1	0	1
1	0	1	1	1
1	1	1	0	1

Proceeding as before, we first evaluate $(A \vee B)$, then we list the circuit output in the last column. Accordingly, we obtain the switching table given above.

Exercises 6.2

Construct the **switching table** for each of the following switching circuits:
1.

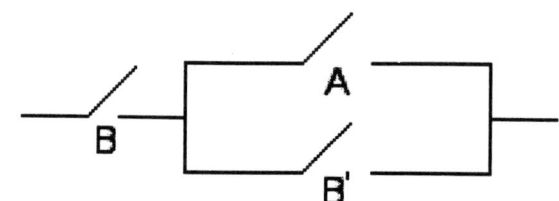

2.

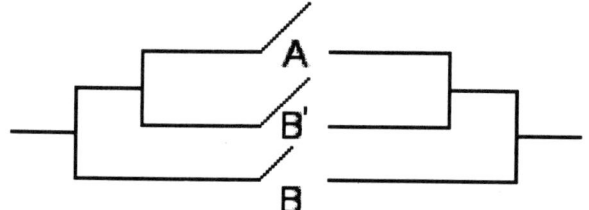

3.

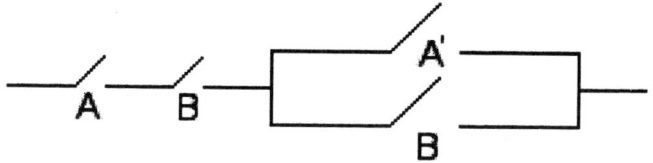

4.

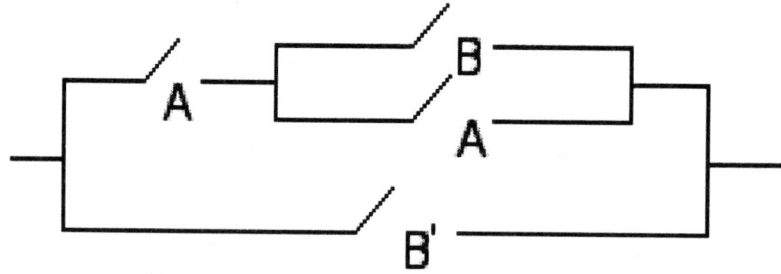

5.

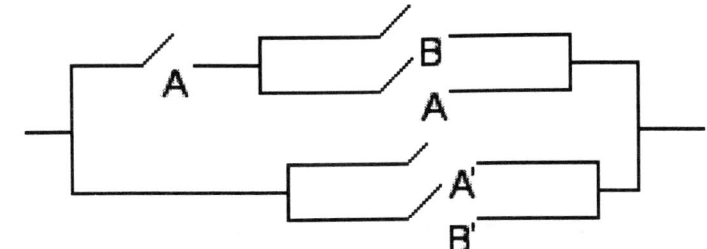

6.

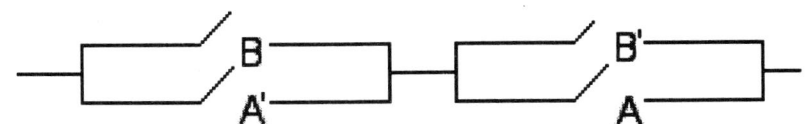

7.

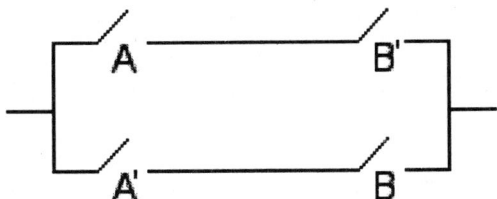

8.

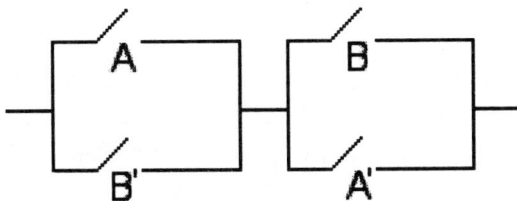

6.3 Boolean Algebra

George Boole (1815 – 64) introduced the so-called Boolean Algebra. A Boolean algebra B is a set containing distinct elements 0 and 1 together with two binary operations, the **sum** denoted by $+$, and the **product** denoted by $*$, and a unary operation called the **complement** denoted by $'$. For all x, y and $z \in B$, the following laws are satisfied.

1. **Associative laws:**

   (i) $x + (y + z) = (x + y) + z$ (ii) $x(yz) = (xy)z$

2. **Commutative laws:**

   (i) $x + y = y + x$ (ii) $xy = yx$

3. **Distributive laws:**

   (i) $x + yz = (x + y)(x + z)$ (ii) $x(y + z) = xy + xz$

4. **Identity laws:**

   (i) $x + 0 = x$ (ii) $x * 1 = x$

5. **Complement laws:**

   (i) $x + x' = 1$ (ii) $x * x' = 0$

The following truth tables illustrate the evaluation of the Boolean expressions that will be discussed later. Given the Boolean algebra defined by the set $B = \{0, 1\}$, and the operation $+$, $*$ and $'$, the following truth tables hold.

+	0	1
0	0	1
1	1	1

*	0	1
0	0	0
1	0	1

x	x'
0	1
1	0

A great attention should be paid to the fact that $1 + 1 = 1$ in Boolean algebra as shown in the table above. This is in fact similar to the parallel connection of two switches A and B. If A is ON, or B is ON, the current flows.

It is also important to note that the binary operation $+$ is similar to the disjunction $\vee$, and the binary operation $*$ is similar to the conjunction $\wedge$ as explained in the previous section.

Accordingly, given the Boolean algebra B, where x and $y \in B$, the following truth tables hold.

x	y	$x+y$
0	0	0
0	1	1
1	0	1
1	1	1

x	y	$x*y$
0	0	0
0	1	0
1	0	0
1	1	1

6.3.1 Boolean Expressions:

A Boolean expression is any expression that contains Boolean variables, and one or more of the operations $+$, $*$ and $'$ and a parenthesis. The expressions

$$xy' + z, x'(y + z)$$

and

$$xy' + yz' + zx$$

are Boolean expressions where x, y and z are Boolean variables. In the following examples, we illustrate the evaluation of Boolean expressions.

It should be noted that in evaluating the Boolean expressions, we first evaluate expressions in parenthesis, then we evaluate complements, products and finally we evaluate sums.

Example 1. Construct the truth table for the Boolean expression E defined by:

$$E = xy' + x'y$$

Solution. We notice that the Boolean expression E contains two variables x and y, therefore the number of rows of the truth table is $2^2 = 4$. Following the tables discussed above we find

x	y	x'	y'	xy'	$x'y$	E
0	0	1	1	0	0	0
0	1	1	0	0	1	1
1	0	0	1	1	0	1
1	1	0	0	0	0	0

Example 2. Construct the truth table for the Boolean expression E defined by:

$$E = x(x' + y) + y'$$

Solution. Following the tables discussed above we find

x	y	x'	y'	$x' + y$	$x(x' + y)$	E
0	0	1	1	1	0	1
0	1	1	0	1	0	0
1	0	0	1	0	0	1
1	1	0	0	1	1	1

Example 3. Construct the truth table for the Boolean expression E defined by:

$$E = xyz' + x'yz + xy'z$$

Solution. In this example, the Boolean expression E contains three variables x, y and z, therefore the number of rows is $2^3 = 8$. Following the tables discussed above we find

x	y	z	x'	y'	z'	xyz'	$x'yz$	$xy'z$	E
0	0	0	1	1	1	0	0	0	0
0	0	1	1	1	0	0	0	0	0
0	1	0	1	0	1	0	0	0	0
0	1	1	1	0	0	0	1	0	1
1	0	0	0	1	1	0	0	0	0
1	0	1	0	1	0	0	0	1	1
1	1	0	0	0	1	1	0	0	1
1	1	1	0	0	0	0	0	0	0

Exercises 6.3.1

Construct the truth table for the following Boolean expressions:

1. $E = (x + y) + x'$

2. $E = (x' + y) + y'$

3. $E = x(x' + y) + x$

4. $E = x'(x + y') + x$

5. $E = xy' + yz'$

6. $E = xy' + yz' + zx'$

7. $E = xy + yz + zx$

8. $E = x(y + z) + z(x + y) + y(z + x)$

9. $E = x(y + z') + z(x + y') + y(z + x')$

10. $E = (x + y + z') + (z + x + y') + (y + z + x')$

6.3.2   The Minimal-Sum Form:

In this section we will focus our study on the algebraic method to find the minimal form of any Boolean expression written in the form of a sum of terms. A well-known method called the Karnaugh map can handle the minimization process to any Boolean expression. The algebraic approach and the Karnaugh map will be used to discuss how a minimal form can be found to any Boolean expression. It should be noted here that the minimal form is a simpler form of the given Boolean expression that will reduce the size of calculations and provide the same column that results when constructing a truth table. In other words, it is possible sometimes to reduce the number of terms of any Boolean expression by an equivalent Boolean expression of a minimal number of terms in a sum form.

The Minimal-Sum Form by Algebra

In this method, we use the following important identities:

$$x + x' = 1$$

$$xx' = 0$$

It should be noted that these two identities can be easily derived by constructing the truth table for the left side of each identity. The minimal-sum form for a Boolean expression can be obtained as follows:

1. For a Boolean expression with an even number of terms:
In this case, we use the method of factoring by grouping, where we group terms that have the same common factors. By factoring each group and using the identities given above we obtain the minimal-sum form. This can be illustrated by the following examples.

Example 4. Find the minimal-sum form for the Boolean expression:

$$E = xy + xy' + x'y + x'y'$$

Solution. Grouping the first two terms and the last two terms, and by factoring each group we find

$$E = x(y + y') + x'(y + y')$$

and this is equivalent to

$$E = x + x'$$

so that

$$E = 1$$

upon using $y + y' = 1$ and $x + x' = 1$.

Example 5. Find the minimal-sum form for the Boolean expression:

$$E = xyz + xy'z + x'yz' + x'y'z'$$

Solution. Grouping the first two terms and the last two terms, and by factoring each group we find

$$E = xz(y + y') + x'z'(y + y')$$

and this is equivalent to

$$E = xz + x'z'$$

upon using $y + y' = 1$.

Example 6. Find the minimal-sum form for the Boolean expression:

$$E = xyz' + xy'z' + x'yz' + x'y'z'$$

Solution. Grouping the first two terms and the last two terms, and by factoring each group we find

$$E = xz'(y + y') + x'z'(y + y')$$

and this is equivalent to

$$E = xz' + x'z'$$

so that

$$E = z'(x + x')$$

which is equivalent to

$$E = z'$$

upon using $x + x' = 1$.

Example 7. Find the minimal-sum form for the Boolean expression:

$$E = xyz + xyz' + xy'z' + xy'z$$

Solution. Proceeding as before, we find

$$E = xy(z + z') + xy'(z + z')$$

and this is equivalent to

$$E = xy + xy'$$

so that

$$E = x$$

upon using $y + y' = 1$.

2. For a Boolean expression with an odd number of terms:
In this case, we first add an additional term to the odd-numbered expression. The additional term must be a term selected from the expression E under discussion and should have two variables, such as x, y, or z, in common with the odd-numbered term. By factoring the resulting expression of even number of terms and by using the identities given above we obtain the minimal-sum form. This can be illustrated by the following examples.

Example 8. Find the minimal-sum form for the Boolean expression:

$$E = x'yz' + x'y'z' + xy'z'$$

Solution. We first notice that the second term and the third term have two variables in common, namely y' and z', hence we add the term $x'y'z'$ to E to obtain

$$E = x'yz' + x'y'z' + xy'z' + x'y'z'$$

Grouping the first two terms and the last two terms, and by factoring each group we find

$$E = x'z'(y + y') + y'z'(x + x')$$

and this is equivalent to

$$E = x'z' + y'z'$$

Example 9. Find the minimal-sum form for the Boolean expression:

$$E = xyz + xyz' + x'yz' + x'y'z' + xy'z'$$

Solution. Grouping the first two terms and the second two terms we find

$$E = xy(z + z') + x'z'(y + y') + xy'z'$$

We notice that two terms of the expression E have two variables in common with the last term of E, namely xyz' and $x'y'z'$. For example if we add the term xyz' to the last expression E and factor we find

$$E = xy(z + z') + x'z'(y + y') + xz'(y + y')$$

and this is equivalent to

$$E = xy + x'z' + xz'$$

Factoring again and using the identity we obtain

$$E = xy + z'$$

On the other hand, if we add the term $x'y'z'$ instead of xyz' to E we find

$$E = xy(z + z') + x'z'(y + y') + y'z'(x + x')$$

or equivalently

$$E = xy + x'z' + y'z'$$

It should be noted here that two minimal forms are obtained in this example. This clearly indicates that the minimal-sum form is not unique.

6.3.3 Karnaugh Map

Karnaugh map can be described as a graphing technique that provides a pictorial method that groups together Boolean terms with common factors for minimizing Boolean expressions. Recall that Boolean algebra was used before to obtain a minimal form for Boolean expressions. The Karnaugh map is a diagram that can also be used to simplify Boolean expressions, because it provides a straightforward technique for minimizing Boolean expressions.

In this section we will focus our work on expressions of three Boolean variables, where Karnaugh map can be constructed in the same way for expressions of more variables.

3-Variable Karnaugh Map

The n-variable Karnaugh map consists of 2^n cells. This means that the 3-variable Karnaugh map consists of 8 cells as shown by Figure 6.4 below:

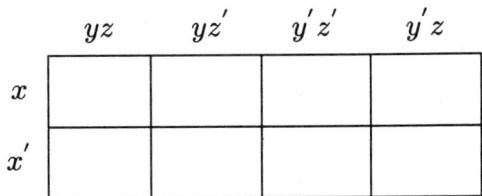

Figure 6.4

The Karnaugh map properties and the minimization process are as follows:
1. Plot 1 in the Karnaugh map for each Boolean term in the cell where the product of the variables at the beginning of the row and at the top of the column corresponds to the product of variables of the given term. For example, the terms xyz and $x'y'z$ are plotted at the leftmost and at the rightmost as shown in Figure 6.5 below:

	yz	yz'	$y'z'$	$y'z$
x	1			
x'				1

Figure 6.5

2. Each cell in a Karnaugh map differs from its adjacent cells by only one variable. This means that two cells are adjacent if they have a common border line. Moreover, each cell has three adjacent cells. It is important to notice that the leftmost cell in each row in Karnaugh map is considered adjacent to the rightmost cell of the same row. This means that the cells xyz and $xy'z$ are adjacent. Similarly, the cells $x'yz$ and $x'y'z$ are adjacent. However, the cells xyz and $x'y'z$ are not adjacent because these cells differ in two variables and do not have a common border line.

3. To find a minimal form, we should consider the following:

(i) We first construct basic rectangles or loops that must contain 2^n cells where $n = 0, 1, 2, 3$.

(ii) For $n = 0$, we obtain a loop of a single cell that cannot be minimized.

(iii) For $n = 1$, a loop of two adjacent cells is obtained. One independent variable can be eliminated and the Boolean expression can be minimized to one term of two variables. This can be shown in Figure 6.6 where the expressions $E = xyz + xyz'$ and $E_1 = xy'z' + x'y'z'$ can be minimized to $E = xy$ and $E_1 = y'z'$ respectively.

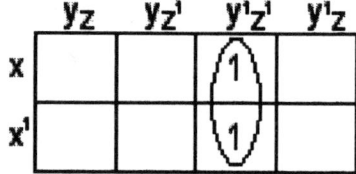

Figure 6.6

(iv) For $n = 2$, a loop of 4 adjacent cells is obtained. Two independent variables can be eliminated and the Boolean expression can be minimized to one term of one variable only. Notice that in this case, the loop consists of a 1×4 cells or a 2×2 cells. Moreover, the loops in any case must be square or rectangular, and cannot be L-shaped or diagonal. Figure 6.7 shows examples of four cells that can be minimized to one variable only.

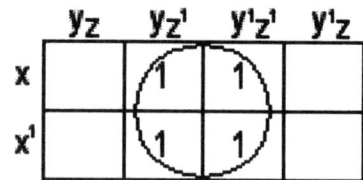

Figure 6.7a,b

In the loops in Figure 6.7 a,b, the minimal form is $E = x$, and for the second loop the minimal form is $E_1 = z'$. However, the minimal form for the loop in Figure 6.7c is $E = z$.

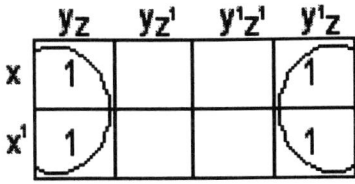

Figure 6.7c

(v) For $n = 3$, the loop consists of eight cells in the form of a 2×4 rectangle. In this case $E = 1$.

(vi) Loops may overlap, especially if the number of terms is $3, 5$ or 7. For example, if the number of cells is 3 as shown by Figure 6.8 below, this can be represented by two overlapping 1×2 rectangles or by two 1×2 and 2×1 rectangles respectively.

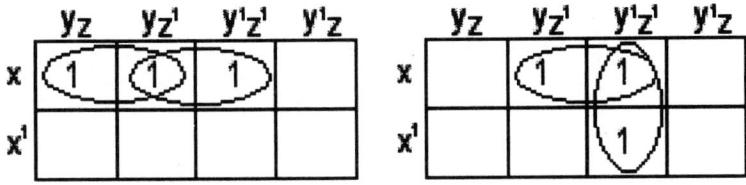

Figure 6.8

For specific cases of 5 cells as shown by Figure 6.9 below, the minimal forms are $E = x + yz'$ and $E_1 = y + x'z'$.

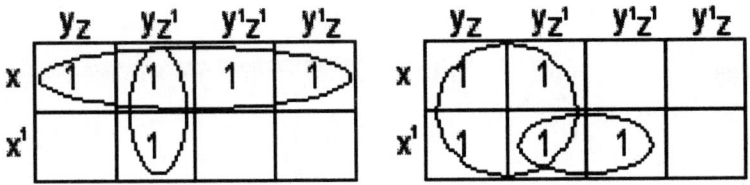

Figure 6.9

(vii) It is important to note that the minimal form of a Boolean expression may not be unique. It is clear from Figure 6.10, that overlapped loops can be made in two distinct ways to give the two minimal forms:

$$E = xy + y'z' + x'y'$$

and

$$E = xy + xz' + x'y'$$

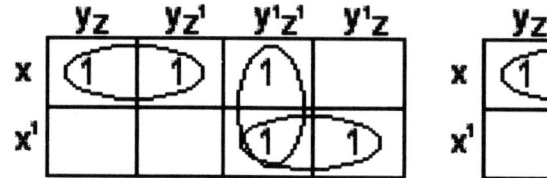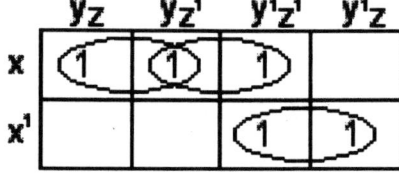

Figure 6.10

Exercises 6.3.2

In Exercises 1 – 6, evaluate the Boolean expressions without using truth tables.

1. $E = (x + x') + (y + y') + (z + z')$

2. $E = x(y + y') + x'(z + z')$

3. $E = (x + x' + y) + (z + z' + y')$

4. $E = (x + x' + xx')$

5. $E = xx'(y + y' + x + x')$

6. $E = x(y + y') + x'(z + z') + xx'$

In Exercises 7 – 14, find the minimal-sum form for the following Boolean expressions by using algebra.

7. $E = xy + xy'$

8. $E = yz + yz'$

9. $E = xy + xy' + x'z + x'z$

10. $E = x'yz + x'yz' + x'y'z' + x'y'z$

11. $E = xy'z' + xy'z + x'y'z' + x'y'z$

12. $E = x'yz + x'y'z + xyz' + xy'z'$

13. $E = xyz + xy'z + xy'z' + xyz'$

14. $E = xyz + xyz' + x'yz + x'yz'$

In Exercises 15 – 20, find the minimal-sum form for the following Boolean expressions by using algebra.

15. $E = xyz + xy'z + xy'z'$

16. $E = xy'z + xy'z' + xyz'$

17. $E = xy'z + x'y'z + x'yz$

18. $E = xyz + xy'z + xy'z' + x'y'z' + x'yz'$

19. $E = xyz' + xy'z + xy'z' + x'y'z' + x'yz'$

20. $E = xyz + xy'z' + xyz' + x'y'z' + x'yz'$

In Exercises 21 – 24, find the minimal-sum form for the following Boolean expressions.

21. $E = xyz + xyz' + x'yz + x'yz' + xy'z' + x'y'z'$

22. $E = xy'z + xyz' + xy'z' + x'y'z + x'yz' + x'y'z'$

23. $E = xyz + xyz' + xy'z' + x'yz + x'yz' + x'y'z' + xy'z + x'y'z$

24. $E = xyz + xyz' + xy'z' + x'yz + x'yz' + x'y'z' + xy'z$

25. Solve exercises 12–22 by using Karnaugh map.

26. Show that the minimal form of the Boolean expression represented by the following Karnaugh map is not unique:

	yz	yz'	$y'z'$	$y'z$
x	1	1	1	1
x'		1	1	

27. Repeat Exercises 26 for the map:

	yz	yz'	$y'z'$	$y'z$
x	1	1	1	
x'	1	1	1	1

28. Repeat Exercises 27 for the map:

	yz	yz'	$y'z'$	$y'z$
x		1	1	1
x'	1	1		

29. Repeat Exercise 28 for the map:

	yz	yz'	$y'z'$	$y'z$
x	1	1	1	1
x'	1			1

30. Show that there is only a unique minimal form for the following Boolean expression represented by the following map:

	yz	yz'	$y'z'$	$y'z$
x	1	1		
x'	1	1		1

Chapter 7

Combinatorics

7.1 Introduction

In this chapter, we will focus our study on counting techniques. The work will be directed towards:

(1) counting the number of possible choices, ways or committees, that we can form from an experiment that consists of several activities, where each activity occurs number of times or number of ways. The following are examples of this type of experiments:

(i)A dealer has red or green cars only, where each type of cars can be found with an automatic or manual transmission. It is useful here to find the number of possible choices a buyer has.

(ii)A committee consists exactly of 1 man and 1 woman. How many distinct committees we can form from 3 men and 2 women.

(iii) An auto license plates have three letters followed by three digits. How many different license plates can be made if repetitions of letters and digits are allowed.

(2) counting the number of different arrangements of r objects that can be formed from n distinct objects. As will be seen later, the number of arrangements can be obtained if the order of the objects matters and is significant on the arrangements or not. This can be clearly seen when discussing the arrangements that can be made from the digits 1, 2, and 3. To form 2-digit numbers we obtain the following 6 ordered arrangements:

$$12, 13, 23$$
$$21, 31, 32$$

whereas we obtain the following 3 unordered arrangements

$$1 \times 2, 1 \times 3, 2 \times 3$$

if a 2-digit product is required where the order has no effect in producing different arrangements.

As stated earlier, we will examine in this chapter all different types of arrangements, and will focus our concern on techniques that count the number of resulting arrangements or choices.

7.2   Fundamental Counting Principle

Suppose an experiment of k distinct activities is to be performed. If the first activity can be performed in n_1 ways, the second activity can be performed in n_2 ways, and so on. The experiment is performed such that each selection is independent of the others, then the number of possible ways to perform this experiment is given by

$$\text{number of possible ways} = n_1 \times n_2 \times n_3 \times \cdots \times n_k$$

This principle is called **The Fundamental Counting Principle**.

Let us consider the example introduced in Section 7.1. If a dealer has red or green cars only, where each kind can be found with automatic or manual transmission. It is obvious that activity 1, the selection of a car according to its color, can be performed in 2 ways. Activity 2, the selection of a car according to its transmission, can be performed in 2 ways. Accordingly, the number of choices available by the buyer is given by

$$\text{number of possible choices} \quad = \quad 2 \times 2 = 4$$

Another example we introduced before in Section 7.1 is about the number of possible committees we can form from 3 men and 2 women if a committee consists exactly of 1 man and 1 woman. It is clear that activity 1, the selection of a man, can be made in 3 possible ways. On the other hand, activity 2, the selection of 1 woman, can be made in 2 ways. We therefore conclude that

$$\text{number of possible committees} \quad = \quad 3 \times 2 = 6$$

The fundamental counting principle will be further illustrated by discussing the following examples.

Example 1. A license plate is to be made of 3 letters followed by 2 digits. How many different license plates we can make if letters and digits can be repeated.

Solution. It is obvious that we have 5 distinct activities that will be performed, namely 3 letters followed by 2 digits. It is thus clear that there are 26 possible choices for each of the first three activities and 10 ways for each of the last two activities. Accordingly,

$$\text{number of plates} \quad = \quad 26 \times 26 \times 26 \times 10 \times 10 = 1757600$$

Example 2. Use Example 1 to find how many different license plates we can make if repetitions in letters cannot be used whereas digits can be repeated.

Solution. We can fill the first place by 26 ways. The second place is then can be filled by 25 ways because if A was selected as shown by the table then A cannot be repeated. In the same manner, if A and B were selected for a specific plate, then we can fill the third place by 24 ways. The selection of digits remains the same as in Example 1.

$$\text{number of plates} \quad = \quad 26 \times 25 \times 24 \times 10 \times 10 = 1560000$$

Example 3. Use Example 1 to find how many different license plates we can make if repetitions in letters and digits are not allowed.

Solution. We can fill the first place by 26 ways. The second place is then can be filled by 25 ways because if A was selected, then A cannot be repeated. In the same manner, we can fill the third place by 24 ways. Similarly, the fourth place and the fifth place can be filled by 10 ways and 9 ways respectively.

$$\text{number of plates} \quad = \quad 26 \times 25 \times 24 \times 10 \times 9 = 1404000$$

Example 4. Jack has 5 pairs of trousers, 7 shirts and 3 pairs of boots. How many different outfits can he wear.

Solution. There are three activities, where the first one can be made by 5 choices, the second by 7 ways and the third by 3 selections. This gives

$$\text{number of different outfits} \quad = \quad 5 \times 7 \times 3 = 105$$

Example 5. How many three-digit numbers are odd.

Solution. The hundreds can be filled by 9 digits from 1 to 9. The ten's place can be filled by 10 digits from 0 to 9. However, in the one's place we can use only 5 digits, namely $1, 3, 5, 7, 9$, because we are seeking odd numbers only. Therefore, we find

$$\text{number of odd three-digit numbers} \ = \ 9 \times 10 \times 5 = 450$$

As a result, we also have 450 even three-digit numbers.

Example 6. A restaurant has a lunch special which consists of 4 kinds of sandwiches, 2 soups, 3 deserts, and 6 kinds of drinks. How many lunch specials can be made if only one of each kind is to be selected.

Solution. It is now obvious that we get

$$\text{number of lunch specials} \ = \ 4 \times 2 \times 3 \times 6 = 144$$

Example 7. A coin is tossed 6 times. How many possible outcomes are there.

Solution. The experiment consists of 6 distinct tossing activities. Each tossing results in 2 choices, head or tail. Consequently,

$$\text{number of possible outcomes} \ = \ 2 \times 2 \times 2 \times 2 \times 2 \times 2 = 64$$

Example 8. A coin is tossed 4 times and a die is rolled 3 times. How many possible outcomes are there.

Solution. The experiment consists of 7 distinct activities. Each tossing results in 2 choices, head or tail and each rolling gives 6 possible outcomes. Consequently,

$$\text{number of possible outcomes} \ = \ 2 \times 2 \times 2 \times 2 \times 6 \times 6 \times 6 = 3456$$

Exercises 7.2

1. A dealer has red, green or blue cars. Each type of cars can be found with automatic or manual transmission. How many possible choices a buyer has?

2. A dealer has red, green, black or white car. Each type of cars can be found with automatic or manual transmission. How many possible choices a buyer has?

3. In a restaurant, there are 3 kinds of soup, 4 kinds of steak and 2 choices for desert. How many possible meals a customer has?

4. A committee of a man and a woman is to be selected from 5 men and 7 women. How many distinct committees can be formed?

5. A student has to select 1 Math course, 1 Science course and 1 English course. The schedule includes 3 distinct Math courses, 2 distinct Science courses and 4 distinct English courses. How many possible selections the student can make?

6. A license plate is to be made of 3 letters followed by 3 digits. How many different license plates we can make if letters and digits can be repeated.

7. A license plate is to be made of 3 letters followed by 3 digits. How many different license plates we can make if digits cannot be repeated.

8. A license plate is to be made of 3 letters followed by 3 digits. How many different license plates we can make if letters and digits cannot be repeated.

9. A coin is tossed 8 times. How many possible outcomes are there?

10. Rolling a die 4 times. How many possible outcomes are there?

11. A coin is tossed 5 times and a die is rolled 4 times. How many possible outcomes are there?

12. A five person committee is to select a chair, secretary and a treasurer. How many selections can be made.

13. A license plate is to be made of 3 letters followed by 3 digits. How many different license plates we can make if any plate cannot start with the letter U and if the letter O is not used anywhere. Moreover, no license plate can have a three digit number greater than 899.

14. A password to a computer consists of 5 characters, 3 letters followed by 2 digits from 1 to 9. How many different passwords we can make if:

(i) repetitions of letters and digits are allowed,

(ii) only repetitions of letters are not allowed.

15. A restaurant has 3 kinds of sandwiches, 4 soups, 3 deserts and 5 drinks. A meal is selected from one of each. How many meals are possible.

16. A test consists of 10 True/False questions. Assuming that all questions are answered, how many ways the test can be answered.

17. How many four-digit numbers are odd?

7.3 The Pigeonhole Principle

The first version of the pigeonhole principle was written by the German mathematician Johann Dirichlet in 1834. If a flock of n pigeons are placed into k pigeonholes, $n > k$, then the pigeonhole principle states that there must be at least one pigeonhole with at least two pigeons in it. The pigeonhole principle applies to other objects besides pigeons and pigeonholes. For example, if we want to store 7 balls into 5 boxes, then there must be at least one box with at least two balls in it.

Theorem 1: The Pigeonhole Principle

If $k + 1$ objects are placed into k boxes, then there is at least one box that contains two or more objects.

This can be illustrated by considering 6 balls to be placed in 5 boxes. Then there is at least one box that contains two or more objects.

However, if we have 12 balls and 5 boxes, then the theorem of the pigeonhole principle should be generalized as shown by the following generalized theorem.

Theorem 2: The Generalized Pigeonhole Principle

If n objects are placed into k boxes, then there is at least one box that contains at least $\lceil \frac{n}{k} \rceil$ objects.

Note the $\lceil x \rceil$ is the ceiling function that denotes the smallest integer larger than or equal to x. For example, $\lceil 2.4 \rceil = 3$, $\lceil -2.4 \rceil = -2$, $\lceil 3 \rceil = 3$ and so on. In what follows we study some examples where the generalized pigeonhole principle will be applied

Example 1. Given 102 people. Find the number of people born in the same month.

Solution. It is obvious that $n = 102$ and $k = 12$. Using the generalized pigeonhole principle gives:

number of people born in the same month $= \lceil \frac{102}{12} \rceil = \lceil 8.5 \rceil = 9$

Example 2. Given a group of 438 people. Find how many people with the same birthday

Solution. It is obvious that $n = 438$ and $k = 365$. Using the generalized pigeonhole principle gives:

number of people with the same birthday $= \lceil \frac{438}{365} \rceil = \lceil 1.2 \rceil = 2$

Example 3. Given the set

$$A = \{1, 2, 3, 4, 5, 6\}$$

If 4 integers are selected from A. Show that we can find 2 integers with its sum equal to 7

Solution. We first group the integers of A into the subsets that include pairs of integers with sum equal to 7. In other words, we form the subsets $\{1,6\}, \{2,5\}, \{3,4\}$. It is obvious that $n = 4$ and $k = 3$. Using the generalized pigeonhole principle gives:

number of integers with sum equal to $7 = \lceil \frac{4}{3} \rceil = \lceil 1.33 \rceil = 2$

Example 4. Given 52 integers selected inclusively from the integers between 1 and 100. Show that 2 of the selected integers are consecutive

Solution. We first group the integers from 1 and 100 into 50 pairs of consecutive integers $\{1,2\}, \{3,4\}, \{5,6\}, \cdots, \{99,100\}$. It is obvious that $n = 52$ and $k = 50$. Using the generalized pigeonhole principle gives:

number of integers $= \lceil \frac{52}{50} \rceil = 2$

Example 5. What is the minimum number of students required in a class so that at least seven students will receive the same grade, if there are five possible grades A, B, C, D, and F.

Solution. It is obvious that $n = x$ and $k = 5$. Using the generalized pigeonhole principle gives:

$$\lceil \frac{x}{5} \rceil = 7$$

This in turn gives

$$x = 5 \times 6 + 1 = 31$$

Example 6. What is the minimum number of students to be enrolled in a university so that at least 60 students come from the same state

Solution. It is obvious that $n = x$ and $k = 50$ states. Using the generalized pigeonhole principle gives:

$$\lceil \frac{x}{50} \rceil = 60$$

This in turn gives

$$x = 50 \times 59 + 1 = 2951$$

Exercises 7.3

1. Given 2921 people. Find the number of people born in the same month

2. Given 3006 people. Find how many people with the same birthday

3. Given the set
$$A = \{1, 2, 3, 4, 5, 6, 7, 8\}$$
If 5 integers are selected from A, show that we can find 2 integers with its sum equal to 9

4. Given the set
$$A = \{2, 3, 4, 5, 6, 7, 8, 9, 10, 11\}$$
If 6 integers are selected from A, show that we can find 2 integers with its sum equal to 13

5. Given a group of 169. Show that there exists at least 7 people with the same last initial

6. Given a group of 143. Show that there exists at least 6 people with the same first initial

7. If you have black, white, red, blue, and green socks in a drawer. How many socks you must choose to get two of the same color

8. If you have black, white, red, blue, yellow, and green socks in a drawer. How many socks you must choose to get two of the same color

9. Given a group of 9 people. Show that there must be at least 5 of the same gender

10. 12 integers are randomly chosen. Show that two of these integers must have the same unit digit

11. Given 162 integers selected inclusively from the integers between 1 and 300. Show that 2 of the selected integers are consecutive

12. What is the minimum number of students required in a class so that at least eight students will receive the same grade, if there are five possible grades A, B, C, D, and F

13. What is the minimum number of people so that at least three people born on the same day of the week

14. What is the minimum number of people so that at least three people born on the same month

15. What is the minimum number of integers we must pick so that at least two of these integers have the same remainder when divided by 5

7.4 Permutations

Given the digits 1, 2 and 3. How many possible two-digit numbers we can form? It is clear that the numbers, or the arrangements, that can be formed are:

$$12, 13, 23$$

$$21, 31, 32$$

We therefore conclude that the number of arrangements is 6. This can be stated by saying that the number of arranging 3 objects where 2 objects are taken at a time is 6. This is usually denoted by:

$$3P2 = 6$$

In other words, the 2-permutations of 3 objects is 6. It is important to note that in permutations, the order of the objects in the resulting arrangement is important. Accordingly, we say that the 6 arrangements obtained above are ordered arrangements or **permutations**. This can be easily seen by noting that the arrangement 23 is different than the arrangement 32. This means that changing the order of the objects in any arrangement results in a different choice.

The formula that will be used to evaluate $3P2$ can be derived from the fundamental principle of counting where place 1 is performed in 3 choices, and hence place 2 is performed in 2 choices. Accordingly,

$$\begin{aligned} 3P2 &\equiv 2 \text{ consecutive decreasing factors that start with } 3 \\ &= 3 \cdot 2 = 6 \end{aligned}$$

In the same manner, we can evaluate $5P3$ as follows:

$$\begin{aligned} 5P3 &\equiv 3 \text{ consecutive decreasing factors that start with } 5 \\ &= 5 \cdot 4 \cdot 3 = 60 \end{aligned}$$

The general formula for permutations is therefore given by

$$nPr = n(n-1)(n-2)\cdots(n-r+1), \quad 1 \le r \le n$$

that gives the number of permutations of n objects arranged by taking r objects at a time.

The following examples explain the evaluation process of permutations.

Example 1. Evaluate $6P2$, $9P3$, $10P4$

Solution.

$$
\begin{aligned}
6P2 &= 6 \cdot 5 = 30 \\
9P3 &= 9 \cdot 8 \cdot 7 = 504 \\
10P4 &= 10 \cdot 9 \cdot 8 \cdot 7 = 5040
\end{aligned}
$$

Example 2. Express $nP2$ and $nP3$ in terms of n

Solution.

$$
\begin{aligned}
nP2 &\equiv 2 \text{ consecutive decreasing factors starting with } n \\
    &= n(n-1) \\
nP3 &\equiv 3 \text{ consecutive decreasing factors starting with } n \\
    &= n(n-1)(n-2)
\end{aligned}
$$

Example 3. Find n if $nP2 = 72$

Solution.

$$
\begin{aligned}
nP2 &= 72 \\
n(n-1) &= 72 \\
n^2 - n - 72 &= 0 \\
(n-9)(n+8) &= 0 \\
n &= 9
\end{aligned}
$$

Notice that n represents a positive number of objects, hence we did not accept the negative value of n.

Example 4. Find n if $nP2 + 2n = 110$

Solution.

$$
\begin{aligned}
nP2 + 2n &= 110 \\
n(n-1) + 2n &= 110 \\
n^2 + n - 110 &= 0 \\
(n-10)(n+11) &= 0 \\
n &= 10
\end{aligned}
$$

7.4.1 Factorial Notation

It is important to note the following permutations:

$$3P3 = 3 \cdot 2 \cdot 1$$
$$4P4 = 4 \cdot 3 \cdot 2 \cdot 1$$
$$5P5 = 5 \cdot 4 \cdot 3 \cdot 2 \cdot 1$$

We notice that if $n = r$ in nPr, we multiply all consecutive decreasing factors that start with n and end with 1. This case, nPn, is usually denoted by $n!$, (read n factorial). In the following, we list the first few values of $n!$

$$0! = 1$$
$$1! = 1$$
$$2! = 2 \cdot 1 = 2$$
$$3! = 3 \cdot 2 \cdot 1 = 6$$
$$4! = 4 \cdot 3 \cdot 2 \cdot 1 = 24$$
$$5! = 5 \cdot 4 \cdot 3 \cdot 2 \cdot 1 = 120$$
$$6! = 6 \cdot 5 \cdot 4 \cdot 3 \cdot 2 \cdot 1 = 720$$
$$7! = 7 \cdot 6 \cdot 5 \cdot 4 \cdot 3 \cdot 2 \cdot 1 = 5040$$

It should be noted here that $0! = 1$ will be proved mathematically below.

Based on the definition of the factorial notation we discussed above, the following general formula for nPr can be written in terms of factorial notations given by:

$$nPr = \frac{n!}{(n-r)!}$$

This alternative expression for nPr can be easily proved by using the left hand side. Thus we find

$$
\begin{aligned}
LHS &= \frac{n!}{(n-r)!} \\
&= \frac{n(n-1)\cdots(n-r+1)(n-r)\cdots 3 \cdot 2 \cdot 1}{(n-r)\cdots 3 \cdot 2 \cdot 1} \\
&= n(n-1)\cdots(n-r+1) \\
&= nPr
\end{aligned}
$$

It is useful to note that this alternative formula gives the following relationship:

$$
\begin{aligned}
nP0 &= \frac{n!}{(n-0)!} \\
&= \frac{n!}{n!} \\
&= 1, \text{for every } n \geq 0
\end{aligned}
$$

and this also gives

$$0! = 1$$

Important formula

Another important formula for $(n+1)!$, that has many applications, can be easily derived. The formula is given by

$$(n+1)! = (n+1)n!$$

To prove this formula, we recall that

$$
\begin{aligned}
(n+1)! &= (n+1)n(n-1)\cdots 3 \times 2 \times 1 \\
&= (n+1)\{n(n-1)\cdots 3 \times 2 \times 1\} \\
&= (n+1)n!
\end{aligned}
$$

Consequently, we find

$$(n+2)! = (n+2)(n+1)! = (n+2)(n+1)n!$$

The formula given above simplifies the computational work. For example,

$$6! = 6 \times 5! = 6 \times 5 \times 4!$$

and so on.

In the following examples, we will illustrate the factorial concept.

Example 5. Evaluate

$$\frac{6!}{4!\,2!\,0!}$$

Solution.

$$\frac{6!}{4!\,2!\,0!} = \frac{6 \cdot 5 \cdot 4!}{(4!)(2 \cdot 1)(1)} = 15$$

Example 6. Evaluate

$$\frac{6!}{3!} - 10\left(\frac{4!}{2!}\right)$$

Solution.

$$\frac{6!}{3!} - 10\left(\frac{4!}{2!}\right) = \frac{6 \cdot 5 \cdot 4 \cdot 3!}{3!} - 10\left(\frac{4 \cdot 3 \cdot 2!}{2!}\right)$$

$$= 120 - 120 = 0$$

Example 7. Find n if $nP2 + nP0 = 29 + 2(0!)$

Solution.

$$\begin{aligned}
nP2 + nP0 &= 29 + 2(0!) \\
n(n-1) + 1 &= 29 + 2 \\
n^2 - n - 30 &= 0 \\
(n-6)(n+5) &= 0
\end{aligned}$$

so that

$$n = 6$$

Example 8. Find n if $nP2 + 2n = 5! - 10$

Solution.

$$\begin{aligned}
nP2 + 2n &= 120 - 10 \\
n(n-1) + 2n &= 110 \\
n^2 + n - 110 &= 0 \\
(n-10)(n+11) &= 0
\end{aligned}$$

so that

$$n = 10$$

Example 9. Use mathematical induction to show that

$$n! \geq 2^{n-1}, \text{ for } n \geq 1$$

Solution.
(i) For $n = 1, 1! = 1, 2^{1-1} = 2^0 = 1$

Thus the statement is true for $n = 1$.

(ii) Assume that the statement is true for $n = k$, i.e. assume that

$$k! \geq 2^{k-1}, k \geq 1$$

(iii) Show that the statement is true for $n = k + 1$, i.e. show that

$$(k + 1)! \geq 2^k$$

$$
\begin{aligned}
\text{LHS} = (k+1)! &= (k+1) \times k! \\
&\geq (k+1) \times 2^{k-1}, \text{ by using induction in (ii)} \\
&\geq 2 \times 2^{k-1}, \text{ because } k+1 \geq 2, \text{ for } k \geq 1 \\
&\geq 2^k
\end{aligned}
$$

Thus the statement is true for $n = k+1$, and hence it is true for all integers $n \geq 1$.

Exercises 7.4

1. Evaluate the following permutations:

   (a) $4P3$ (b) $7P3$ (c) $9P4$ (d) $8P0$

2. Evaluate the following expressions:

   (a) $(0!)(1!)(2!)$ (b) $6! - 5! + 4!$

   (c) $\dfrac{2! + 0!}{2! - 0!}$ (d) $\dfrac{2P0 + 0!}{3P0 + 0!}$

3. Evaluate the following expressions:

   (a) $\dfrac{7!}{5! \, 3! \, 1!}$ (b) $\dfrac{8!}{5! \, 2! \, 0!}$

   (c) $\dfrac{4! + 2!}{4P2 + 0!}$ (d) $\dfrac{5! + 2! + 1!}{4! + 3(3!) - 0!}$

4. Evaluate $\dfrac{7P4}{4!}$

5. Evaluate $\dfrac{8P5}{5!}$

6. Evaluate $4P0 - 4P1 + 4P2 - 4P3 + 4P4$

7. Evaluate $5P0 - 5P1 + 5P2 - 5P3 + 5P4 - 5P5$

8. Evaluate $\frac{6P0+6P3+6P6}{4P0+4P2+4P4}$

9. Show that $(1+1)! = 1! + 1!$, but $(2+1)! \neq 2! + 1!$

10 Show that $(2 \times 1)! = 2! \times 1!$, but $(3 \times 2)! \neq 3! \times 2!$

11. Find n if $nP2 = 56$

12. Find n if $nP2 + 3! = 4! + 12$

13. Find n if $nP2 = 5n + 4! + 3! + 10(0!)$

14. Find n if $nP3 = n^3 - 5! - 13$

15. Find n if $nP3 = n^3 - 5! + 4!$

16. Find n if $10(nP1) + 2(nP2) = nP3$

17. Find n if $nP2 + nP3 = 81n$

Prove each of the statements in 18–24 by using mathematical induction:

18. $n! > n^2$, $n \geq 4$

19. $n! > n^3$, $n \geq 6$

20. $n^n \geq n!$, $n \geq 1$

21. $n! > 2^n$, $n \geq 4$

22. $(2n)! < 2^{n-1}(n!)^2$, $n \geq 5$

23. $\dfrac{1}{2!} + \dfrac{2}{3!} + \dfrac{3}{4!} + \cdots + \dfrac{n}{(n+1)!} = 1 - \dfrac{1}{(n+1)!}, n \geq 1$

24. $1(1!) + 2(2!) + 3(3!) + \cdots + n(n!) = (n+1)! - 1, n \geq 1$

7.5 Combinations

Given the digits 1, 2, and 3. How many possible products we can form where each product consists of the multiplication of any two distinct digits? It is clear that the products are:

$$1 \times 2, 1 \times 3, 2 \times 3$$

We note here that changing the order will not produce a different product. The products 2×3 and 3×2 are the same. The resulting arrangements are called unordered arrangements or combinations. Accordingly, the order has no effect on the result. The combinations of n objects, where each arrangement contains r objects is denoted by nCr or $\binom{n}{r}$. This means that $6C3$ is the number of ways that we can arrange 6 objects taken 3 at a time where order has no effect on the formation of the arrangements. Other combinations can be interpreted in a similar way.

We can easily show that nCr is related to nPr by the following formula

$$nCr = \frac{nPr}{r!}$$

or equivalently

$$nCr = \frac{n!}{r!(n-r)!}$$

that will be used in this section for evaluation purposes. To illustrate the evaluation process of any combination, we discuss the following examples.

Example 1. Evaluate $6C3, 7C4, 10C2$

Solution.

$$6C3 = \frac{6P3}{3!} = \frac{6 \cdot 5 \cdot 4}{3 \cdot 2 \cdot 1} = 20$$

$$7C4 = \frac{7P4}{4!} = \frac{7 \cdot 6 \cdot 5 \cdot 4}{4 \cdot 3 \cdot 2 \cdot 1} = 35$$

$$10C2 = \frac{10P2}{2!} = \frac{10 \cdot 9}{2 \cdot 1} = 45$$

Example 2. Evaluate

$$\binom{4}{0} - \binom{4}{1} + \binom{4}{2} - \binom{4}{3} + \binom{4}{4}$$

Solution.

$$\binom{4}{0} - \binom{4}{1} + \binom{4}{2} - \binom{4}{3} + \binom{4}{4}$$
$$= \frac{4P0}{0!} - \frac{4P1}{1!} + \frac{4P2}{2!} - \frac{4P3}{3!} + \frac{4P4}{4!}$$
$$= \frac{1}{1} - \frac{4}{1} + \frac{12}{2} - \frac{24}{6} + \frac{24}{24} = 0$$

Example 3. Evaluate $7C2$ and $7C5$. Use the result to derive a conclusion.

Solution.

$$7C2 = \frac{7P2}{2!} = \frac{7 \cdot 6}{2 \cdot 1} = 21$$

$$7C5 = \frac{7P5}{5!} = \frac{7 \cdot 6 \cdot 5 \cdot 4 \cdot 3}{5 \cdot 4 \cdot 3 \cdot 2 \cdot 1} = 21$$

The conclusion that can be made here is that although $7P5 \neq 7P2$, but $7C5 = 7C2$. This can be stated as a formula given by

$$nCr = nC(n - r)$$

This formula is mostly used if r, in nCr, is large, and in particular if $r > \frac{n}{2}$. For example, $8C6 = 8C2$, $70C67 = 70C3$, and so on. This formula simplifies the computational work as shown by the following example.

Example 4. Use the last formula to evaluate $200C198, 1000C999$

Solution.

$$200C198 = 200C2 = \frac{200P2}{2!} = \frac{200 \times 199}{2!} = 19900$$

$$1000C999 = 1000C1 = \frac{1000P1}{1!} = \frac{1000}{1} = 1000$$

Example 5. Evaluate

$$\binom{5}{0} + \binom{5}{1} + \binom{5}{2} + \binom{5}{3} + \binom{5}{4} + \binom{5}{5}$$

Solution.

$$\binom{5}{0} + \binom{5}{1} + \binom{5}{2} + \binom{5}{3} + \binom{5}{4} + \binom{5}{5}$$

$$= \binom{5}{0} + \binom{5}{1} + \binom{5}{2} + \binom{5}{2} + \binom{5}{1} + \binom{5}{0}$$

$$= 1 + 5 + 10 + 10 + 5 + 1 = 2^5 = 32$$

Example 6. Find n if

$$nC2 = 28$$

Solution.

$$
\begin{aligned}
nC2 &= 28 \\
\frac{nP2}{2!} &= 28 \\
nP2 &= 56 \\
n^2 - n - 56 &= 0 \\
(n-8)(n+7) &= 0 \\
n &= 8
\end{aligned}
$$

Example 7. Find n if

$$nC2 + nP2 = 10C2$$

Solution.

$$
\begin{aligned}
nC2 + nP2 &= 10C2 \\
\frac{nP2}{2!} + nP2 &= \frac{10P2}{2!} \\
\frac{n(n-1)}{2!} + n(n-1) &= 45 \\
n^2 - n - 30 &= 0 \\
(n-6)(n+5) &= 0 \\
n &= 6
\end{aligned}
$$

Example 8. Find n if

$$nP3 + 12(nC3) = 126n$$

Solution.

$$
\begin{aligned}
nP3 + 12(nC3) &= 126n \\
n(n-1)(n-2) + 12\left(\frac{n(n-1)(n-2)}{3!}\right) &= 126n \\
3n(n-1)(n-2) &= 126n
\end{aligned}
$$

Dividing both sides by n we find

$$
\begin{aligned}
n^2 - 3n - 40 &= 0 \\
(n-8)(n+5) &= 0 \\
n &= 8
\end{aligned}
$$

Exercises 7.5

1. Evaluate the following combinations:

   (a) $8C2$ (b) $10C3$ (c) $6C4$ (d) $4C0$

2. Evaluate the following combinations:

   (a) $100C98$ (b) $500C499$

   (c) $200C197$ (d) $1000C998$

3. Evaluate the following expression:
$$\binom{4}{0} + \binom{4}{1} + \binom{4}{2} + \binom{4}{3} + \binom{4}{4}$$

4. Evaluate the following expression:
   $$5C0 - 5C1 + 5C2 - 5C3 + 5C4 - 5C5$$

5. Show that
$$\binom{6}{0} + \binom{6}{1} + \binom{6}{2} + \binom{6}{3} + \binom{6}{4} + \binom{6}{5} + \binom{6}{6} = 2^6$$

6. Find x if $xP2 = \frac{4P0 + 4P1 + 4P2 + 4P3 + 19(0!)}{4C0 - 4C1 + 4C2 - 4C3 + 3(0!)}$

7. Find n if $nC2 = 45$

8. Find n if $nC2 + 3! = 4! - 3$

9. Find n if $nC2 + nP2 = 5! - 36$

10. Find n if $nC2 = nP2 - 4! - 4$

11. Find n if $nC2 + nC3 = 8n$

12. Find n if $nP3 - n(nC2) = 14n$

13. Find n if $nP3 + 18(nC3) = 80n$

14. Find n if $nP3 - n(nC2) = 27n$

15. Find n if $16(nC2) - nP3 - nP2 = 7n$

16. Find n if $(nP2) \times (nC2) = \frac{25}{2}n^2$

17. Find n if $nP3 + nC3 = 5n^2$

18. Find n if $nP2 + nC3 = 8n + 2!$

7.6   Binomial Theorem

The **Binomial Theorem** introduces a useful technique to find the expansion of two terms raised to a positive integer power n. Examples of two terms raised to a positive integer power are $(x+y)^4$, $(x-y)^5$, and $(2x+3y)^6$. The theorem in its simplest form is given by

$$(x+y)^n = \binom{n}{0}x^n y^0 + \binom{n}{1}x^{n-1}y^1 + \binom{n}{2}x^{n-2}y^2 + \cdots + \binom{n}{n}x^0 y^n$$

The following remarks can be drawn here:

1. The expansion of $(x + y)^n$ consists of $(n + 1)$ terms. This means, for example, that the expansion of $(x + y)^6$ gives 7 terms.

2. To determine the expansion, it is recommended first to evaluate the binomial coefficients

$$\binom{n}{0}, \binom{n}{1}, \binom{n}{2}, \cdots, \binom{n}{n}$$

3. The first term of the expansion has the form $x^n y^0$ and the last term has the form $x^0 y^n$.

4. The variables in the successive terms can be easily obtained from the first term by decreasing the power of x by 1 for each term and end with x^0, and by increasing the power of y by 1 for each term and end with y^n. The following scheme summarizes the use of the Binomial theorem for $(x + y)^n$.

$$\text{coefficients} \rightarrow \quad \binom{n}{0} \quad \binom{n}{1} \quad \binom{n}{2} \quad \cdots \quad \binom{n}{n-1} \quad \binom{n}{n}$$

$$\downarrow \quad\quad \downarrow \quad\quad \downarrow \quad\quad \downarrow \quad\quad \downarrow \quad\quad \downarrow$$

$$\text{variables} \rightarrow \quad x^n y^0 \quad x^{n-1}y^1 \quad x^{n-2}y^2 \quad \cdots \quad x^1 y^{n-1} \quad x^0 y^n$$

5. In expanding $(x - y)^n$, we follow the same procedure as before except that the resulting terms alternate in sign, where the first term will be given a positive sign, the second term will be given a negative sign, and so on.

   The following examples illustrate the technique of the binomial theorem.

Example 1. Find the expansion of $(x + y)^4$

Solution.

$$\text{binomial coefficients} \rightarrow \quad \binom{4}{0} \quad \binom{4}{1} \quad \binom{4}{2} \quad \binom{4}{3} \quad \binom{4}{4}$$

$$\downarrow \quad \downarrow \quad \downarrow \quad \downarrow \quad \downarrow$$

$$\text{variables} \rightarrow \quad x^4 y^0 \quad x^3 y^1 \quad x^2 y^2 \quad x^1 y^3 \quad x^0 y^4$$

Accordingly, we find

$$(x + y)^4 = x^4 + 4x^3 y + 6x^2 y^2 + 4xy^3 + y^4$$

obtained upon evaluating the binomial coefficients, multiplying the coefficients by the variables and adding the resulting terms.

Example 2. Find the expansion of $(x - y)^5$

Solution.

$$\text{binomial coefficients} \rightarrow \quad \binom{5}{0} \quad \binom{5}{1} \quad \binom{5}{2} \quad \binom{5}{3} \quad \binom{5}{4} \quad \binom{5}{5}$$

$$\downarrow \quad \downarrow \quad \downarrow \quad \downarrow \quad \downarrow \quad \downarrow$$

$$\text{variables} \rightarrow \quad x^5 y^0 \quad x^4 y^1 \quad x^3 y^2 \quad x^2 y^3 \quad x^1 y^4 \quad x^0 y^5$$

It is worth noting that the resulting terms alternate in sign as discussed before. Accordingly, we find

$$(x - y)^5 = x^5 - 5x^4 y + 10x^3 y^2 - 10x^2 y^3 + 5xy^4 - y^5$$

obtained upon evaluating the binomial coefficients, multiplying the coefficients by the variables and adding the resulting terms.

Example 3. Find the expansion of $(x + 2y)^4$

Solution.

$$\text{coefficients} \rightarrow \quad \binom{4}{0} \quad \binom{4}{1} \quad \binom{4}{2} \quad \binom{4}{3} \quad \binom{4}{4}$$

$$\downarrow \quad \downarrow \quad \downarrow \quad \downarrow \quad \downarrow$$

$$\text{variables} \rightarrow \quad x^4 (2y)^0 \quad x^3 (2y)^1 \quad x^2 (2y)^2 \quad x^1 (2y)^3 \quad x^0 (2y)^4$$

Therefore, we obtain

$$
\begin{aligned}
(x+2y)^4 &= x^4 + 4x^3(2y) + 6x^2(4y^2) + 4x(8y^3) + (16y^4)\\
&= x^4 + 8x^3y + 24x^2y^2 + 32x^3y + 16y^4
\end{aligned}
$$

obtained upon evaluating the binomial coefficients, multiplying the coefficients by the variables and adding the resulting terms.

Example 4. Find the expansion of $(2x - 3y)^5$

Solution. The binomial coefficients and the variables are shown by

$$
\binom{5}{0}\quad \binom{5}{1}\quad \binom{5}{2}\quad \binom{5}{3}\quad \binom{5}{4}\quad \binom{5}{5}
$$

$$
\downarrow\qquad \downarrow\qquad \downarrow\qquad \downarrow\qquad \downarrow\qquad \downarrow
$$

$$
(2x)^5\quad (2x)^4(3y)^1\quad (2x)^3(3y)^2\quad (2x)^2(3y)^3\quad (2x)^1(3y)^4\quad (3y)^5
$$

It is worth noting that the resulting terms alternate in sign as discussed before. Accordingly, we find

$$
\begin{aligned}
(2x-3y)^5 &= 32x^5 - 5(16x^4)(3y) + 10(8x^3)(9y^2)\\
&\quad - 10(4x^2)(27y^3) + 5(2x)(81y^4) - (243y^5)\\
&= 32x^5 - 240x^4y + 720x^3y^2 - 1080x^2y^3 + 810xy^4 - 243y^5
\end{aligned}
$$

obtained by following the procedure discussed before.

Example 5. Find the first four terms in the expansion of $(x+y)^{10}$

Solution. The complete expansion of $(x+y)^{10}$ consists of 11 distinct terms. However, it is required here to find the first four terms. Hence, we find

$$
\text{Binomial coefficients} \rightarrow \quad \binom{10}{0}\quad \binom{10}{1}\quad \binom{10}{2}\quad \binom{10}{3}\quad \cdots
$$

$$
\downarrow\quad \downarrow\quad \downarrow\quad \downarrow\quad \downarrow
$$

$$
\text{Variables} \rightarrow \quad x^{10}y^0\quad x^9y^1\quad x^8y^2\quad x^7y^3\quad \cdots
$$

Accordingly, we find

$$
(x+y)^{10} = x^{10} + 10x^9y + 45x^8y^2 + 120x^7y^3 + \cdots
$$

7.6.1   Binomial Coefficients

It is interesting to note that the term containing $x^{n-k}y^k$ in the expansion of $(x+y)^n$ is of the form $\binom{n}{k}x^{n-k}y^k$. This means that y has the power k and the coefficient of the term containing $x^{n-k}y^k$ is $\binom{n}{k}$. We can use this fact in the following examples.

Example 6. Find the coefficient of the term containing x^7y^4 in the expansion of $(x+y)^{11}$

Solution. As stated before the coefficient of the term containing x^7y^4 is $\binom{11}{4}$, by noting that $k=4$. Therefore, the coefficient is

$$\binom{11}{4} = \frac{11P4}{4!} = \frac{11 \cdot 10 \cdot 9 \cdot 8}{4 \cdot 3 \cdot 2 \cdot 1} = 330$$

Example 7. Find the coefficient of the term containing x^6y^k in the expansion of $(x+y)^{10}$

Solution. It is clear that $6+k=10$. This gives $k=4$. As a result, the coefficient is

$$\binom{10}{4} = 210$$

Example 8. One term in the expansion of $(x+y)^n$ contains x^ry^2. The coefficient of this term is 45. Find n and r.

Solution. It is obvious that $n=r+2$. Using this for the binomial coefficient we find

$$\binom{r+2}{2} = 45$$
$$\frac{(r+2)(r+1)}{2!} = 45$$
$$r^2 + 3r - 88 = 0$$
$$(r-8)(r+11) = 0$$
$$r = 8$$

7.6.2   Pascal's Triangle

We close this section by introducing an alternative method to determine the binomial coefficients of the terms of the expansion of $(x + y)^n$. It is useful to notice that

$$(x + y)^0 = 1$$
$$(x + y)^1 = x + y$$
$$(x + y)^2 = x^2 + 2xy + y^2$$
$$(x + y)^3 = x^3 + 3x^2y + 3xy^2 + y^3$$
$$(x + y)^4 = x^4 + 4x^3y + 6x^2y^2 + 4xy^3 + x^4$$
$$(x + y)^5 = x^5 + 5x^4y + 10x^3y^2 + 10x^2y^3 + 5xy^4 + x^5$$
$$(x + y)^6 = x^6 + 6x^5y + 15x^4y^2 + 20x^3y^3 + 15x^2y^4 + 6xy^5 + y^6$$

$$\vdots$$

$$(x + y)^n = x^n + \frac{n}{1!}x^{n-1}y + \frac{n(n - 1)}{2!}x^{n-2}y^2 + \frac{n(n - 1)(n - 2)}{3!}x^{n-3}y^3 + \cdots$$

We can easily conclude the following:

1. The sum of powers of x and y of each term is equal to the power n.

2. A pattern can be established from the coefficients of the terms. This pattern can be written in a triangular format as shown below:

$$
\begin{array}{ccccccccccccc}
 & & & & & & 1 & & & & & & \\
 & & & & & 1 & & 1 & & & & & \\
 & & & & 1 & & 2 & & 1 & & & & \\
 & & & 1 & & 3 & & 3 & & 1 & & & \\
 & & 1 & & 4 & & 6 & & 4 & & 1 & & \\
 & 1 & & 5 & & 10 & & 10 & & 5 & & 1 & \\
1 & & 6 & & 15 & & 20 & & 15 & & 6 & & 1 \\
\end{array}
$$

$$\vdots$$

$$1 \quad \frac{n}{1!} \quad \frac{n(n - 1)}{2!} \quad \frac{n(n - 1)(n - 2)}{3!} \quad \cdots \qquad 1$$

This triangular pattern is called Pascal's triangle. We can easily observe that:

(i) The first row is 1.
(ii) The numbers on the edges of each row begins with 1 and ends in 1.
(iii) Each interior number other than the first and the last is obtained by adding the two coefficients above it.
(iv) The sum of the numbers in each row is a power of 2.
(v) Each row gives the coefficients for binomial coefficients. The $x^{n-k}y^k$ terms are obtained in a regular way as presented before.

Example 9. Expand $(x+3)^5$ using Pascal's triangle.

Solution.

$$
\begin{aligned}
(x+3)^5 &= x^5 + 5(x)^4(3) + 10(x)^3(3)^2 + 10(x)^2(3)^3 + 5(x)(3)^4 + (3)^5 \\
&= x^5 + 15x^4 + 90x^3 + 270x^2 + 405x + 243
\end{aligned}
$$

Exercises 7.6

1. Find the expansion of $(x+y)^6$

2. Find the expansion of $(x-y)^6$

3. Find the expansion of $(3x+y)^5$

4. Find the expansion of $(2x-y)^5$

5. Find the expansion of $(3x+2y)^5$

6. Find the expansion of $(3x-2y)^6$

7. Find the expansion of $(x^2+y)^5$

8. Find the expansion of $(\frac{1}{2}x+y)^4$

9. Find the first four terms of the expansion of $(x+y)^{16}$

10. Find the first five terms of the expansion of $(x+3y)^{20}$

11. Find the first four terms of the expansion of $(x-2y)^{36}$

12. Find the coefficient of the term containing x^8y^9 in the expansion of $(x+y)^{17}$

13. Find the coefficient of the term containing x^4y^6 in the expansion of $(x-y)^{10}$

14. Find the coefficient of the term containing $x^9 y^{11}$ in the expansion of $(x-y)^{20}$

15. One term in the expansion of $(x+y)^n$ contains $x^4 y^8$. Find n and the coefficient of the term

16. One term in the expansion of $(x+y)^n$ contains $x^6 y^9$. Find n and the coefficient of the term

17. One term in the expansion of $(x+y)^{14}$ contains $x^9 y^k$. Find k and the coefficient of the term

18. One term in the expansion of $(x+y)^{18}$ contains $x^{12} y^k$. Find k and the coefficient of the term

19. One term in the expansion of $(x+y)^n$ contains $x^r y^2$. The coefficient of the term containing $x^r y^2$ is 28. Find n and r

20. One term in the expansion of $(x+y)^n$ contains $x^r y^3$. The coefficient of the term containing $x^r y^3$ is 20. Find n and r

21. Use Pascal's triangle to expand $(x+1)^6$

22. Use Pascal's triangle to expand $(x^2+2)^3$

23. Use the binomial theorem to the expression $(1+3)^n$ to show that

$$\binom{n}{0}3^0 + \binom{n}{1}3^1 + \binom{n}{2}3^2 + \cdots + \binom{n}{n}3^n = 4^n, n \geq 0$$

24. Use the binomial theorem to the expression $(5-1)^n$ to show that

$$\binom{n}{0}5^n - \binom{n}{1}5^{n-1} + \binom{n}{2}5^{n-2} - \cdots + (-1)^n\binom{n}{n}5^0 = 4^n, n \geq 0$$

25. Use Exercises 23 and 24 to evaluate

$$\frac{\binom{n}{0}3^0 + \binom{n}{1}3^1 + \binom{n}{2}3^2 + \cdots + \binom{n}{n}3^n}{\binom{n}{0}5^n - \binom{n}{1}5^{n-1} + \binom{n}{2}5^{n-2} - \cdots + (-1)^n\binom{n}{n}5^0}$$

26. Evaluate the following

$$\frac{\binom{n}{0}5^0 + \binom{n}{1}5^1 + \binom{n}{2}5^2 + \cdots + \binom{n}{n}5^n}{\binom{n}{0}4^n - \binom{n}{1}4^{n-1} + \binom{n}{2}4^{n-2} - \cdots + (-1)^n\binom{n}{n}4^0}$$

Chapter 8

Probability

8.1 Finite Probability

8.1.1 Sample Space

In this section, we will study the outcomes when a certain experiment is performed. An **experiment** is an operation that produces a set of outcomes that can be observed and listed. Examples of experiments are tossing a coin, rolling a die, drawing a ball from a bag that contains several colored balls, a survey conducted on female and male students about two specific courses, and drawing a card from a standard deck of cards.

A **sample space** S is defined to be the set of all possible outcomes when performing an experiment. The sample space S is identical to the universal set U that contains all elements under discussion. To study the idea of probability, it is useful to discuss how we can list the outcomes of an experiment, or the results of a survey conducted, that form the elements of the sample space S. The concept of listing the outcomes of an experiment will be illustrated by discussing the following examples.

Example 1. Tossing one coin for one time, write the sample space S of all possible outcomes of this experiment.

Solution. A coin has two faces marked as Head and Tail symbolized by H and T respectively. Therefore, the sample space of this experiment is given by:

$$S = \{H, T\}$$

Example 2. A family has one child. Write the sample space S of all possibilities of the sex of the child.

Solution. The child may be a Girl or a Boy symbolized by G or B respectively. Therefore, the sample space of this experiment is given by:

$$S = \{G, B\}$$

Example 3. Rolling a die once. Write the sample space S of all possible outcomes of this experiment.

Solution. It is well-known that a die has 6 faces marked from 1 to 6. Therefore, the sample space of the rolling experiment is given by:

$$S = \{1, 2, 3, 4, 5, 6\}$$

Example 4. A bag contains three red balls and four green balls. A ball is drawn at random. Write the space of all possible outcomes of the drawing.

Solution. We denote the balls of the bag by $R_1, R_2, R_3, G_1, G_2, G_3$, and G_4. Hence, we find

$$S = \{R_1, R_2, R_3, G_1, G_2, G_3, G_4\}$$

We next consider how to construct a sample space S if an experiment is repeated twice. We point out that S can be obtained by two ways as will be discussed by the following examples.

Example 5. Tossing one coin twice, write the space S of all possible outcomes.

Solution. As stated above, S can be obtained by two ways, namely:

1. The Chart Method:

Recall that tossing a coin once results in

$$S_1 = \{H, T\}$$

and when the coin is tossed for the second time we find

$$S_2 = \{H, T\}$$

To combine both results, we list the elements of S_1 in a column of a chart and the elements of S_2 in the row of that chart. Consequently, we combine each element of the column with each element of the row. The procedure discussed above is shown in the chart below.

$S_1 \setminus S_2$	H	T
H	HH	HT
T	TH	TT

Accordingly, the sample space of all possible outcomes of tossing a coin twice, or equivalently the sample space S of tossing two coins one time, is given by:

$$S = \{HH, HT, TH, TT\}$$

On the other hand, the space S can also be determined by:

2. The Tree Diagram:

In this representation, two branches are drawn from a start point of a tree to represent S_1. We then continue by drawing two other branches representing S_2 for every branch of S_1. The procedure discussed above is shown in the figure below.

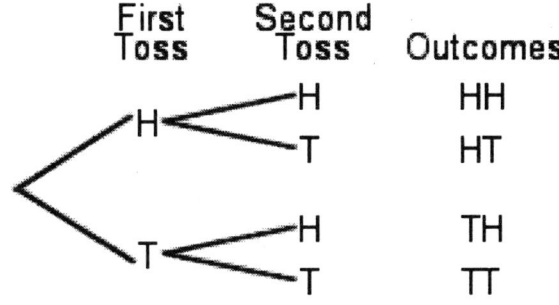

It is clear from the tree diagram that the outcomes are represented by four branches, hence the sample space S is given by

$$S = \{HH, HT, TH, TT\}$$

Example 6. Write the sample space that results from rolling a die followed by tossing a coin.

Solution. We first use the chart method.

1. The chart method
Rolling a die once gives

$$S_1 = \{1, 2, 3, 4, 5, 6\}$$

and tossing a coin gives

$$S_2 = \{H, T\}$$

Combining the results by using the chart method we obtain

$S_1 \setminus S_2$	H	T
1	1H	1T
2	2H	2T
3	3H	3T
4	4H	4T
5	5H	5T
6	6H	6T

The sample space has 12 elements given by:

$$S = \{1H, 1T, 2H, 2T, \cdots, 6H, 6T\}$$

2. The tree diagram:

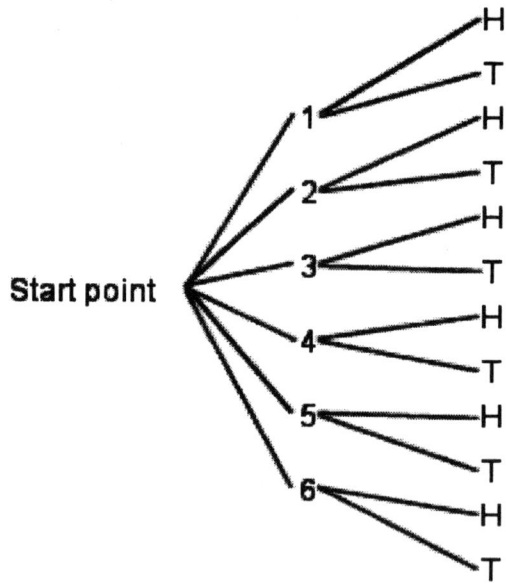

We first draw 6 branches from the start point, each branch represents one element of S_1, given by

$$S_1 = \{1, 2, 3, 4, 5, 6\}$$

We then draw 2 branches, each represents one element of S_2, for each branch drawn before, where S_2 is given by

$$S_2 = \{H, T\}$$

Notice that we obtain 12 branches as shown by the tree diagram below. The branches give the sample space S obtained before and given by

$$S = \{1H, 1T, 2H, 2T, \cdots, 6H, 6T\}$$

The procedure is illustrated by the tree diagram shown before.

Example 7. Write the sample space that results from rolling two dice.

Solution. Rolling each die independently gives

$$S_1 = \{1, 2, 3, 4, 5, 6\}, \; S_2 = \{1, 2, 3, 4, 5, 6\}$$

$S_1 \setminus S_2$	1	2	3	4	5	6
1	(1,1)	(1,2)				
2	(2,1)		(2,3)			
3	(3,1)	(3,2)				(3,6)
4				(4,4)		
5			(5,3)		(5,5)	
6		(6,2)		(6,4)		(6,6)

It should be noted here that the tree diagram is not practical. For this reason, the chart method is used as shown above. It is clear that the sample space consists of 36 elements, each is an ordered pair, given by:

$$S = \{(1, 1), (1, 2), (1, 3), \cdots, (6, 5), (6, 6)\}$$

In the figure shown above we listed few elements of the space S. Other elements can be listed.

8.1.2 Events

In Chapter 5, we have studied that for each set S, we can construct 2^n subsets of S, where n is the number of elements of the set S. It should be noted here that, in this chapter, the subset will be called **event**. Therefore, an **event** E is a subset of a sample space S. The elements of E are usually determined by using given specifications or descriptions. This means that an event E is obtained from the space S by listing only the elements of S that justify the given condition. Accordingly, to determine any event E, it is advised to list first the space S, then we derive the elements of E from S. This can be easily explained by considering the following examples.

Example 8. Rolling a die once. Write the event that an even number occurs.

Solution. The sample space S that results from rolling a die is:

$$S = \{1, 2, 3, 4, 5, 6\}$$

The event E consists of all even numbers of S, hence we have

$$E = \{2, 4, 6\}$$

Example 9. Rolling a die once. Write the event P that a prime number occurs.

Solution. The sample space S that results from rolling a die is:

$$S = \{1, 2, 3, 4, 5, 6\}$$

The event P consists of all prime numbers of S, hence we have

$$P = \{2, 3, 5\}$$

Example 10. Rolling a die once. Write the event F that a number greater than 7 occurs.

Solution. The sample space that results from rolling a die is:

$$S = \{1, 2, 3, 4, 5, 6\}$$

To list the elements of F, it is clear that S has no element > 7. In other words, the event F is simply the empty set ϕ that has been discussed before, hence we write

$$F = \phi$$

Example 11. Tossing two coins. Write the following events:

(a) A is an event that at least one H occurs.

(b) B is an event that at most one H occurs.

(c) C is an event that H occurs first.

(d) D is an event that the two outcomes are identical.

Solution. Using Example 5 discussed before, we find

$$S = \{HH, HT, TH, TT\}$$

(a) The given description of the elements of A is that at least one H occurs. This means that A contains elements of S where number of heads is ≥ 1. Hence, we find

$$A = \{HH, HT, TH\}$$

(b) The event B that number of heads is ≤ 1, i.e. number of heads $= 0, 1$, therefore we find

$$B = \{HT, TH, TT\}$$

(c) The event C contains the elements of S that starts with H, hence we find

$$C = \{HH, HT\}$$

(d) The event D contains the elements where both outcomes are the same, hence we find

$$D = \{HH, TT\}$$

Example 12. Rolling two dice. Write the following events:

(a) A is an event that both dice show the same number.

(b) B is an event that the sum of the outcomes is ≥ 10.

(c) C is an event that one die only (not both) shows the number 4.

(d) D is an event that the sum of the two outcomes is ≥ 14.

Solution. In Example 7, we have shown that the sample space consists of 36 elements, each is an ordered pair, given by:

$$S = \{(1, 1), (1, 2), (1, 3), \cdots, (6, 5), (6, 6)\}$$

The elements of the events A, B, and C are shown by the following table.

$S_1 \setminus S_2$	1	2	3	4	5	6
1	a			c		
2		a		c		
3			a	c		
4	c	c	c	a	c	b c
5				c	a b	b
6				c b	b	a b

(a) In this event, the two components of each ordered pair are the same. Therefore, we find

$$A = \{(1, 1), (2, 2), (3, 3), (4, 4), (5, 5), (6, 6)\}$$

Notice that we denoted the elements of A by lowercase letters a.

(b) The event B contains outcomes (x, y) such that $x + y \geq 10$. Using the table shown above we find

$$B = \{(4, 6), (5, 5), (5, 6), (6, 4), (6, 5), (6, 6)\}$$

Notice that we denoted the elements of B by lowercase letters b.

(c) The event C contains all elements of the form $(x, 4)$ or $(4, y)$ where $x \neq 4$ and $y \neq 4$. Accordingly, we obtain

$$C = \{(4, 1), (4, 2), (4, 3), (4, 5), (4, 6), (1, 4), (2, 4), (3, 4), (5, 4), (6, 4)\}$$

Notice that we denoted the elements of C by lowercase letters c.

(d) We can easily observe that the largest sum of any two shows is 12. This results from the outcome at the rightmost corner $(6, 6)$. This gives:

$$D = \phi$$

Example 13. A family has three children. Write the following events:

(a) A is an event that the family has exactly two girls.

(b) B is an event that the first two children are boys.

(c) C is an event that the family has at most one girl.

(d) D is an event that all children are of the same sex.

Solution. The three sample spaces for one child only are given

$$S_1 = \{G, B\}, \ S_2 = \{G, B\}, \ S_3 = \{G, B\}$$

Combining the results for the first two children we find

$$S_4 = \{GG, GB, BG, BB\}$$

obtained by using the chart method. To combine S_4 with S_3, we use the chart shown below.

$S_3 \setminus S_4$	G	B
GG	GGG	GGB
GB	GBG	GBB
BG	BGG	BGB
BB	BBG	BBB

The sample space S in this case is

$$S = \{GGG, GGB, GBG, GBB, BGG, BGB, BBG, BBB\}$$

(a) The event A consists of the elements of S that have exactly two girls, hence we have
$$A = \{GGB, GBG, BGG\}$$

(b) The event B contains the last two elements of S, hence we obtain
$$B = \{BBG, BBB\}$$

(c) The given description of the event C means that the number of girls is 0 or 1. Therefore, we find
$$S = \{GBB, BGB, BBG, BBB\}$$

(d) The event D that consists of all children of the same sex is given by:
$$S = \{GGG, BBB\}$$

8.1.3   Number of Elements of a Sequence

Given the sequence of numbers by

$$a, a + d, a + 2d, a + 3d, \cdots, a + d(n - 1)$$

where
first term $a_1 = a$,
last term $a_n = a + d(n - 1)$
common difference $= d$
This means that in a sequence of numbers with a fixed common difference d, the number of terms of this sequence can be obtained by using the formula

$$n = \frac{\text{last term} - \text{first term} + \text{common difference}}{\text{common difference}} = \frac{a_n - a_1 + d}{d}$$

This formula can be used to determine the number of terms in a sample space of integers of a given description. The following two examples illustrate the idea.

Example 14. How many three-digit integers are from the sequence

$$104, 105, 106, \cdots, 900$$

Solution. It is clear that
first term $a_1 = 104$
last term $a_n = 900$
common difference $d = 1$
Using the formula stated before we find

$$n = \frac{900 - 104 + 1}{1} = 797$$

Example 15. How many three-digit integers are from the sequence

$$100, 101, 102, \cdots, 998, 999$$

are divisible by 6?

Solution. The sequence of integers that are divisible by 6 is

$$102, 108, 114, \cdots, 996$$

first term $a_1 = 102$
last term $a_n = 996$
common difference $d = 6$
This gives

$$n = \frac{996 - 102 + 6}{6} = 150$$

8.1.4   Probability of an Event

So far, we have discussed how we can list the elements of a sample space S that contains all possible outcomes if an experiment is performed. In addition, we have discussed how we can derive specific subsets of S, called events, that satisfy given specifications. It remains now to introduce the concept of the probability that an event E will occur. Given a sample space S, we denote the number of elements of S by $n(S)$, and the number of elements of an event E of S by $n(E)$. The **probability** that E will occur, denoted by $P(E)$, is defined by the formula:

$$P(E) = \frac{n(E)}{n(S)}$$

This means that to determine $P(E)$, we first determine $n(S)$ and $n(E)$ simply by listing the elements of S and the elements of E as discussed before. We then apply the formula $P(E)$ defined above. To illustrate the idea of calculating $P(E)$, we discuss the following examples.

Example 16. A die is rolled one time. Find the probability of the following events:

(a) A is an event that a number ≥ 5 will occur.

(b) B is an event that an odd number will occur.

(c) C is an event that an odd prime number will occur.

(d) D is an event that a number ≥ 8 will occur.

(e) E is an event that a number ≤ 6 will occur.

Solution. Rolling a die for one time gives the space:

$$S = \{1, 2, 3, 4, 5, 6\}, \quad n(S) = 6$$

(a) The event A is defined by

$$A = \{5, 6\}, \quad n(A) = 2$$

Hence, we find

$$P(A) = \frac{n(A)}{n(S)} = \frac{2}{6} = \frac{1}{3}$$

(b) The event B is defined by

$$B = \{1, 3, 5\}, \quad n(B) = 3$$

Hence we find

$$P(B) = \frac{n(B)}{n(S)} = \frac{3}{6} = \frac{1}{2}$$

(c) Recall that 1 is an odd number but not a prime number. The event C is defined by

$$C = \{3, 5\}, \quad n(C) = 2$$

Hence we find

$$P(C) = \frac{n(C)}{n(S)} = \frac{2}{6} = \frac{1}{3}$$

(d) A regular die has only six faces marked from 1 to 6. Accordingly, it is impossible that a number ≥ 8 will occur. Hence we find

$$D = \phi, n(D) = 0$$

$$P(D) = \frac{n(D)}{n(S)} = 0$$

(e) Each number marked on a die is ≤ 6. Therefore, we find

$$E = \{1, 2, 3, 4, 5, 6\}$$

This means that

$$E = S$$

$$P(E) = \frac{n(E)}{n(S)} = 1$$

The last two events D and E are called the impossible and the certain events.

Example 17. Rolling two dice. Find the probability of the following events:

(a) A is an event that both dice show the same number.

(b) B is an event that the sum of the two outcomes is ≤ 3.

   (c) C is an event that the sum is ≥ 9.

(d) D is an event that the sum of the two shows is exactly 5.

Solution. As shown before the sample space contains 36 elements.

$S_1 \setminus S_2$	1	2	3	4	5	6
1	a b	b		d		
2	b	a	d			
3		d	a			c
4	d			a	c	c
5				c	a c	c
6			c	c	c	a c

For simplicity reasons, we used the lowercase letters a, b, c and d to indicate the elements of the events A, B, C and D respectively.

(a) As shown in the table we have

$$A = \{(1,1), (2,2), (3,3), (4,4), (5,5), (6,6)\}, \quad n(A) = 6$$

Therefore,

$$P(A) = \frac{n(A)}{n(S)} = \frac{6}{36} = \frac{1}{6}$$

(b) As shown in the table we have

$$B = \{(1,1), (1,2), (2,1)\}, \quad n(B) = 3$$

Therefore,

$$P(B) = \frac{n(B)}{n(S)} = \frac{3}{36} = \frac{1}{12}$$

(c) We use the table and count the elements indicated by c, hence we find

$$n(C) = 10$$

Therefore,

$$P(C) = \frac{n(C)}{n(S)} = \frac{10}{36} = \frac{5}{18}$$

(d) We use the table to find

$$D = \{(1,4), (2,3), (3,2), (4,1)\}, \quad n(D) = 4$$

Therefore,

$$P(D) = \frac{n(D)}{n(S)} = \frac{4}{36} = \frac{1}{9}$$

Example 18. A bag contains 4 red balls and 3 green balls. A ball is selected at random from the bag. Find the probability of the following events:

(a) R is an event that the ball selected is red.

(b) G is an event that the ball selected is green.

(c) B is an event that the ball selected is blue.

Solution. The sample space S contains 7 balls defined by:

$$S = \{R_1, R_2, R_3, R_4, G_1, G_2, G_3\}, \quad n(S) = 7$$

(a) The event R is given by

$$R = \{R_1, R_2, R_3, R_4\}, \quad n(R) = 4$$

Therefore,

$$P(R) \quad = \quad \frac{n(R)}{n(S)} = \frac{4}{7}$$

(b) The event G is given by

$$G = \{G_1, G_2, G_3\}, \quad n(G) = 3$$

Therefore,

$$P(G) \quad = \quad \frac{n(G)}{n(S)} = \frac{3}{7}$$

(c) The bag does not contain any blue ball, hence

$$B = \phi, \quad n(B) = 0$$

Therefore,

$$P(B) \quad = \quad \frac{n(B)}{n(S)} = 0$$

Example 19. Find the probability that a randomly selected three-digit integer is divisible by 8 from all three-digit integers

Solution. The sample space of all three-digit integers is given by the sequence

$$100, 101, 102, \cdots, 998, 999$$

The number of integers of this sequence is

$$n = \frac{999 - 100 + 1}{1} = 900$$

The event E of all three-digit integers divisible by 8 is given by

$$104, 112, 120, , \cdots, 984, 992$$

The number of integers of the last sequence is given by

$$n = \frac{992 - 104 + 8}{8} = 112$$

Therefore, the probability is given by

$$P(E) = \frac{n(E)}{n(S)} = \frac{112}{900} = \frac{28}{225}$$

Exercises 8.1

1. In the octal system, find the probabilities of the following events:

(a) an odd digit is used.

(b) an even prime digit is used.

(c) the digit 8 is used.

2. One die is rolled. Find the probability of the following events:

(a) the die shows an even number > 3

(b) the die shows a number x such that $1 < x < 5$

(c) the die shows a number y where y is at most 4

3. A bag contains 7 green balls and 4 blue balls. A ball is selected at random. Find the probability that the ball drawn is:

(a) a green ball

(b) a blue ball

(c) a red ball

4. A factory produces specific items where it was found that 6.5 % of the produced items are defect. An item is selected at random. Find the probability that this item is not defect.

5. A pond has 250 bass fish, 200 perch and 350 trout. A fish is caught at random. Find the probability that the fish caught is:

(a) a trout

(b) a perch

(c) a catfish

6. The final grades of a math section of 45 students were as follows:
10 students got A
15 students got B
12 students got C
8 students got D.
A student is selected at random. Find the probability that this is

(a) an A student

(b) a D student

(c)an F students

7. A survey has been conducted over 800 students. The data has been summarized in the following table:

	Math M	English E	History H
Girls G	80	202	130
Boys B	78	160	150

A student is selected at random. Find the probability that the student is:

(a) a girl

(b) taking a Math course

(c) taking a History course

(d) taking an English course

8. Tossing a coin twice. Find the probability of the following events:

(a) A is the event that exactly one tail occurs

(b) B is the event that at least one tail occurs

(c) C is the event that exactly two heads occur

9. Two dice are rolled. Find the probability of the following events:

(a) A is the event that the sum of the two shows is exactly 10

(b) B is the event that the sum of the two shows is at least 10

(c) C is the event that the sum of the two shows is a prime number ≤ 5

10. Two dice are rolled. Find the probability of the following events:

(a)A is the event that both shows are primes < 5

(b) B is the event that the product of the two shows is ≤ 36

(c)C is the event that the sum of the two shows is ≥ 40

11. Two dice are rolled. Find the probability of the following events:

(a) A is the event that the sum of the two shows is exactly 7

(b) B is the event that the sum of the two shows is ≥ 7

(c) C is the event that one die only shows 5

12. Two dice are rolled. Find the probability of the following events:

(a) A is the event that the two shows are primes ≥ 5

(b) B is the event that the product of the two shows is a multiple of 5

(c) C is the event that the two shows are even numbers

13. A family has 3 children. Find the probability of the following events:

(a) A is the event that at least 2 children are boys

(b) B is the event that the second child is a girl

(c) C is the event that at most 1 child is a girl

14. A family has 3 children. Find the probability of the following events:

(a) A is the event that the last 2 children are girls

(b) B is the event that the family has exactly 2 girls

(c) C is the event that the first two children are not of the same sex

15. A card is drawn from a standard deck of cards. Find the probability that the card drawn is:

(a) a red card

(b) a diamond card

(c) a black picture

(d) a red queen

(e) a diamond queen

16. A card is drawn from a standard deck of cards. Find the probability that the card dawn is:

(a) a 6

(b) a black 6

(c) a king

(d) a queen

(e) a black card x such that $3 \le x \le 6$

17. A die is rolled three times. Find the probability that all three shows are the same

18. A die is rolled three times. Find the probability that the sum of the three shows is ≤ 18

19. A sample space S of a specific experiment contains 112 elements. If the probability of an event E is $P(E) = \frac{1}{8}$. Find $n(E)$

20. If the probability of an event E, where $n(E) = 231$, of a sample space S is $P(E) = \frac{3}{7}$. Find $n(S)$

21. A number is selected from a sequence of consecutive integers from 0 to 99. Find the probability of the following events:

(a) A is the event that the number is even

(b) B is the event that the number ends in 6

(c) C is the event that the nonzero number is divisible by 6

22. A number is selected from a sequence of numbers 50 to 199. Find the probability of the following events:

(a) A is the event that the number is even

(b) B is the event that the number ends in 5

(c) C is the event that the number is divisible by 5

23. Find the probability of a three-digit number that is divisible by 7 from the sequence 100 to 999

24. Find the probability of a four-digit number that is divisible by 9 from the sequence 1000 to 1949

25. A number is selected from a sequence

$$0, 2, 4, 6, \cdots, 248, 250$$

Find the probability that the number ends in 4

8.2 Rules of Probability

In the preceding section, the probability that an event E will occur has been discussed. The event E was defined as a subset of S with elements satisfying a given specification. Concerning E as a subset of S, the following conclusions will be drawn:

1. The event E may be an **impossible** event, hence $E = \phi$, where $n(\phi) = 0$, and therefore

$$P(\phi) = 0$$

2. The event E may be a **certain** event that includes all elements of the sample space S. In this case $E = S$, and therefore

$$P(E) = \frac{n(S)}{n(S)} = 1$$

3. Frequently, the event E is a proper subset of S that contains some elements of S, hence $n(E) < n(S)$, and therefore

$$P(E) = \frac{n(E)}{n(S)} < 1$$

This means that if the event E is not an impossible event, or is not a certain event, then the probability is a positive fraction < 1.

The three possible cases of an event E indicate the following formal conclusion:

$$0 \leq P(E) \leq 1$$

It seems now reasonable to introduce two important formulas that are usually used in probability spaces. Recall that the concepts of the complement of a set, denoted by A^c, the union of two sets A and B, denoted by $A \cup B$, and the intersection of two sets A and B, denoted by $A \cap B$ were discussed. Accordingly, it is important now to derive and use the probability formulas related to these concepts. The formulas will be derived as follows:
1. The complementary event denoted by E^c:
We recall from our study of the set theory that

$$E \cup E^c = S$$

since E and E^c are disjoint sets, we find

$$n(E) + n(E^c) = n(S)$$

Dividing both sides of the last equation by $n(S)$ gives

$$\frac{n(E)}{n(S)} + \frac{n(E^c)}{n(S)} = \frac{n(S)}{n(S)}$$

and this gives the formula

$$P(E) + P(E^c) = 1$$

or equivalently

$$P(E^c) = 1 - P(E)$$

2. The events $A \cup B$ and $A \cap B$:
We can easily show that, for two overlapped sets A and B, the number of

elements of the union and the number of elements of the intersection of two sets are related by

$$n(A \cup B) = n(A) + n(B) - n(A \cap B)$$

where A and B are events derived from a sample space S. Dividing the last equation by $n(S)$ gives

$$\frac{n(A \cup B)}{n(S)} = \frac{n(A)}{n(S)} + \frac{n(B)}{n(S)} - \frac{n(A \cap B)}{n(S)}$$

and this gives the formula

$$P(A \cup B) = P(A) + P(B) - P(A \cap B)$$

It is worth noting here that if $A \cap B = \phi$, the two disjoint subsets A and B are usually called in probability as **mutually exclusive events**. In this case, we find

$$P(A \cap B) = 0$$

and hence

$$P(A \cup B) = P(A) + P(B)$$

The two basic formulas discussed above will be illustrated by discussing the following examples.

Example 1. Given $P(A) = 0.12$, find $P(A^c)$

Solution. Using the formulas given above, we find

$$
\begin{aligned}
P(A^c) &= 1 - P(A) \\
&= 1 - 0.12 = 0.88
\end{aligned}
$$

Example 2. Given $P(A) = 0.32$, $P(B) = 0.56$, $P(A \cap B) = 0.18$. Find $P(A \cup B)$

Solution. Using the formulas given above, we find

$$
\begin{aligned}
P(A \cup B) &= P(A) + P(B) - P(A \cap B) \\
&= 0.32 + 0.56 - 0.18 = 0.7
\end{aligned}
$$

Example 3. Given $P(A) = 0.47$, $P(B) = 0.56$, $P(A \cup B) = 0.72$. Find $P(A \cap B)$

Solution. Using the formulas given above, we find

$$
\begin{aligned}
P(A \cup B) &= P(A) + P(B) - P(A \cap B) \\
0.72 &= 0.47 + 0.56 - P(A \cap B)
\end{aligned}
$$

so that

$$P(A \cap B) = 0.31$$

Example 4. Given $P(A) = 0.42$, $P(B) = 0.38$, A and B are mutually exclusive events. Find $P(A \cup B)$

Solution. Using the formulas given above, we find

$$
\begin{aligned}
P(A \cup B) &= P(A) + P(B) - P(A \cap B) \\
&= 0.42 + 0.38 - 0 = 0.8
\end{aligned}
$$

Example 5. Rolling two dice. The event A that both dice show the same number. The event B that the sum of the two dice is ≥ 9. Find the following probabilities:

(a) $P(A)$ (b) $P(A^c)$ (c) $P(B^c)$

(d) $P(A \cap B)$ (e) $P(A \cup B)$

Solution. We denote the elements of the events A and B by the lowercase letters a and b as shown by the following figure.

$S_1 \setminus S_2$	1	2	3	4	5	6
1	a					
2		a				
3			a			b
4				a	b	b
5				b	a b	b
6			b	b	b	a b

(a) $n(A) = 6$, hence $P(A) = \frac{1}{6}$

(b) $P(A^c) = 1 - P(A) = \frac{5}{6}$

(c) $n(B) = 10$, hence $P(B^c) = 1 - \frac{10}{36} = \frac{13}{18}$

(d) $n(A \cap B) = 2$, hence $P(A \cap B) = \frac{2}{36}$

e) $P(A \cup B) = P(A) + P(B) - P(A \cap B) = \frac{1}{6} + \frac{10}{36} - \frac{2}{36} = \frac{7}{18}$

Example 6. Rolling two dice. A is the event that both dice show the same number, B is the event that the sum of the two dice is exactly 7, and C is the event that at least one die shows 6. Find the following probabilities:

(a) $P(A^c)$

   (b) $P(A \cap B)$

(c) $P(B \cap C)$

(d) $P(A \cup B)$

(e) $P(B \cup C)$

Solution. We denote the elements of the events A, B and C by the lowercase letters a, b and c as shown by the following table.

$S_1 \setminus S_2$	1	2	3	4	5	6
1	a					b c
2		a			b	c
3			a	b		c
4			b	a		c
5		b			a	c
6	b c	c	c	c	c	a c

(a) $n(A) = 6$, hence $P(A^c) = 1 - P(A) = \frac{5}{6}$

(b) $n(A \cap B) = 0$, hence $P(A \cap B) = 0$

(c) $n(B \cap C) = 2$, hence $P(B \cap C) = \frac{1}{18}$

(d) $P(A \cup B) = P(A) + P(B) - P(A \cap B) = \frac{6}{36} + \frac{6}{36} - 0 = \frac{1}{3}$

(e) $P(B \cup C) = P(B) + P(C) - P(B \cap C) = \frac{6}{36} + \frac{11}{36} - \frac{2}{36} = \frac{5}{12}$

Exercises 8.2

1. Find $P(A^c)$ if $P(A) = 0.58$

2. Find $P(A)$ if $P(A^c) = 0.64$

In Exercises 3 – 7, find $P(A \cup B)$ from the information given:

3. $P(A) = 0.36$, $P(B) = 0.57$, $P(A \cap B) = 0.23$

4. $P(A) = 0.42$, $P(B) = 0.36$, $P(A \cap B) = 0.22$

5. $P(A) = 0.23$, $P(B) = 0.46$, A and B are mutually exclusive

6. $P(A^c) = 0.38$, $P(B^c) = 0.87$, A and B are mutually exclusive

7. $P(A) = \frac{1}{3}$, $P(B) = \frac{1}{12}$, $A \cap B = \phi$

In Exercises 8 – 12, determine $P(A \cap B)$ from the information given:

8. $P(A) = 0.71$, $P(B) = 0.39$, $P(A \cup B) = 0.78$

9. $P(A) = 0.6$, $P(B) = 0.3$, $P(A \cup B) = 0.75$

10. $P(A) = \frac{3}{4}$, $P(B) = \frac{1}{12}$, $P(A \cup B) = \frac{2}{3}$

11. $P(A) = \frac{2}{3}$, $P(B) = \frac{1}{3}$, $A \cup B = S$

12. $P(A^c) = \frac{1}{4}$, $P(B^c) = \frac{5}{6}$, $P((A \cup B)^c) = \frac{1}{6}$

13. Given $P(A) = 0.3$, $P(A \cup B) = 0.7$, $P(A \cap B) = 0.2$, find $P(B^c)$

14. Given $P(B^c) = \frac{2}{3}$, $P(A \cup B) = \frac{5}{6}$, $A \cap B = \phi$, find $P(A^c)$

15. Tossing a coin twice. A is the event that at least one Head occurs. B is the event that at most one Head occurs. Determine the following probabilities:

(a) $P(A^c)$

(b) $P(A \cup B)$

(c) $P(A \cap B)$

(d) $P((A \cap B)^c)$

16. A family has two children. A is the event that the second child is a Girl. B is the event that the family has at least one Boy. Determine the following probabilities:

(a) $P(A^c)$

(b) $P(A \cup B)$

(c) $P(A \cap B)$

(d) $P((A \cap B)^c)$

17. A family has three children. Given the following events:

(a) A is the event that the family has at least 2 Boys

(b) B is the event that the second child is a Girl

(c) C is the event that all children are of the same sex

Determine the following probabilities:

(a) $P(C^c)$

(b) $P(A \cap B)$

(c) $P(A \cup B)$

(d) $P((C \cap B)^c)$

18. The following table contains information about a survey carried on several persons about orange juice and apple juice:

	Orange O	Apple A
Female F	12	17
Male M	11	10

Determine the following probabilities:

(a) $P(M)$

(b) $P(F \cap O)$

(c) $P(M \cap A)$

(d) $P(F \cup O)$

(e) $P(M \cup A)$

19. Rolling two dice:
A is an event that the sum of the two outcomes is an odd prime
B is an event that the first die shows an odd prime
C is an event that one die only (not both) shows a 3

(a) Find $P(A^c)$

(b) Find $P(A \cap B)$

(c) Find $P(A \cup B)$

(d) Find $P(B \cap C)$

(e) Find $P(B \cup C)$

20. Rolling two dice:
A is an event that the sum of the two outcomes is 6 or 8
B is an event that the first die is a prime number
.C is an event that at least one die (may be both) shows a 4

(a) Find $P(A^c)$

(b) Find $P(A \cap B)$

(c) Find $P(A \cup B)$

(d) Find $P(B \cap C)$

(e) Find $P(B \cup C)$

21. Rolling two dice:
A is an event that the sum of the outcomes is ≤ 5
B is an event that the first die is a multiple of 3
C is an event that the product of the outcomes is a prime number

(a) Find $P(B^c)$

(b) Find $P(A \cap B)$

(c) Find $P(A \cup B)$

(d) Find $P(B \cap C)$

(e) Find $P(B \cup C)$

22. Rolling two dice:
A is an event that the sum of the outcomes is ≤ 4

B is an event that the first die is a prime number < 5
C is an event that the first die shows a 5 or 6

(a) Find $P(B^c)$

(b) Find $P(A \cap B)$

(c) Find $P(A \cup B)$

(d) Find $P(B \cap C)$

(e) Find $P(B \cup C)$

23. Students in a specific course are interviewed about their grades and gender. The following table shows the obtained results:

	A	B	C
Female F	14	21	22
Male M	12	25	26

A student is selected at random

(a) Find the probability that the student grade is not a C

(b) Find the probability that the student is a female and grade is B

(c) Find the probability that the student is a female or grade is B

(d) Find the probability that the students is a male and grade is C

(e) Find the probability that the student is a male or grade is C

24. Students in a specific course are interviewed about their grades and gender. The following table shows the obtained results:

	A	B	C	D
Male M	2	6	11	1
Female F	3	9	16	2

A student is selected at random

(a) Find the probability that the student grade is not a D

(b) Find the probability that the student is a female and grade is C

(c) Find the probability that the student is a female or grade is C

(d) Find the probability that the students is a male and grade is B

(e) Find the probability that the student is a male or grade is B

8.3 Conditional Probability

In this section we will discuss the probability of an event A that is affected by a condition given in another event B. This can be illustrated by discussing the following problem.

Suppose that a die is tossed once. It is required to determine the following:

1. Find the probability that a 4 turns up.

2. Find the probability that a 4 turns up if it is known that the outcomes are ≥ 3

To determine the probabilities, we first list the elements of the sample space S by:

$$S = \{1, 2, 3, 4, 5, 6\}$$

1. The event A that a 4 turns up is expressed by

$$A = \{4\}$$

Accordingly, we find

$$P(A) = \frac{1}{6}$$

2. It seems that we are asking the same question. However, a condition here is given that will reduce the number of elements of the sample space. The given condition states that we should focus our attention on all outcomes that are ≥ 3. Using this condition, we obtain the reduced sample space S_1 defined by

$$S_1 = \{3, 4, 5, 6\}$$

with

$$A = \{4\}$$

Consequently, the probability in this case, where a condition is given, is given by

$$P(A) = \frac{n(A)}{n(S_1)} = \frac{1}{4}$$

It is clear that the given condition affected the result and a different answer was obtained. This is normal since we are seeking the occurrence of a 4 among a reduced sample space that is controlled by a described condition.

In the following we present another example that is worth of discussing. Suppose that a bag contains four red balls and five green balls.

A ball is selected at random. It is required to determine the following probabilities:

1. Find the probability that the ball drawn is red.

2. A second ball is drawn. Find the probability that the second ball drawn is red if the first drawn ball is not replaced in the bag.

As discussed before, the sample space that best describes the contents of the bag is given by:

$$S = \{R_1, R_2, R_3, R_4, G_1, G_2, G_3, G_4, G_5\}$$

1. The event B that a red ball is selected is expressed by

$$B = \{R_1, R_2, R_3, R_4\}$$

Accordingly, we find

$$P(B) = \frac{4}{9}$$

It should be noted here that the probability will remain the same if the ball is replaced in the bag. In this case the sample space is not affected.

However, a condition is given in that the first ball drawn is red and not replaced in the bag. As discussed before, this condition will reduce number of elements of the sample space.

2. Using the given condition, we obtain the reduced sample space S_1 defined by

$$S_1 = \{R_2, R_3, R_4, G_1, G_2, G_3, G_4, G_5\}$$

with

$$B_1 = \{R_2, R_3, R_4\}$$

Consequently, the probability in this case, where a condition is given, is given by

$$P(B) = \frac{n(B_1)}{n(S_1)} = \frac{3}{8}$$

The examples discussed above show how the probability is affected by the condition added in the problem. However, a formula can be derived, and will be easily used, that will determine the probability in case a condition is given.

The **conditional probability** of an event A in a sample space S, given that another

event B has occurred, is denoted by $P(A|B)$, read as "probability of A if B" or "probability of A given the event B". The conditional probability $P(A|B)$ is defined by

$$P(A|B) = \frac{P(A \cap B)}{P(B)}, \ P(B) \neq 0$$

As will be seen from the following examples, the formula simplifies the calculations work.

Example 1. Tossing a coin once. Find the probability that a 4 turns up if it is known that the outcomes are ≥ 4.

Solution. Tossing a coin once gives the sample space S by

$$S = \{1, 2, 3, 4, 5, 6\}$$

Notice that two events are described. The event A that a 4 turns up, hence we have

$$A = \{4\}$$

In addition, a statement that follows the preposition "if" describes the condition of the event B. The condition given describes B in that it contains all outcomes that are ≥ 3. Consequently, in set notation, we find

$$B = \{4, 5, 6\}$$

so that

$$A \cap B = \{4\}$$

Using the formula for the conditional probability given above, we find

$$P(A|B) = \frac{P(A \cap B)}{P(B)}$$
$$= \frac{1/6}{3/6} = \frac{1}{3}$$

Example 2. A family has two children. Find the probability that both children are girls if it is known that the first child is a girl.

Solution. The sample space S for this family is given by

$$S = \{GG, GB, BG, BB\}$$

Notice that two events are described. The event A that both children are girls, hence we have

$$A = \{GG\}$$

The condition that follows the preposition "if" defines the event B in that it consists of all elements of S where the first child is a girl. hence, we find

$$B = \{GB, GG\}$$

so that

$$A \cap B = \{GG\}$$

Using the formula for the conditional probability given above, we find

$$P(A|B) = \frac{P(A \cap B)}{P(B)}$$
$$= \frac{1/4}{2/4} = \frac{1}{2}$$

Example 3. Rolling two dice. A is the event that the two outcomes are the same. B is the event that the sum of the outcomes is > 8. Find

(a) $P(A|B)$

(b) $P(B|A)$

Solution. The sample space S consists of 36 squares as shown by the table below. The elements of the events A and B are symbolized by the lowercase letters a and b. Besides, we can easily observe that the events A and B contains 6 elements and 10 elements respectively.

$S_1 \setminus S_2$	1	2	3	4	5	6
1	a					
2		a				
3			a			b
4				a	b	b
5				b	a b	b
6			b	b	b	a b

However, the common elements between A and B are 2 elements. This means that

$$A \cap B = \{(5,5),(6,6)\}$$

Using the formula for the conditional probability given above, we find

(a) $P(A|B) = \dfrac{P(A \cap B)}{P(B)} = \dfrac{2/36}{10/36} = \dfrac{1}{5}$

(b) $P(B|A) = \dfrac{P(B \cap A)}{P(A)} = \dfrac{2/36}{6/36} = \dfrac{1}{3}$

Theorem.

Although $A \cap B = B \cap A$, but in general

$$P(A|B) \neq P(B|A)$$

Finally, we consider the following example.

Example 4. A survey has been conducted on a math section. The following table shows the results of the survey.

	Junior J	Senior R
Female F	9	8
Male M	11	7

Find

(a) $P(F)$

(b) $P(F \cap R)$

(c) $P(M \cup J)$

(d) $P(F|R)$

(e) $P(M|J)$

Solution. Using the table shown above, we find the sample space S has 35 elements distributed as 17 females and 18 males. On the other hand, S is distributed as 20 juniors and 15 seniors.

(a) The event F contains 17 elements, therefore we find

$$P(F) = \frac{17}{35}$$

(b) The event $F \cap R$ contains 8 elements, therefore we find

$$P(F \cap R) = \frac{8}{35}$$

(c) The event $M \cup J$ contains 27 elements, therefore we find

$$P(M \cup J) = \frac{27}{35}$$

$$(d) P(F|R) = \frac{P(F \cap R)}{P(R)} = \frac{8}{15}$$

$$e P(M|J) = \frac{P(M \cap J)}{P(J)} = \frac{11}{20}$$

Exercises 8.3

1. A die is rolled once. Find the probability that the out come is :

(a) a 4

(b) a 4 if the outcomes are even

2. Two coins are tossed. Find the probability that the outcome is:

(a) exactly one head occurs

(b) exactly one head occurs if at least one head occurs

3. One card is selected from a deck of cards. Find the probability that the outcome is:

(a) a diamond card

(b) a diamond card if the selected card is a red card

4. A family has three children. Find the probability that this family has:

(a) three girls

(b) three girls if this family has at least 2 girls

5. Two dice are rolled. A is the event that the sum of the outcomes is exactly 7. B is the event that at least one die shows a 3(may be both). Find

(a) $P(A|B)$

(b) $P(B|A)$

6. Two dice are rolled. A is the event that the sum is ≤ 5. B is the event that the two outcomes are the same. Find

(a) $P(A|B)$

(b) $P(B|A)$

7. Two dice are rolled. A is the event that the sum is ≤ 5. B is the event that the sum is ≥ 10. Find

(a) $P(A|B)$

(b) $P(B|A)$

8. Two hundred persons are interviewed about their preferred programs. The collected data is given by the table:

	News N	Classics C	Talk T
Female F	20	32	50
Male M	25	58	15

Find

(a) $P(F)$

(b) $P(C)$

(c) $P(F \cup T)$

(d) $P(M \cap N)$

(e) $P(M|T)$

9. One hundred students in a specific course are interviewed about their grades. The following table shows the results obtained:

	A	B	C
Female F	14	11	22
Male M	12	15	26

Find

(a) $P(A^c)$

(b) $P(A \cup M)$

(c) $P(F \cap C)$

(d) $P(M|B)$

(e) $P(B|F)$

10. Three hundred math and chemistry major students are interviewed about the core courses they registered. The results are given by the table:

	English E	History H	Art A
Math M	40	55	50
Chemistry C	45	65	45

Find

(a) $P((E \cap M)^c)$

(b) $P((C \cup H)^c)$

(c) $P(M|A)$

(d) $P(H|C)$

(e) $P(E|H)$

11. The following table shows the number of voters in a small town:

	Republican R	Democratic D	Independent I
Male M	184	187	129
Female F	229	131	140

A person is selected at random. Answer the following:

(a) Find the probability that this person is a male and a republican.

(b) Find the probability that this person is a male or a democratic.

(c) Find the probability that this person is not independent.

(d) Find the probability that this person is a female if the person is a democratic.

(e) If the person is independent, find the probability that the person is a male.

12.   The following table shows the grades of students in three different courses:

	A	B	C
Math M	10	15	25
History H	22	28	30
English E	21	27	22

(a) Find $P(M \cap A)$

(b) Find $P(H \cup B)$

(c) Find $P(E \cup C)^c$

(d) Find $P(C|H)$

(e) Find $P(H|C)$

13. Rolling two dice:
A is an event that the sum of the two outcomes is an odd prime
B is an event that the first die shows an odd prime
C is an event that one die only (not both) shows a 3

(a) Find $P(A^c)$

(b) Find $P(A \cap B)$

(c) Find $P(A \cup B)$

(d) Find $P(B|C)$

(e) Find $P(A|B)$

14. Rolling two dice:
A is an event that the sum of the two outcomes is 6 or 8
B is an event that the first die is a prime number
C is an event that at least one die (may be both) shows a 4

(a) Find $P(A^c)$

(b) Find $P(A \cap B)$

(c) Find $P(A \cup B)$

(d) Find $P(B|C)$

(e) Find $P(C|B)$

15. Rolling two dice:
A is an event that the sum of the outcomes is ≤ 5
B is an event that the first die is a multiple of 3
C is an event that the product of the outcomes is a prime number

(a) Find $P(B^c)$

(b) Find $P(A \cap B)$

(c) Find $P(A \cup B)$

(d) Find $P(B|C)$

(e) Find $P(A|B)$

16. Rolling two dice:
A is an event that the sum of the outcomes is ≤ 4
B is an event that the first die is a prime number < 5
C is an event that the first die shows a 5 or 6

(a) Find $P(B^c)$

(b) Find $P(A \cap B)$

(c) Find $P(A \cup B)$

(d) Find $P(B|C)$

(e) Find $P(A|B)$

17. Students in a specific course are interviewed about their grades and gender. The following table shows the obtained results:

	A	B	C
Female F	14	21	22
Male M	12	25	26

A student is selected at random

(a) Find the probability that the student grade is not a C

(b) Find the probability that the student is a female and grade is B

(c) Find the probability that the student is a female or grade is B

(d) Find the probability that the students is a male if grade is C

(e) If the grade is a B, find the probability that the student is a female

18. Students in a specific course are interviewed about their grades and gender. The following table shows the obtained results:

	A	B	C	D
Male M	2	6	11	1
Female F	3	9	16	2

A student is selected at random

(a) Find the probability that the student grade is not a D

(b) Find the probability that the student is a female and grade is C

(c) Find the probability that the student is a female or grade is C

(d) Find the probability that the students is a male if grade is B

(e) If the grade is A, find the probability that the student is a female

19. Students in a specific course are interviewed about their grades and gender. The following table shows the obtained results:

	A	B	C	D
Female F	2	7	13	3
Male M	4	9	17	5

A student is selected at random

(a) Find the probability that the student grade is not B

(b) Find the probability that the student is a female and grade is C

(c) Find the probability that the student is a female or grade is C

(d) Find the probability that the students is a female if grade is A

(e) If the student is a male, find the probability that the grade is D

20. Students in a specific course are interviewed about their grades and gender. The following table shows the obtained results:

	A	B	C	D	E
Female F	4	12	13	8	1
Male M	3	7	11	9	2

A student is selected at random

(a) Find the probability that the student grade is not A

(b) Find the probability that the student is a female and grade is B

(c) Find the probability that the student is a male or grade is C

(d) Find the probability that the students is a male if grade is D

(e) If the student is a female, find the probability that the grade is E

8.4   Repeated Experiments

In this section we study an experiment that consists of two or more of smaller experiments. Such an experiment is called a **repeated experiment** or **compound experiment**. Example of repeated experiment is drawing a marble twice from a bag that contains 4 red marbles and 5 green marbles. Another example of a repeated experiment is rolling one die followed by tossing a coin twice.

It is useful to use a tree diagram to represent the repeated experiments. Moreover, the tree diagram enables us to compute the probabilities of the smaller events. As presented before, we normally start with the root of a tree, we then first draw one branch for each possible event, such as drawing a red marble. Second, for each branch constructed in the first step, we draw one branch for each event involved in the second trial of the experiment. We continue in this way. The probability of each event is obtained by multiplying probabilities of the branches of each path that gives a specific event. The following illustrative examples explain the concept of repeated experiments.

Example 1. A bag contains 5 red and 4 green marbles. Two marbles are selected one at a time. Draw a proper tree diagram and show the probability of each path, provided that:
(a) the marble drawn is replaced in the bag
(b) the marble drawn is not replaced in the bag

Solution.

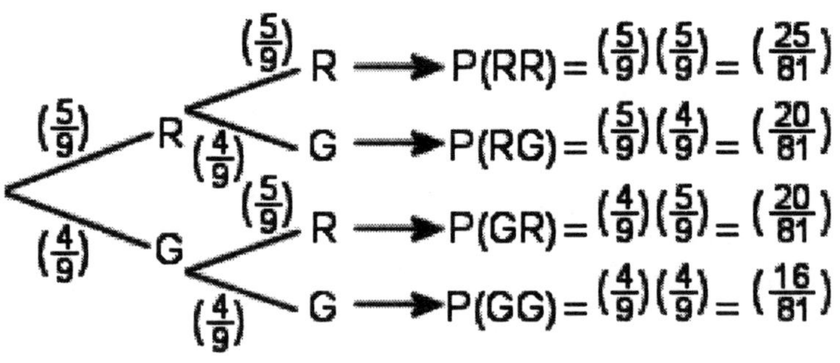

$$P(RR) = \left(\tfrac{5}{9}\right)\left(\tfrac{5}{9}\right) = \left(\tfrac{25}{81}\right)$$

$$P(RG) = \left(\tfrac{5}{9}\right)\left(\tfrac{4}{9}\right) = \left(\tfrac{20}{81}\right)$$

$$P(GR) = \left(\tfrac{4}{9}\right)\left(\tfrac{5}{9}\right) = \left(\tfrac{20}{81}\right)$$

$$P(GG) = \left(\tfrac{4}{9}\right)\left(\tfrac{4}{9}\right) = \left(\tfrac{16}{81}\right)$$

Figure 8.1

(a) Notice that the sample space consists of 9 marbles. Drawing a marble for the first time indicates that

$$P(R) = \frac{5}{9}, \; P(G) = \frac{4}{9}$$

This is shown by the first two branches in the figure 8.1. Notice that the drawn marble is replaced, therefore there is no change in the sample space or the number of red and green marbles. For each branch constructed, we draw two branches to show the second drawn marble. The probability of each branch is the same as before.

As indicated before, we obtain the probability of each path by multiplying the probabilities of the branches of that path. In other words, we find

$$P(RR) \;\; = \;\; \left(\frac{5}{9}\right)\left(\frac{5}{9}\right) = \frac{25}{81}$$

$$P(RG) \;\; = \;\; \left(\frac{5}{9}\right)\left(\frac{4}{9}\right) = \frac{20}{81}$$

$$P(GR) \;\; = \;\; \left(\frac{4}{9}\right)\left(\frac{5}{9}\right) = \frac{20}{81}$$

$$P(RR) \;\; = \;\; \left(\frac{4}{9}\right)\left(\frac{4}{9}\right) = \frac{16}{81}$$

(b) Notice here that the selected marble is not replaced back. Although there is no change in the sample space in the first drawing, but the number of red or green marbles will change, because the experiment is conducted without replacement. As a result, the sample space will change as well.

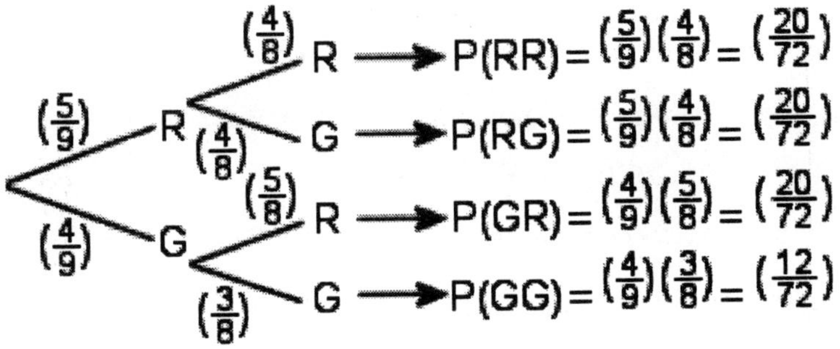

Figure 8.2

If the first marble selected was red and was not replaced, then $n(R) = n(G) = 4$, and therefore $n(S) = 8$. This means that the probability of each branch of the second trial, as shown by the figure 8.2, is

$$P(R) = \frac{4}{8}, \; P(G) = \frac{4}{8}$$

However, if the first marble selected was green and was not replaced, then $n(R) = 5, (G) = 3$, and therefore $n(S) = 8$. This means that the probability of each branch of the second trial, as shown by the figure, is

$$P(R) = \frac{5}{8}, \; P(G) = \frac{3}{8}$$

As discussed above, we obtain the probability of each path by multiplying the probabilities of the branches of that path. In other words, we find

$$P(RR) \;=\; \left(\frac{5}{9}\right)\left(\frac{4}{8}\right) = \frac{20}{72}$$

$$P(RG) \;=\; \left(\frac{5}{9}\right)\left(\frac{4}{8}\right) = \frac{20}{72}$$

$$P(GR) \;=\; \left(\frac{4}{9}\right)\left(\frac{5}{8}\right) = \frac{20}{72}$$

$$P(RR) \;=\; \left(\frac{4}{9}\right)\left(\frac{3}{8}\right) = \frac{12}{72}$$

Example 2. Use Example 1, where the two marbles were selected with replacement, to find the probabilities of the following events:

(a) A is an event that the two marbles selected are of different colors
(b) B is an event that the two marbles selected are of same color
(c) C is an event that at least one of the two marbles selected is red

Solution. Using the tree diagram shown above in Figure 8.2 we find

(a) P(A)= P(RG)+P(GR) = $\frac{20}{81} + \frac{20}{81} = \frac{40}{81}$

(b) P(B)= P(RR)+P(GG) = $\frac{25}{81} + \frac{16}{81} = \frac{41}{81}$

(c) P(C)= P(RR) +P(RG)+P(GR)= $\frac{25}{81} + \frac{20}{81} + \frac{20}{81} = \frac{65}{81}$

Example 3. Use Example 1, where the two marbles were selected without replacement, to find the probabilities of the following events:

(a) A is an event that the two marbles selected are of different colors
(b) B is an event that the two marbles selected are of same color
(c) C is an event that at least one of the two marbles selected is red

Solution. Using the tree diagram shown above in Figure 8.2 we find

(a) P(A)= P(RG)+P(GR) $= \frac{20}{72} + \frac{20}{72} = \frac{5}{9}$

(b) P(B)= P(RR)+P(GG) $= \frac{20}{72} + \frac{12}{72} = \frac{4}{9}$

(c) P(C)= P(RR) +P(RG)+P(GR)$= \frac{20}{72} + \frac{20}{72} + \frac{20}{72} = \frac{5}{6}$

Exercises 8.4

1. A bag contains 7 red and 3 green marbles. Two marbles are selected at random one at a time. The drawn marble is replaced in the bag. Use a tree diagram to find the probabilities of the following events:
(a) A is an event that the two marbles selected are of different colors
(b) B is an event that the two marbles selected are of same color
(c) C is an event that at least one of the two marbles selected is green

2. Repeat Exercise 1 if the drawn is conducted without replacement.

3. A bag contains 5 red, 3 blue and 2 green marbles. Two marbles are selected at random one at a time. The drawn marble is replaced in the bag. Use a tree diagram to find the probabilities of the following events:
(a) A is an event that the two marbles selected are of different colors
(b) B is an event that the two marbles selected are of same color
(c) C is an event that exactly one of the two marbles is red

4. Repeat Exercise 3 if the drawn is conducted without replacement.

5. A bag contains 6 red and 4 green marbles. Three marbles are selected at random one at a time. The drawn marble is replaced. Use a tree diagram to find the probabilities of the following events:
(a) A is an event that the three marbles selected are of same colors
(b) B is an event that exactly two marbles of the three drawn are red
(c) C is an event that exactly two marbles of the three drawn are green

6. Repeat Exercise 5 if the drawn is conducted without replacement.

7. In a class of 50 students, 30 students are female students and 20 are male. It was found that 40% of the female and 30% of the male students got an A in a test, others got a non A (NA) in the same test. Find the probability of

the students with an A grade.

8. Rolling a die three times. A success is considered if the die shows 5 or 6, and a failure if the die shows 1,2,3 or 4. Find the probability of exactly two successes.

Chapter 9

Functions and Algorithms

9.1 Functions

The area A of a square depends on the length x of the square. The equation $A = x^2$ is a rule that assigns for every value of x exactly one value of A. In this case, where the rule $A = x^2$ connects x to A, we say that A is a function of x.

The notation $f(x)$, read " f of x" is normally used and denotes the value f that results for a given value of x. Accordingly, a function f is a rule that assigns to each value of x, usually belongs to a set called the **domain**, one value $f(x)$ in another set called the **range**.

To determine the value of $f(x)$ for a given value of the variable x, we simply substitute the given value of x in the given rule to evaluate the corresponding value of $f(x)$. This can be illustrated by the following examples.

Example 1. Given the function

$$f(x) = x^2 + 3x$$

Evaluate $f(0), f(-2), f(2)$

Solution.

$$
\begin{aligned}
f(x) &= x^2 + 3x \\
f(0) &= 0 + 0 = 0 \\
f(-2) &= (-2)^2 - 3 \times 2 = -2 \\
f(2) &= 2^2 + 3 \times 2 = 10
\end{aligned}
$$

Example 2. Given the function

$$f(x) = \sqrt{x-1}$$

Evaluate $f(1), f(5), f(0)$

Solution.

$$
\begin{aligned}
f(x) &= \sqrt{x-1} \\
f(1) &= \sqrt{1-1} = 0 \\
f(5) &= \sqrt{5-1} = 2 \\
f(0) &= \sqrt{0-1} = \sqrt{-1}, \text{(undefined)}
\end{aligned}
$$

It is useful to note here that $f(x)$ is not defined at $x = 0$, because there is no square root to a negative real number in the real number system. This occurred because $x = 0$ is not a value that will assign a real value for $f(x)$. As will be discussed later, $x = 0$ is not an element of the domain, the set of all values of x that define $f(x)$. The domain in this example is identical to the coins that will enable you to make a phone call, namely nickels, dimes and quarters. Other types of currency are not acceptable by a public telephone, hence a call is not possible without using nickels, dimes or quarters.

It is worth noting here that any given rule for $f(x)$ will not guarantee the determination of a corresponding value of $f(x)$ for every value of x. Using a computer to evaluate $f(0)$ in Example 2 will result in an error message.

Example 3. Given the function

$$f(x) = \frac{1}{x^2 - x}$$

Evaluate $f(2), f(3), f(1)$

Solution.

$$
\begin{aligned}
f(x) &= \frac{1}{x^2 - x} \\
f(2) &= \frac{1}{2^2 - 2} = \frac{1}{2} \\
f(3) &= \frac{1}{3^2 - 3} = \frac{1}{6} \\
f(0) &= \frac{1}{1^2 - 1} = \frac{1}{0}, \text{(undefined)}
\end{aligned}
$$

We note here that $f(1)$ is not defined in this example, because we cannot divide by 0. This means that $x = 1$ is not a value that will define a real value for $f(x)$.

An important conclusion can be made here is that the square root of a negative real number and division by 0 are not defined operations in the real number system, and hence $f(x)$ cannot be evaluated for this value of x. This conclusion must be considered in computer programming as well.

It is useful to note that the absolute value of a number x, usually denoted by $|x|$, is defined by

$$|x| = \begin{cases} x & \text{for } x \geq 0 \\ -x & \text{for } x < 0 \end{cases}$$

This means that $|-4| = 4$, and $|4| = 4$. We now turn to the following example.

Example 4. Given the function

$$f(x) = |x| + 1$$

Evaluate $f(-3), f(0), f(3)$

Solution.

$$\begin{aligned} f(x) &= |x| + 1 \\ f(-3) &= |-3| + 1 = 3 + 1 = 4 \\ f(0) &= |0| + 1 = 0 + 1 = 1 \\ f(3) &= |3| + 1 = 3 + 1 = 4 \end{aligned}$$

It is important to note here that a function $f(x)$ is not always defined by a one piece rule. Sometimes, a function is defined by two or more pieces. A useful example in this regard is the driving speed of a car. The speed function is defined by 30 mph in regular streets, 20 mph near schools and greater than or equal 45 mph and less than or equal to 65 mph in highways. This example is usually expressed by the following expression:

$$\text{speed} = \begin{cases} 30\,\text{mph} & \text{in regular streets} \\ 20\,\text{mph} & \text{in school zones} \\ \geq 45\,\text{mph and} \leq 65\,\text{mph} & \text{in highways} \end{cases}$$

Other examples can be found in many applications such as the grading system of a course, where grade A will be given to scores greater than or equal to 90, B will be given to scores from 80 to 89 an so on.

It is helpful to note that in functions defined by pieces like the speed example introduced before, each piece of the function is defined for corresponding values of x that are usually given. To evaluate the function for a given value of x, we have to determine the related interval that contains the given value of x. We then substitute the given value of x in the corresponding piece of $f(x)$. This may be illustrated by the following examples.

Example 5. Given the function

$$f(x) = \begin{cases} 3x + 1 & \text{for } -2 \leq x < 0 \\ x^2 + x + 1 & \text{for } 0 \leq x \leq 2 \end{cases}$$

Evaluate $f(-1), f(0), f(1), f(2), f(3)$

Solution. It is easily observed that this function is defined by 2 pieces, where the first piece $3x + 1$ will only be used for all values of x that belong to the interval $-2 \leq x < 0$. The second piece given by $x^2 + x + 1$ will only be used for all values of x that belong to the interval $0 \leq x \leq 2$. Since $x = 1$ is contained in the first interval of the x values, we therefore use the first piece of $f(x)$. In a similar discussion, we use the second piece for $x = 0, 1$, and 2. For $x = 3$, it is clear that this value of x does not belong to any interval of determination of the function, hence there is no rule that will assign $f(3)$. The previous discussion can be illustrated as follows:

$$\begin{aligned} f(-1) &= 3(-1) + 1 = -2 \\ f(0) &= 0^2 + 0 + 1 = 1 \\ f(1) &= 1^2 + 1 + 1 = 3 \\ f(2) &= 2^2 + 2 + 1 = 7 \\ f(3) &\quad \text{is undefined} \end{aligned}$$

Example 6. Given the function

$$f(x) = \begin{cases} 3x + 1 & \text{for } x \leq 0 \\ 3x^2 + 1 & \text{for } x > 0 \end{cases}$$

Evaluate $f(-2), f(0), f(2)$

Solution. It is clear that $x = -2$ belongs to the interval of the first piece, hence the first rule will be used for evaluating $f(-2)$. On the other hand, $x = 0$ and $x = 2$ belong to the second interval of the second piece. Accordingly, we find

$$
\begin{aligned}
f(-2) &= 3(-2) + 1 = -5 \\
f(0) &= 3 \times 0^2 + 1 = 1 \\
f(2) &= 3 \times 2^2 + 1 = 13
\end{aligned}
$$

Example 7. Given the function

$$
f(x) = \begin{cases}
-x + 1 & \text{for } x < 0 \\
3 & \text{for } x = 0 \\
x + 1 & \text{for } x > 0
\end{cases}
$$

Evaluate $f(1), f(-1), f(0), f(-4), f(4)$

Solution. We note here that this function consists of three pieces. The first piece is applicable for negative real numbers, the second piece is applicable only for one value of x namely $x = 0$, and the third piece is applicable for positive real numbers. Accordingly, we find

$$
\begin{aligned}
f(1) &= 1 + 1 = 2 \\
f(-1) &= -(-1) + 1 = 2 \\
f(0) &= 3 \\
f(-4) &= -(-4) + 1 = 5 \\
f(4) &= 4 + 1 = 5
\end{aligned}
$$

Example 8. Given the function

$$
f(x) = \begin{cases}
kx + 3 & \text{for } -5 \leq x < 0 \\
x^2 - 3 & \text{for } 0 \leq x \leq 5
\end{cases}
$$

Find k such that $f(-3) = f(3)$

Solution. The first rule should be used to evaluate $f(-3)$. The second rule should be used to evaluate $f(3)$. Accordingly, we find

$$\begin{aligned} f(-3) &= f(3) \\ -3k + 3 &= 6 \\ k &= -1 \end{aligned}$$

Example 9. Given the function

$$f(x) = \begin{cases} x^2 + ax + 6 & \text{for } x < 0 \\ x^3 - ax^2 + bx & \text{for } x \geq 0 \end{cases}$$

Find a and b such that $f(-1) = f(1)$, and $f(-2) = f(2)$.

Solution. For $f(-1), f(-2)$ we use the first rule and for $f(1), f(2)$ we use the second rule. This gives

$$\begin{aligned} 1 - a + 6 &= 1 - a + b \\ 4 - 2a + 6 &= 8 - 4a + 2b \end{aligned}$$

Solving these two equations we find

$$a = 5, b = 6$$

Example 10. Find a and b such that $f(-1) = f(1)$, and $f(-4) = f(4)$ where

$$f(x) = \begin{cases} x^2 + ax + 4 & \text{for } -6 \leq x < 0 \\ ax + b & \text{for } 0 \leq x < 3 \\ ax^3 + bx - 60 & \text{for } 3 \leq x \leq 6 \end{cases}$$

Solution. For $f(-1), f(-4)$, the first rule is used, and for $f(1)$, the second rule is used, and for $f(4)$, the third rule is used. This gives

$$\begin{aligned} 1 - a + 4 &= a + b \\ 16 - 4a + 4 &= 64a + 4b - 60 \end{aligned}$$

Solving these two equations we find

$$a = 1, b = 3$$

Exercises 9.1

1. Given $f(x) = 3x^2 + 2x - 1$. Evaluate $f(-1), f(0), f(1)$

2. Given $f(x) = x^3 + 2x - 4$. Evaluate $f(-2), f(0), f(2)$

3. Given $f(x) = |x - 1|$. Evaluate $f(-3), f(0), f(3)$

4. Given $f(x) = |3x - 5|$. Evaluate $f(-2), f(0), f(2)$

5. Given $g(x) = \sqrt{x + 1}$. Evaluate $g(0), g(3), g(8)$

6. Given $g(x) = \sqrt{x^2 + 1}$. Evaluate $g(0), g(-2), g(\sqrt{15})$

7. Given $h(x) = \sqrt{3 - x}$. Evaluate $h(0), h(-1), h(3)$

8. Given $f(x) = \dfrac{x + 1}{x - 1}$. Evaluate $f(0), f(2), f(3)$

9. Given $f(x) = \dfrac{x + 2}{x^2 + 2}$. Evaluate $f(0), f(-2), f(3)$

10. Given $f(x) = \dfrac{x^2 - 1}{x^2 + 1}$. Evaluate $f(0), f(1), f(-1)$

11. Given $f(x) = \dfrac{x + 1}{x^2 - 5x + 6}$. Evaluate $f(0), f(1), f(-1)$

12. Given $f(x) = \begin{cases} x^2 + 1 & \text{for } -4 \le x < 0 \\ x^2 - 1 & \text{for } 0 \le x \le 4 \end{cases}$

Evaluate $f(-2), f(0), f(3)$

13. Given $f(x) = \begin{cases} x + 1 & \text{for } x < 0 \\ 2x - 3 & \text{for } x \ge 0 \end{cases}$

Evaluate $f(-1), f(0), f(2)$

14. Given $f(x) = \begin{cases} x^3 + 1 & \text{for } x < 0 \\ x^2 & \text{for } x \ge 0 \end{cases}$

Evaluate $f(-1), f(0), f(2)$

15. Given $f(x) = \begin{cases} x + 3 & \text{for } x < 0 \\ 4 & \text{for } x = 0 \\ x^3 + 1 & \text{for } x > 0 \end{cases}$

Evaluate $\dfrac{f(-2) - f(0) + f(2)}{|f(-9)|}$

16. Given $f(x) = \begin{cases} x + 4 & \text{for } x < -2 \\ 2x & \text{for } -2 \le x \le 2 \\ x + 4 & \text{for } x > 2 \end{cases}$

Evaluate $f(-2), f(-4), f(4)$

17. Given $f(x) = \begin{cases} kx + 2 & \text{for } -3 \le x < 0 \\ x^2 - 6 & \text{for } 0 \le x \le 3 \end{cases}$

Find k such that $f(-2) = f(2)$

18. Given $f(x) = \begin{cases} kx^2 - 2x + 1 & \text{for } -2 \le x < 0 \\ x^2 + x - 6 & \text{for } 0 \le x \le 2 \end{cases}$

Find k such that $f(-1) = f(1)$

19. Given $f(x) = \begin{cases} kx^3 + x - 4 & \text{for } x < 0 \\ x^2 + kx - 20 & \text{for } x \ge 0 \end{cases}$

Find k such that $f(-2) = f(2)$

20. Given $f(x) = x^2 - 7x + 6$. Find k if $f(k) = 0$

21. Given $f(x) = x^3 - x^2 - 4x + 4$. Find k if $f(k) = 0$

22. Given $f(x) = 1 + x|x|$. Find k if $f(k) = 5$

23. Given $f(x) = \sqrt{x^2 + 3x - 1}$. Find k if $f(k) = 3$

24. Given the function
$$f(x) = \begin{cases} x^2 - bx + 12 & \text{for } -4 \le x \le 0 \\ ax + b & \text{for } 0 < x \le 2 \\ bx^3 + 2bx + 51 & \text{for } 2 < x \le 4 \end{cases}$$

Find a and b such that $f(-1) = f(1)$, $f(-3) = f(3)$

25. Given the function
$$f(x) = \begin{cases} x^2 + bx + 17 & \text{for } -8 \le x < 0 \\ 49 - bx^2 & \text{for } 0 < x < 4 \\ 6x + a & \text{for } 4 \le x \le 8 \end{cases}$$

Find a and b such that $f(-4) = f(4)$, $f(3) = f(5)$

26. Given the function
$$f(x) = \begin{cases} x^2 - bx + 3a - 27 & \text{for } -4 \le x \le 0 \\ ax + a + 14b & \text{for } 0 < x \le 2 \\ bx^3 + 2bx + 51 & \text{for } 2 < x \le 4 \end{cases}$$

Find a and b such that $f(-1) = f(1)$, and $f(-3) = f(3)$.

27. Given the function
$$f(x) = \begin{cases} x^2 - bx + 5a + 38 & \text{for } -4 \le x \le 0 \\ ax + a + 12b & \text{for } 0 < x \le 2 \\ bx^3 + 2bx - 83 & \text{for } 2 < x \le 4 \end{cases}$$

Find a and b such that $f(-2) = f(2)$, and $f(-3) = f(3)$.

28. Given the function
$$f(x) = \begin{cases} x^2 - bx + 3a + 44 & \text{for } -4 \le x \le 0 \\ ax + 25a + 14b & \text{for } 0 < x \le 2 \\ bx^3 + 2bx - 4 & \text{for } 2 < x \le 4 \end{cases}$$

Find a and b such that $f(-2) = f(2)$, and $f(-3) = f(3)$.

29. Given the function

$$f(x) = \begin{cases} ax + b & \text{for } -6 \leq x \leq -3 \\ bx + a & \text{for } -3 < x \leq 0 \\ ax^2 - bx + 4 & \text{for } 0 < x \leq 4 \end{cases}$$

Find a and b such that $f(-4) = f(-2)$ and $f(-1) = f(2)$

30. Given the function

$$f(x) = \begin{cases} ax + b & \text{for } -4 \leq x < 0 \\ 4bx^3 - 3a & \text{for } 0 \leq x \leq 2 \\ ax^2 - bx - 3 & \text{for } 2 < x \leq 4 \end{cases}$$

Find a and b such that $f(2) = f(-2)$ and $f(1) = f(3)$

9.2   Domain of a Function

We discussed before that a function $f(x)$ is not always defined for all real values of x. The public phone accepts only specific coins to make phone connections. In Example 2 of Section 8.1, it was impossible to determine the square root of a negative quantity. In Example 3 of the previous section, the fractional function is not determined for any value of x that makes the denominator 0. In Example 4 of Section 8.1, the function was defined for specific values of x defined by the intervals specified. Accordingly, a function is defined only for specific values of x called the **domain** of $f(x)$. The **domain** of a function $f(x)$ is defined as the set of all real values of x for which $f(x)$ is determined. To find the domain of $f(x)$, we consider the following classes of functions:

1. **Polynomials:** for polynomials of the form

$$f(x) = a_n x^n + a_{n-1} x^{n-1} + \cdots + a_1 x + a_0$$

with no specified restrictions on the values of x, the domain is all real numbers. If specific restrictions were imposed on the values of x, the restrictions should be considered in determining the domain. This may be explained by the following example.

Example 1. Find the domain of each of the following functions:

(a) $f(x) = x^3 - x^2 + 1$

(b) $f(x) = x^4 + x^2 + 4$

(c) $f(x) = x^2 + 2x + 1$ for $-3 \leq x \leq 3$

Solution.

(a) The domain of the polynomial $f(x) = x^3 - x^2 + 1$ is all real numbers. The domain is thus given by: $D = \{x \mid x \in R\}$

(b) The domain of the polynomial $f(x) = x^4 + x^2 + 4$ is all real numbers. The domain is thus given by: $D = \{x \mid x \in R\}$

(c) A restriction is given here on the values of x to be used. Accordingly, the domain of $f(x) = x^2 + 2x + 1$ is given by all real values of x of the interval $-3 \leq x \leq 3$. This means that $f(3)$ is defined whereas $f(4)$ is undefined because $x = 4$ does not belong to the domain of validity.

2. Absolute value functions: the absolute value function denoted by $f(x) = |x|$ is defined by

$$f(x) = |x| = \begin{cases} x & \text{for } x \geq 0 \\ -x & \text{for } x < 0 \end{cases}$$

We can easily observe from the definition that $|x|$ can be determined for any real number x, hence the domain is all real numbers $(-\infty, \infty)$. This will be illustrated by the following example.

Example 2. Find the domain of each of the following functions:

(a) $f(x) = |x - 5|$

(b) $g(x) = |5x + 3| - 6x$, for $-2 \leq x \leq 2$

Solution.

(a) The domain of $f(x)$ is all real numbers, $-\infty < x < \infty$

(b) The domain of $g(x)$ is given by the specified interval $-2 \leq x \leq 2$

3. Rational functions: a rational function is a function of the form

$$f(x) = \frac{P(x)}{Q(x)}$$

where $P(x)$ and $Q(x)$ are polynomials. The domain of a rational function is all real values of x that do not produce 0 in the denominator. In other words, the domain of a rational function can be obtained by solving the non-equality

$$Q(x) \neq 0$$

Solving the non-equality gives the excluded values of x. This indicates that the domain is all real numbers excluding all values of x that result from solving the non-equality $Q(x) \neq 0$.

The domain of rational functions can be illustrated by using the illustrative example.

Example 3. Find the domain of each of the following:

(a) $f(x) = \dfrac{2}{x-1}$

(b) $f(x) = \dfrac{x+1}{x^2 - 5x + 4}$

(c) $f(x) = \dfrac{x-1}{x^2 + 1}$

Solution.

(a) The denominator $Q(x) = x - 1$
To find the excluded values of x, we therefore solve

$$
\begin{aligned}
x - 1 &\neq 0 \\
x &\neq 1
\end{aligned}
$$

Hence, the domain is all real numbers excluding $x = 1$. This may also be expressed in a format set by

$$\text{Domain} = \{x | x \in R, x \neq 1\}$$

where R represents all real numbers.

(b) The denominator $Q(x) = x^2 - 5x + 4$
To find the excluded values of x, we therefore solve

$$
\begin{aligned}
x^2 - 5x + 4 &\neq 0 \\
(x - 1)(x - 4) &\neq 0 \\
x &\neq 1, 4
\end{aligned}
$$

Hence, the domain is all real numbers excluding $x = 1, 4$. This may also be expressed in a format set by

$$\text{Domain} = \{x | x \in R, x \neq 1, 4\}$$

(c) The denominator $Q(x) = x^2 + 1$
To find the excluded values of x, we therefore solve

$$x^2 + 1 \neq 0$$

We note here that x^2 is always positive for negative and positive values of x. As a result, there is no real x such that $x^2 + 1 = 0$. Therefore, the domain is the set of all real numbers R. This may also be expressed in a format set by

$$\text{Domain} = \{x | x \in R\}$$

4. Square root functions: a square root function is a function of the form

$$f(x) = \sqrt{g(x)}$$

where $g(x)$ is a polynomial, absolute value function or a rational function. In this text we will focus our study on the case where $g(x)$ is a polynomial. As discussed before, the square root function $f(x)$ is defined only if $g(x) \geq 0$. Consequently, the domain of a square root function is determined by solving the inequality

$$g(x) \geq 0$$

The interval that results from solving the above inequality will give the domain of the function. This technique will be explained by discussing the following illustrative example.

Example 4. Find the domain of the following functions:

(a) $f(x) = \sqrt{3x - 6}$

(b) $f(x) = \sqrt{18 - 6x}$

(c) $f(x) = \sqrt{x^2 + 4}$

Solution.

(a) To determine the domain, we solve the inequality

$$3x - 6 \geq 0$$
$$x \geq 2$$

The domain is thus defined by all real numbers x such that

$$x \geq 2$$

or equivalently, the domain is given by the interval

$$2 \leq x < \infty$$

(b) To determine the domain, we solve the inequality

$$
\begin{aligned}
81 - 6x &\geq 0 \\
-6x &\geq -18 \\
x &\leq 3
\end{aligned}
$$

The domain is all real numbers x such that $x \leq 3$, or equivalently, the domain is given by the interval

$$-\infty < x \leq 3$$

(c) In this example, we note that $x^2 + 4$ is always positive for all real values of x. Accordingly, the domain of $f(x)$ is all real numbers.

5. Piecewise Functions(Functions defined in pieces): functions of this type are usually given in pieces (rules) where each piece is defined for a given distinct interval of validity. It is clear here that the domain is the union of these specified distinct intervals. The domain of this type of functions will be illustrated by discussing the following example.

Example 5. Find the domain of the following functions:

(a) $f(x) = \begin{cases} x + 1 & \text{for } -3 \leq x < 0 \\ x^2 - 1 & \text{for } 0 \leq x \leq 3 \end{cases}$

(b) $f(x) = \begin{cases} 3x + 1 & \text{for } x < 0 \\ x^2 + 1 & \text{for } x > 0 \end{cases}$

(c) $f(x) = \begin{cases} x^3 + 1 & \text{for } x < 3 \\ x^2 + 1 & \text{for } x \geq 3 \end{cases}$

Solution.

(a) The set

$$\{x \mid -3 \le x < 0\}$$

and the set

$$\{x \mid 0 \le x \le 3\}$$

are the specified intervals that will define $f(x)$ for piece 1 and piece 2 respectively. The union of these two intervals defined by

$$D = \{x \mid -3 \le x \le 3\}$$

gives the domain D of this function.

(b) The set

$$\{x \mid x < 0\}$$

and the set

$$\{x \mid x > 0\}$$

are the specified intervals that will define $f(x)$ for piece 1 and piece 2 respectively. We note here that $x = 0$ does not belong to any interval. Therefore, the domain D of this function is the set of all real numbers excluding $x = 0$. This can be expressed in a set form by

$$D = \{x \mid x \in R, x \ne 0\}$$

or by

$$D = \{x \mid -\infty < x < 0 \text{ or } 0 < x < \infty\}$$

(c) It is clear here that the union of the two intervals is the set of all real numbers. Hence, the domain D is the set of all real numbers. This can be expressed in a format set by

$$D = \{x \mid x \in R\} \quad \text{or} \quad D = \{x \mid -\infty < x < \infty\}$$

In what follows we discuss the domain of functions that include rational and square root functions.

Example 6. Find the domain of each of the following functions:

(a) $f(x) = \dfrac{1}{x^2 - 2x}$

(b) $g(x) = \sqrt{x-2} + \dfrac{1}{x-3}$

(c) $h(x) = \dfrac{1}{\sqrt{x-2}}$

Solution.

(a) To find the excluded values of x, we therefore solve

$$x^2 - 2x \neq 0$$
$$x(x-2) \neq 0$$
$$x \neq 0, 2$$

Hence, the domain is all real numbers excluding $x = 0, 2$. This may also be expressed in a format set by

$$\text{Domain} = \{x | x \in R, x \neq 0, 2\}$$

where R represents all real numbers.

(b) We should consider first the domain of the square root to define the interval of validity of this part. We then consider the domain of the denominator to find the excluded value of x from the interval obtained before. Using the square root, we solve the inequality

$$x - 2 \geq 0$$
$$x \geq 2$$

that defines the interval of validity of the square root. We next consider the denominator, therefore we solve

$$x - 3 \neq 0$$
$$x \neq 3$$

This means that the domain D is given by the interval $x \geq 2$ excluding $x = 3$. In a format set, this can be expressed by

$$D = \{x \, | \, 2 \leq x < 3 \text{ or } 3 < x < \infty\}$$

(c) Proceeding as before, we consider the square root to define the interval, whereas the denominator gives the excluded values of x from this interval. This means that we solve the inequality

$$x - 2 \geq 0$$
$$x \geq 2$$

The denominator gives

$$x - 2 \neq 0$$
$$x \neq 2$$

This means that the domain D is given by the interval $x \geq 2$ excluding $x = 2$. In other words, the domain D is the interval $x > 2$. In a format set, this can be expressed by

$$D = \{x \mid x > 2\}$$

We point out here that the **range** of a function, the values of $f(x)$ that result from using the domain, will not be discussed in this text.

Exercises 9.2

Find the domain of the following functions:

1. $f(x) = x + 1$

2. $f(x) = 2x^3 + 3x - 1, \ -2 \leq x \leq 4$

3. $f(x) = \dfrac{1}{x + 1}$

4. $f(x) = \dfrac{1}{x^2 + x}$

5. $f(x) = \dfrac{1}{x^2 + 16}$

6. $f(x) = \sqrt{-x}$

7. $f(x) = \dfrac{x + 1}{x^2 + 6x}$

8. $f(x) = \dfrac{3}{x^2 - 4x}$

9. $f(x) = \dfrac{x^2 + 1}{x^2 - 4}$

10. $f(x) = \dfrac{|x|}{x}$

11. $f(x) = \dfrac{2}{x^2 + 6x - 7}$

12. $f(x) = \sqrt{-x}$

13. $f(x) = \sqrt{12 - 3x}$

14. $f(x) = x + |x - 2|$

15. $f(x) = \dfrac{\sqrt{x - 6}}{x - 14}$

16. $f(x) = \dfrac{\sqrt{x - 1}}{x^2 - 5x + 6}$

17. $f(x) = \dfrac{1}{x^2 - 6x - 7}$

18. $f(x) = \dfrac{\sqrt{x}}{x^2 - 2x}$

19. $f(x) = \dfrac{\sqrt{9 - 3x}}{x - 1}$

20. $f(x) = \dfrac{1}{x^3 + x^2 - 2x}$

21. $f(x) = \begin{cases} x + 1 & \text{for } x \le -1 \\ x^2 + 1 & \text{for } x > -1 \end{cases}$

22. $f(x) = \begin{cases} x^2 + 1 & \text{for } -5 \le x < 0 \\ x^2 - 1 & \text{for } 0 < x \le 5 \end{cases}$

23. $f(x) = \begin{cases} 4 & \text{for } -4 \le x < -1 \\ |x| & \text{for } -1 < x < 1 \\ -4 & \text{for } 1 \le x \le 4 \end{cases}$

24. $f(x) = \begin{cases} x - 1 & \text{for } x < -2 \\ -x^2 + 1 & \text{for } -2 \le x \le 2 \\ x - 5 & \text{for } x > 2 \end{cases}$

9.3 Limits

In Section 9.1, we have discussed how a function can be evaluated for given values of x selected from the domain of that function. It is useful to study the behavior of functions when x becomes very large, or in other words, when x approaches infinity. It is clear that the function cannot be evaluated directly as used before. Instead, we will use a different approach called **limits** to find the behavior of a function $f(x)$ as x approaches infinity. The limits topic is usually introduced in calculus books, hence we skip details. We point out that the limit of a function $f(x)$ as x approaches a constant number a will not be introduced in this text, instead it will be left to a calculus course. In this text, our concern will be on finding the limit of a rational function $f(x)$ as x approaches infinity, denoted by the symbol ∞. This specific type of limits as x approaches ∞ will be discussed below. The notation $x \to \infty$ is read as x approaches ∞. We usually write

$$\lim_{x \to \infty} f(x) = L$$

which means that $f(x)$ gets close to the limit L as x approaches ∞. For example, given the function

$$f(x) = \frac{2x + 2}{x + 1}$$

the value of this function for any given x from its domain can be evaluated by direct substitution. However, the function $f(x)$ becomes very close to 2 when x gets closer to ∞. In this text, we will introduce a simplified technique for determining the limit of a rational function as $x \to \infty$. This technique will be summarized as follows:

1. If the degree of the numerator is less than the degree of the denominator, then the limit of this rational function is 0. This can be easily seen by noting that x is very large, hence the denominator with the larger power dominates and is very large compared to the numerator. Accordingly, the denominator is very small compared to the larger denominator.

2. If the degree of the numerator is greater than the degree of the denominator, then the limit of this rational function is ∞. This can be easily seen by noting that x is very large, hence the numerator with the larger power dominates and is very large compared to the denominator.

3. If the degree of the numerator is equal to the degree of the denominator, then the limit of this rational function is obtained by dividing the leading

coefficients of the numerator and the denominator. Notice that the leading coefficient of any polynomial is the coefficient of x with the highest power.

An alternative approach that works successfully for all cases 1, 2, and 3 that were discussed above will be used as follows. The limit of a rational function can be obtained by retaining the dominant term of the numerator and the dominant term of the denominator. Note that the dominant term in a polynomial as x approaches ∞, is the term with the highest power. We next simplify the result that contains the dominant terms only. Finally, we substitute x by infinity in the simplified form to determine the limit. It should be noted here that

$$\frac{c}{\infty} \to 0$$

where c is a constant.

The following examples will be used to illustrate the limits concept.

Example 1. Find the following limits:

(a) $\lim\limits_{x \to \infty} \dfrac{x^2 + 1}{x^3 + 1}$

(b) $\lim\limits_{x \to \infty} \dfrac{x^2 - 1}{x + 1}$

(c) $\lim\limits_{x \to \infty} \dfrac{2x^2 + 3x + 1}{3x^2 - 3x + 1}$

Solution.

(a) Since the degree of the numerator is less than the degree of the denominator, hence

$$\lim_{x \to \infty} \frac{x^2 + 1}{x^3 + 1} = 0$$

This can also be obtained by retaining the dominant term of the numerator and the dominant term of the denominator and simplifying to obtain

$$\lim_{x \to \infty} \frac{x^2 + 1}{x^3 + 1} = \lim_{x \to \infty} \frac{x^2}{x^3}$$
$$= \lim_{x \to \infty} \frac{1}{x} = 0$$

obtained upon substituting $x = \infty$ in $\frac{1}{x}$.

(b) Since the degree of the numerator is greater than the degree of the denominator, hence

$$\lim_{x \to \infty} \frac{x^2 - 1}{x + 1} = \infty$$

This can also be obtained by retaining the dominant term of the numerator and the dominant term of the denominator and simplifying to obtain

$$\begin{aligned}\lim_{x \to \infty} \frac{x^2 - 1}{x + 1} &= \lim_{x \to \infty} \frac{x^2}{x} \\ &= \lim_{x \to \infty} \frac{x}{1} \\ &= \infty\end{aligned}$$

(c) Since the numerator and the denominator have the same degree, hence the limit is the ratio of the leading coefficients given by

$$\lim_{x \to \infty} \frac{2x^2 + 3x + 1}{3x^2 - 3x + 1} = \frac{2}{3}$$

This can also be obtained by retaining the dominant term of the numerator and the dominant term of the denominator and simplifying to obtain

$$\begin{aligned}\lim_{x \to \infty} \frac{2x^2 + 3x + 1}{3x^2 - 3x + 1} &= \lim_{x \to \infty} \frac{2x^2}{3x^2} \\ &= \lim_{x \to \infty} \frac{2}{3} \\ &= \frac{2}{3}\end{aligned}$$

Exercises 9.3

Find the limits for Exercises 1 – 12

1. $\lim_{x \to \infty} \dfrac{x - 1}{x + 1}$

2. $\lim_{x \to \infty} \dfrac{6x + 4}{3x - 2}$

3. $\lim_{x \to \infty} \dfrac{x^2 + 2x + 1}{x^3 - 2x + 3}$

4. $\lim_{x \to \infty} \dfrac{x^3 - 3x + 1}{x^2 - 2x + 4}$

5. $\lim\limits_{x \to \infty} \dfrac{x^2 + 4}{x - 4}$

6. $\lim\limits_{x \to \infty} \dfrac{x^3 + 5}{x^4 - 5}$

7. $\lim\limits_{x \to \infty} \dfrac{x^3 + x + 1}{x^2 - x + 1}$

8. $\lim\limits_{x \to \infty} \dfrac{x^2 + 4}{x^2 - 4}$

9. $\lim\limits_{x \to \infty} \dfrac{1 - x - x^2}{1 + x + 2x^2}$

10. $\lim\limits_{x \to \infty} \dfrac{x(x + 2)}{(x + 1)^2}$

11. $\lim\limits_{x \to \infty} \dfrac{x(1 - x + x^2)}{(x^2 + 1)(x - 1)}$

12. $\lim\limits_{x \to \infty} \dfrac{x(1 - 2x)}{(1 + x)(1 - x)}$

In Exercises 13 – 16, find the dominant term of each polynomial as $x \to \infty$

13. $f(x) = 3x + 1$

14. $f(x) = 2x^3 + 3x^2 + 1$

15. $f(x) = x^2 + 4x$

16. $f(x) = x^4 + x^3 + 2$

Find the limits for Exercises 17 – 20

17. Find $\lim\limits_{x \to \infty} \dfrac{x^a - x^b}{x^c - x^d}, a < b < c < d$

18. Find $\lim\limits_{x \to \infty} \dfrac{a^x - b^x}{c^x - d^x}, a < b < c < d$

19. Find $\lim\limits_{x \to \infty} \dfrac{1 + 36x^4 - 40x^2}{1 - 9x^4 + 40x^2}$

20. Find $\lim\limits_{x \to \infty} \dfrac{1 + x|2x|}{1 + x|x|}$

9.4 Algorithms and Asymptotic Notations

A computer programmer uses a specified set of instructions, called an **algorithm** to process a considerably large amount of input data. The algorithm constructed includes mathematics operations, call procedures, loops and nested loops, subroutines, arrays and computer concepts that are usually discussed in computer texts. The computer will follow the algorithm constructed to solve the problem under discussion. An algorithm is a sequence of steps that transform set of values as the input and produces other set of values as the output. In addition, an algorithm can be defined as set of rules used for carrying out calculations.

A special attention should be considered in constructing a computer algorithm. For a large volume of input data, an algorithm should be constructed in a way such that the **running time** is reasonable and the **space** memory available exceeds the input size. The running time of an algorithm and the space are functions of the input data to be processed. The running time and the space will be explained as follows:

1. The **running time** needed to run a computer program: it is noticed that the time needed to execute a program depends mainly on the number of executive instructions used by the program. However, the time needed can be minimized by reducing the number of instructions and loops used in the algorithm. This is of course a characteristic of an efficient and experienced programmer.

2. The **space** of a memory: the data and parameters used in any program should take into consideration the space of the memory available by the computer. An efficient programmer is one that uses data and parameters that do not exceed the space memory of the computer.

To study the time and space requirements discussed above, we note first that both concepts depend on the input size that includes number of instructions n, where $n > 0$. In what follows, we will study the asymptotic notations that involves Big-Oh O notation, and its relatives Big-Omega Ω notation and Big-Theta Θ notation. The other notations little-Oh o and little-omega ω will not be addressed in this text.

9.4.1 Big-Oh Notation

We will discuss a notation, denoted by O and read as **Big-Oh**, to study the rate of growth of time and space function $f(n)$, where n is the number of instructions in a program. Big-Oh notation is a mathematical notation used to describe the asymptotic behavior of functions. It provides an asymptotic upper bound for a function.

Let $f(x)$ and $g(x)$ be non-negative, non-decreasing functions of x. We say $f(x) = O(g(x))$ if there exists positive constants M and x_0 such that

$$f(x) \leq Mg(x), \text{for all } x > x_0$$

Figure 9.1 shows the graphs of the functions $f(x)$ and $Mg(x)$. It is clear that the graph of $f(x)$ is bounded above by the curve $Mg(x)$, so that $Mg(x)$ is an upper bound of $f(x)$, for all values of $x \geq x_0$, and M is a positive constant. In other words, the graph of $f(x)$ as shown by Figure 9.1 lies closer to the x-axis and on or below the graph of $Mg(x)$ for large values of x where $x \geq x_0$. This means that $f(x)$ grows no faster than $Mg(x)$, i.e a multiple of $g(x)$, for every $x \geq x_0$.

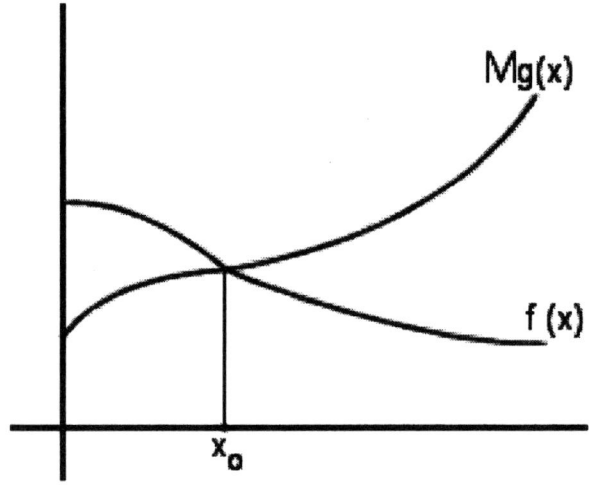

Figure 9.1

For space and time function $f(n)$, we next use the Big-Oh notation for this function. Let $f(n)$ and $g(n)$ be functions whose domains are positive integers. We write

$$f(n) = O\left(g(n)\right),$$

read " f is Big-Oh of g(n)" if there exists constants M and n_0 such that

$$|f(n)| \leq M\,|g(n)|\,, \text{for all } n \geq n_0.$$

This means that if $f(n) = O(g(n))$, then $f(n)$ grows no faster than $g(n)$.

To clearly explain the significant importance of the "Big-Oh" notation, we compare the running time of two algorithms. In the first algorithm, assume that it was required to write a set of 100 positive integers in one row. This can be easily done by using one loop in a computer program that runs from 1 to 100. It is clear in this case that the running time of this program follows the function:

$$f(n) = n$$

where n is the number of times the loop will be executed. Assume that the time required for executing the loop for one time is 1 millisecond, then

$$f(100) = 100 \, \text{milliseconds}$$

is the time needed to execute the loop for 100 times. In the second example, assume that we have to write an array of 100 rows and 100 columns. This means that we have to write the integers 1 to 100 in each row. In this case two loops should be used, an outer loop and an inner loop. It is clear in this case that the running time of this program follows the function:

$$f(n) = n^2$$

where n is the number of times that each of the outer the loop and the inner loop will be executed. Assume that the time required for executing the loop for one time is 1 millisecond, then

$$f(100) = 100^2 = 10000 \, \text{milliseconds}$$

It is easily seen that the algorithm that needs one loop produces better running time compared to the nested loop where two loops were needed. In these examples in particular, we conclude that reducing the number of interior loops in a nested loop will reduce the running time of an algorithm.

We now return to the "Big-Oh" notation that we introduced before. We illustrate the idea first by considering a polynomial such as

$$f(n) = 2n^3 + 4n^2 + 1$$

It is easily observed in this polynomial that the dominant term, that we discussed in the previous section, as $n \to \infty$ is $2n^3$. Ignoring the coefficient of $2n^3$ means that $g(n) = n^3$. This leads to the following "Big-Oh" notation:

$$f(n) = O(n^3)$$

This is consistent with the definition introduced before, where

$$\lim_{n \to \infty} \left| \frac{2n^3 + 4n^2 + 1}{n^3} \right| = 2$$

obtained by using the limits concept discussed in the previous section. Table 9.1 below shows the smallest function $\log_2 n$ at the left increasing to the dominant term $x!$ at the most right.

n	$\log_2 n$	n	$n \log_2 n$	$n^2, n^3, ..$	$2^n, 3^n, ..$	$n!$
8	3	8	24	$64, 512, ..$	$256, 6561, ..$	40320
16	4	16	64	$258, 4096, ..$	$65536, 4.3 \times 10^6, ..$	2.09×10^{13}

Table 9.1

An important conclusion can be made here in that the "Big-Oh" notation is used to capture the most dominant term in a **time** or a **space** function given by $f(n)$. It is easily observed that for polynomials, the most dominant term, as $n \to \infty$, is the leading term of that polynomial. However, the task of determining the most dominant term in other functions is subjected to a rule, called the **common order rule**, that defines the dominant term when more than one term exists. For this purpose, we consider the Table 9.1 that will explain the common order rule. In Table 9.1, we introduced the most important functions that are normally used in a data structure course.

It is important to point out that we should ignore the coefficient of the dominant term. We can easily observe from the table of orders shown above that, for example for $n = 8$, the smallest value is $\log_2 8 = 3$ and the largest value is $8! = 40320$, hence $n!$ dominates every function in the row. The same conclusion can be made for other values of n. The table can be used to make a comparison on the terms included in the time or space function. For example, $n! > 2^n$, and $2^n > n^3$, $n^3 > n\log_2 n$, and so on.

To select the "Big-Oh" notation for any time or space function $f(n)$, we apply the following steps :

1. If $f(n) = C$, where C is a constant, then $f(n) = O(1)$.

2. If $f(n)$ is expressed in a polynomial form defined by:

$$f(n) = a_r n^r + a_{r-1} n^{r-1} + a_{r-2} n^{r-2} + \cdots + a_1 n + a_0$$

then

$$f(n) = O(n^r)$$

where the constant coefficient a_r is ignored. For example, for

$$f(n) = 6n^3 - 7n^4 + 1$$

then

$$f(n) = O(n^4)$$

3. If $f(n)$ is a function that includes two or more of the terms shown in Table 9.1 of common orders, the term that comes at the right is the most dominant term that will be used to define the "Big-Oh" for the function $f(n)$, where other terms will not be considered in the notation.

4. Notice that

$$\log^a n \prec n^b \prec c^n$$

for any a, b, and c, provided that $b > 0$ and $c > 1$, where $\prec$ stands for asymptotically smaller. This indicates that logarithmic functions are asymptotically smaller than polynomial function, and the last one is smaller than exponential functions. This can be seen from Table 9.1, where $\log n \prec n \prec 2^n$.

5. We can also easily show that

$$n \log n > n, n^2 \log n > n^2, n^3 \log n > n^3, \cdots$$

For example, for $f(n) = 4n^3 \log n + n - n^3$, $f(n) = O(n^3 \log n)$. The "Big-Oh notation" will be explained by using the following illustrative examples.

Example 1. Establish the best "Big-Oh" notation for the function

$$f(n) = 12$$

Solution. The given function $f(n)$ is a polynomial that contains only a constant number, hence

$$f(n) = O(1)$$

Example 2. Establish the best "Big-Oh" notation for the function

$$f(n) = 8n^4 + n^3 - 2n^2 + 12$$

Solution. The given function $f(n)$ is a polynomial, where the dominant term is $8n^4$, hence

$$f(n) = O(n^4)$$

Example 3. Establish the best "Big-Oh" notation for the function

$$f(n) = 4n\log_2 n + n^2 + 2^n$$

Solution. The given function $f(n)$ includes 3 different terms, where 2^n occurs at the rightmost as shown by the table of common orders, hence

$$f(n) = O(2^n)$$

Example 4. Establish the best "Big-Oh" notation for the function

$$f(n) = \log_2 n + n + n\log_2 n$$

Solution. The given function $f(n)$ includes 3 different terms, where $n\log_2 n$ occurs at the rightmost as shown by the table of common orders, hence

$$f(n) = O(n\log_2 n)$$

Example 5. Establish the best "Big-Oh" notation for the function

$$f(n) = (n-1)(1 + \log_2 n)$$

Solution. The given function $f(n)$ can be rewritten in the form

$$f(n) = n - 1 + n\log_2 n - \log_2 n$$

includes 4 different terms, where $n\log_2 n$ occurs at the rightmost as shown by the table of common orders, hence

$$f(n) = O(n\log_2 n)$$

Example 6. Establish the best "Big-Oh" notation for the function

$$f(n) = (3 + 2\log_2 n)(n^2 + n)$$

Solution. In this example, the given function can be rewritten as

$$f(n) = 3n^2 + 3n + 2n^2 \log_2 n + 2n \log_2 n$$

This in turn gives

$$f(n) = O(n^2 \log_2 n)$$

9.4.2 Big-Oh for Summation Formulas

To find the Big-Oh notation for summation formulas we use Appendix B, where formulas for summation of terms are given. Recall that some of these formulas are proved by mathematical induction as shown in Chapter 4. The Big-Oh notation for the summation of terms can be explained by the following illustrative examples.

Example 7. Establish the "Big-Oh" notation for the summation

$$1^2 + 2^2 + 3^2 + \cdots + n^2$$

Solution. Using Appendix B, we find that

$$1^2 + 2^2 + 3^2 + \cdots + n^2 = \frac{n(n+1)(2n+1)}{6}$$

This means that the statement is $O(n^3)$

Example 8. Establish the "Big-Oh" notation for the summation

$$1^3 + 3^3 + 5^3 + \cdots + (2n-1)^3$$

Solution. Using Appendix B, we find that

$$1^3 + 3^3 + 5^3 + \cdots + (2n-1)^3 = n^2(2n^2 - 1)$$

This means that the statement is $O(n^4)$

9.4.3 Limits of Functions

In Section 9.3, we examined limits for rational functions where numerator and denominator are polynomials. It is useful to use the concept of dominant term and the common order rule to find limits of the ratio of functions where numerator and denominator any function. Table 9.1 should be used to select the dominant term. Unlike the Big-Oh notation, the coefficients of the dominant terms of the numerator and the denominator should not be ignored. This will be illustrated by the following two examples.

Example 9. Evaluate

$$\lim_{n \to \infty} \frac{2! + n^3 + 18(2^n)}{3! - n^3 + 6(2^n)}$$

Solution. Using Table 9.1, the dominant terms in the numerator and the denominator are $18(2^n)$ and $6(2^n)$.

$$\lim_{n\to\infty} \frac{2! + n^3 + 18(2^n)}{3! - n^3 + 6(2^n)} = \lim_{n\to\infty} \frac{18(2^n)}{6(2^n)} = 3$$

Example 10. Evaluate

$$\lim_{n\to\infty} \frac{4 + n\log_2 n + 4(n!)}{2 + n^2 - 2(n!)}$$

Solution. Using Table 9.1, the dominant terms in the numerator and in the denominator are $4(n!)$ and $-2(n!)$. Consequently we find

$$\lim_{n\to\infty} \frac{4 + n\log_2 n + 4(n!)}{2 + n^2 - 2(n!)} = \lim_{n\to\infty} \frac{4(n!)}{-2(n!)} = -2$$

In closing this subsection, we point out that the concept of function dominance has been discussed so far. It is interesting now to point out to the concept of **analysis of an algorithm.** By algorithm analysis, we mean the process of estimating the time and space needed to execute an algorithm, or in other words, the process of measuring the efficiency of an algorithm. As indicated before, if the program needs long time or requires large space, computer may not execute it. The critical part for a computer program is most often its running time. The process of calculating the running time of any algorithm is usually introduced in a data structure course, because it heavily depends on the loops and the nested loops used among other factors. However, if the time function is $f(n) = O(n)$, this means that the program uses an $O(n)$ algorithm, or simply the algorithm solves the problem in a small fraction of time unit. The $O(n\log_2 n)$ algorithm requires more time if compared with the $O(n)$ algorithm. The $O(n^2)$ algorithm uses more time than the $O(n)$ and the $O(n\log_2 n)$ algorithms. The table of common orders Table 9.1 is a useful tool for time requirement and for comparison reasons.

9.4.4   Big Omega Notation

We will now discuss a new notation, denoted by Ω and read as **Big-Omega**, to describe lower bounds of functions. Let $f(x)$ and $g(x)$ be non-negative, non-decreasing functions of x. We say $f(x) = \Omega(g(x))$ if there exists positive constant m and x_0 such that

$$f(x) \geq mg(x), \text{ for all } x > x_0, \text{ and for some positive } m$$

In other words, we say that $f(x)$ is bounded below by $g(x)$ up to a constant factor. This is almost the same definition as Big Oh, except that $f(x) = \Omega(g(x))$, this makes $g(x)$ a lower bound function, instead of an upper bound function as in the case of $f(x) = O(g(x))$. Figure 9.2 below helps to visualize the relationship between $O(g(x))$ and $\Omega(g(x))$.

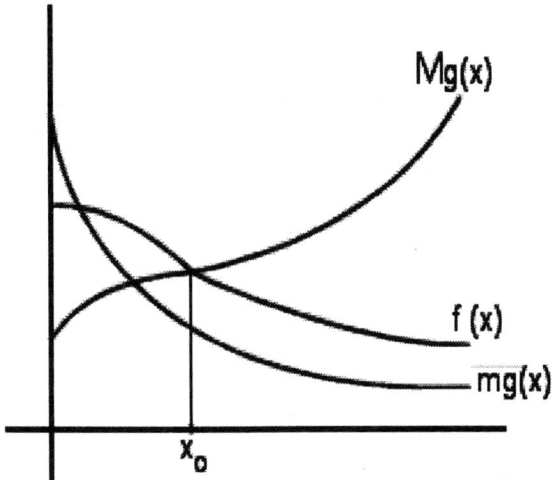

Figure 9.2

9.4.5 Big Theta Notation

We will finally discuss a third notation, denoted by Θ and read as **Big-Theta**, to combine the upper bounds and the lower bounds of functions. The Big-Oh O is an upper bound, whereas the Big-Omega Ω is a lower bound. The Big-Theta Θ combines the O and Ω notations.

Let $f(x)$ and $g(x)$ be non-negative, non-decreasing functions of x. We say $f(x) = \Theta(g(x))$ if there exists positive constants m and M, and x_0 such that

$$mg(x) \leq f(x) \leq Mg(x), \text{for all } x > x_0, \text{ and for some positive } m \text{ and } M$$

The Θ notation only describes the case where the upper and lower bounds of a function are on the same order of magnitude. In other words, the Θ notation is used when the function $f(x)$ can be bounded both from above and below by the same function $g(x)$. This is basically saying that the function, $f(x)$ is bounded both from the top and bottom by the same function $g(x)$.

Thus $f(x) = \Theta(g(x))$ indicates that $f(x) = \Omega(g(x))$ and $f(x) = O(g(x))$. The notations Ω and Θ are often used in computer science but rare in mathematics.

Table 9.2 below shows a comparison between the three notations defined above.

Notation	Intuition	Relation
$f(x) = O(g(x))$	f is bounded **above** by $g(x)$	$f(x) \le Mg(x)$
$f(x) = \Omega(g(x))$	f is bounded **below** by $g(x)$	$f(x) \ge mg(x)$
$f(x) = \Theta(g(x))$	f is bounded **above** and **below** by $g(x)$	$mg(x) \le f(x) \le Mg(x)$

Table 9.2

Exercises 9.4

Establish the best "Big-Oh" form for each of the following exercises.

1. $f(n) = 6$

2. $f(n) = 5!$

3. $f(n) = 4n^2 + 6$

4. $f(n) = 4n^3 - 2n^2 + 6n$

5. $f(n) = (2n + 1)^2$

6. $f(n) = (n + 1)^3$

7. $f(n) = 6n + n\log_2 n$

8. $f(n) = n^2 + \log_2 n$

9. $f(n) = (3n + 2)(2 + \log_2 n)$

10. $f(n) = n\log_2 n + n^2 + 2^n$

11. $f(n) = n^3 + (n + 1)^4$

12. $f(n) = n^2 + 2^n + n!$

13. $f(n) = \dfrac{n^2 + 2}{n^2 + 3}$

14. $f(n) = \dfrac{n^3 + n\log_2 n}{n^2 + 1}$

15. $f(n) = \dfrac{n(n+2)(n-2)}{(n+1)(n-1)}$

16. $f(n) = \dfrac{(n+3)(n-3)}{(n+4)(n-4)}$

17. $f(n) = nP2 + nP3$

18. $f(n) = n(nP2) - nP3$

19. $f(n) = 3(n!) + n^5 + 10$

20. $f(n) = 6n^2 + 2^n + n\log_2 n$

21. $f(n) = 1 \times 2 \times 3 \times 4 \times \cdots \times n$

In Exercises 22–26, Use Appendix B to find the Big-Oh notation for the following statements:

22. $f(n) = 1 + 2 + 3 + 4 + \cdots + n$

23. $1 + 3 + 5 + \cdots + (2n - 1)$

24. $1^2 + 2^2 + 3^2 + \cdots + n^2$

25. $1^2 + 3^2 + 5^2 + \cdots + (2n - 1)^2$

26. $1 + 2 + 2^2 + 2^3 + \cdots + 2^{n-1}$

In Exercises 27–32, evaluate the given limits:

27. $\displaystyle\lim_{n\to\infty} \dfrac{1 + \log_2 n + 2n\log_2 n}{1 - n + n\log_2 n}$

28. $\displaystyle\lim_{n\to\infty} \dfrac{n^2 + n^3 + 4!(n!)}{n + n^2 + 2!(n!)}$

29. $\displaystyle\lim_{n\to\infty} \dfrac{1 + 2n + 3!(2^n)}{1 + n + 2!(2^n)}$

30. $\displaystyle\lim_{n\to\infty} \dfrac{4 + (2^n)}{1 + n!}$

31. $\displaystyle \lim_{n \to \infty} \frac{(1 + 4n^2)(3 + 2\log n)}{(1 + n^2)(5 - \log n)}$

32. $\displaystyle \lim_{n \to \infty} \frac{(4 + 6\log n)(11 - 7n) + 42n\log n}{(4 + 3\log n)(11 - 7n) + 21n\log n}$

In Exercises 33–40, find the Big-Oh notation for each exercise:

33. $f(n) = 600 + 1000!$

34. $f(n) = (7 - 4n\log n)(5n - 13) + 20n^2\log n$

35. $f(n) = (7 - 4\log n)(5n - 13) + 20n\log n$

36. $f(n) = n^2 + n^3 + 5(2^n) + 7(3^n)$

37. $f(n) = (1 + 4!n)(n\log n - 1)$

38. $f(n) = (1 + 4n)(n^2\log n - n^2)$

39. $f(n) = (1 + 4n)(n^2\log n - n^2) - 4n^3\log n$

40. $f(n) = (n + \log n)(n - \log n)$

Chapter 10

Recursion

10.1 Recurrence Relations

In Section 9.1, we have discussed functions where for every value of x of its domain, a corresponding value $f(x)$ is determined. In this section, a distinct relation, called **recurrence relation**, will be presented. A recurrence relation is defined as any equation involving two or more terms of a sequence such as

$$a_{n+1} = 2a_n,\ a_0 = 1,$$
$$a_{n+2} = 5a_{n+1} - 9a_n,\ a_0 = 1, a_1 = 2$$

10.1.1 Origins of Recurrence Relations:

It is useful to study how recurrence relations arise. Recurrence relations are usually derived from equations $a_n = f(n)$ by substituting n by $n+1, n+2, \cdots$. A recurrence relation is then established between terms of $a_j, j \geq n$. Unlike equations, a recurrence relation should be developed with one or more initial values $a_0, a_1, \cdots$ obtained by substituting $n = 0, 1, \cdots$ into the equation under discussion. The approach for deriving recurrence relations with initial values can be illustrated by the following examples.

Example 1. Consider the equation:

$$a_n = Cn!$$

where C is a constant. Construct an equivalent recurrence relation.

Solution. Substituting n by $n+1$ into both sides of the equation we get

$$a_{n+1} = C(n+1)!$$

Dividing the second equation by the first we find

$$\frac{a_{n+1}}{a_n} = \frac{(n+1)!}{n!} = \frac{(n+1)n!}{n!} = (n+1)$$

Therefore we obtain the recurrence relation

$$a_{n+1} = (n+1)a_n, \; a_0 = C, n \geq 0$$

The initial value a_0 is obtained by substituting $n = 0$ into the equation $a_n = Cn!$. Unlike equations, recurrence relations must be defined with one or more initial values.

Example 2. We next consider the equation

$$a_n = 3(2^n)$$

Solution. Substituting n by $n+1$ into both sides of the equation we get

$$a_{n+1} = 3(2^{n+1})$$

Dividing the second equation by the first we find

$$\frac{a_{n+1}}{a_n} = \frac{2^{n+1}}{2^n} = 2$$

Therefore we obtain the recurrence relation

$$a_{n+1} = 2a_n, \; a_0 = 3, n \geq 0$$

Example 3. We next consider the equation

$$a_n = 2^n + 3^n$$

Solution. Notice here that the equation includes two distinct terms 2^n and 3^n. Therefore, we should substitute $n = n+1$ and $n = n+2$ to find a recurrence relation that is equivalent to the given equation. Substituting n by $n+1$ first and then by $n+2$ into both sides of the equation we get

$$\begin{aligned} a_{n+1} &= 2^{n+1} + 3^{n+1} = 2(2^n) + 3(3^n) \\ a_{n+2} &= 2^{n+2} + 3^{n+2} = 4(2^n) + 9(3^n) \end{aligned}$$

To find the recurrence relation between a_{n+2}, a_{n+1} and a_n, we set

$$a_{n+2} = Aa_{n+1} + Ba_n$$

where A and B are constants that will be determined. Substituting the equations obtained before we find

$$4(2^n) + 9(3^n) = A\left(2(2^n) + 3(3^n)\right) + B\left(2^n + 3^n\right)$$

Equating the coefficients of 2^n and 3^n from both sides we obtain the system of equations

$$2A + B = 4, 3A + B = 9$$

Solving this system we find

$$A = 5, B = -6$$

Therefore the solution is

$$a_{n+2} = 5a_{n+1} - 6a_n, \ a_0 = 2, \ a_1 = 5, n \geq 0$$

Example 4. We next consider the equation

$$a_n = 5^n - 3^n$$

Solution. Notice here that the equation includes two distinct terms 5^n and 3^n. Substituting n by $n + 1$ first and then by $n + 2$ into both sides of the equation we get

$$
\begin{aligned}
a_{n+1} &= 5^{n+1} - 3^{n+1} = 5(5^n) - 3(3^n) \\
a_{n+2} &= 5^{n+2} - 3^{n+2} = 25(5^n) - 9(3^n)
\end{aligned}
$$

To find the recurrence relation between a_{n+2}, a_{n+1} and a_n, we set

$$a_{n+2} = Aa_{n+1} + Ba_n$$

where A and B are constants that will be determined. Substituting the equations obtained before we find

$$25(5^n) - 9(3^n) = A\left(5(5^n) - 3(3^n)\right) + B\left(5^n - 3^n\right)$$

Equating the coefficients of 2^n and 3^n from both sides we obtain the system of equations

$$5A + B = 25, 3A + B = 9$$

Solving this system we find

$$A = 8, B = -15$$

Therefore the solution is

$$a_{n+2} = 8a_{n+1} - 15a_n, \ a_0 = 0, \ a_1 = 2, n \geq 0$$

10.1.2   Solving Recurrence Relation:

In this part we will study a method to solve a linear homogeneous recurrence relation with constant coefficients. The order of a recurrence relation is the difference between the highest and the lowest subscripts. The study will be focused on first-order and second-order linear homogeneous recurrence relations.

The First-Order Linear Homogeneous Recurrence Relation

We first start our study on the first-order linear homogeneous recurrence relation given by

$$a_{n+1} = ka_n, a_0 = A$$

where k and A are constants. The given recurrence relation is:
(i) of first-order; the difference between highest and lowest subscripts is 1;
(ii) linear because a_{n+1} and a_n have the first exponent;
(iii) homogeneous because each term has a form of a_n;
(iv) of constant coefficients.

To solve the linear homogeneous recurrence relation with constant coefficients, we assume that $a_n = Cx^n$, where C is a constant. Substituting this assumption into the recurrence relation gives a first degree equation that can be solved for one value of x. The constant C can be determined by using the given initial value $a_0 = A$. This can be explained as follows.

Example 5. Solve the following recurrence relation

$$a_{n+1} = 4a_n, a_0 = 5$$

Solution. We first set $a_n = Cx^n$. Substituting this assumption into the relation gives

$$x^{n+1} = 4x^n$$

which gives

$$x = 4$$

Accordingly we find

$$a_n = C(4)^n$$

Substituting $n = 0$ and using the initial value $a_0 = 5$ we find

$$C = 5$$

We therefore obtain the unique solution of the given recurrence relation

$$a_n = 5(4)^n$$

The Second-Order Linear Homogeneous Recurrence relation

We next consider the recurrence relation

$$a_{n+2} = k_0 a_{n+1} + k_1 a_n$$

with initial values given by

$$a_0 = c_0, a_1 = c_1$$

where k_0, k_1 are constants and c_0, c_1 are given initial values. Notice that we use two initial values for the second order recurrence relation. The given recurrence relation is:

(i) of second order; because the difference between highest and lowest subscripts is 2;

(ii) linear because a_{n+2}, a_{n+1} and a_n have the first exponent;

(iii) homogeneous because each term has a form of a_n;

(iv) of constant coefficients.

To solve the linear homogeneous recurrence relation with constant coefficients, we assume that $a_n = Cx^n$, where C is a constant. Substituting this assumption into the recurrence relation gives a quadratic equation $x^2 - k_0 x - k_1 = 0$ that can be solved for x. Recall that a quadratic equation may give:

(i) two distinct real roots x_1 and x_2;

(ii) two equal real roots $x_1 = x_2$;

(iii) two complex roots.

We will consider the first two cases of real roots only.

Distinct Real Roots:

For two distinct real roots x_1 and x_2, it is normal to set a_n as a linear combination of these two values of x in the form

$$a_n = Ax_1^n + Bx_2^n$$

where the constants A and B can be determined by using the given two initial values. Having determined the constants A and B, the unique solution is thus obtained. The concept of forming a linear combination is used in solving linear ordinary differential equations to obtain a unique solution. This will be illustrated by the following examples.

Example 6. Consider the following recurrence relation

$$a_n - 3a_{n-1} + 2a_{n-2} = 0, n \geq 3, a_1 = 3, a_2 = 5$$

Solution. Let $a_n = Cx^n$, C is a constant. The recurrence relation becomes

$$x^n - 3x^{n-1} + 2x^{n-2} = 0$$

or equivalently

$$x^2 - 3x + 2 = 0$$

This means that $x = 1, 2$. Accordingly, we obtain

$$a_n = A(1)^n + B(2)^n$$

where A and B are constants that will be determined by using the initial values. Substituting $n = 1$ then $n = 2$ we find

$$A + 2B = 3, \; A + 4B = 5$$

This gives

$$A = 1, B = 1$$

This means that

$$a_n = 1 + 2^n$$

Example 7. Consider the following recurrence relation

$$a_n - 8a_{n-1} + 15a_{n-2} = 0, n \geq 2, a_0 = 5, a_1 = 21$$

Solution. Let $a_n = Cx^n$, C is a constant. The recurrence relation becomes

$$x^n - 8x^{n-1} + 15x^{n-2} = 0$$

or equivalently

$$x^2 - 8x + 15 = 0$$

This means that $x = 3, 5$. Accordingly, we obtain

$$a_n = A(3)^n + B(5)^n$$

where A and B are constants that will be determined by using the initial values. Substituting $n = 0$ the $n = 1$ we find

$$A + B = 5, \ 3A + 5B = 21$$

This gives

$$A = 2, B = 3$$

This means that

$$a_n = 2(3^n) + 3(5^n)$$

Equal Real Roots:
In this case a single root is obtained where $x_1 = x_2$. In this case we set a_n in the form

$$a_n = A(x_1)^n + Bn(x_1)^n$$

where the constants A and B are determined by using the given initial values. Notice that we multiplied x_1 by n for the term that involves x_2. The following two examples explain the present approach.

Example 8. Consider the following recurrence relation

$$a_{n+2} - 6a_{n+1} + 9a_n = 0, n \geq 0, a_0 = 1, a_1 = 6$$

Solution. Let $a_n = Cx^n, C$ is a constant. The recurrence relation becomes

$$x^2 - 6x + 9 = 0$$

This means that $x = 3, 3$. Accordingly, we obtain

$$a_n = A(3)^n + Bn(3)^n$$

where A and B. Substituting $n = 0$ then $n = 1$, and using the initial values we find

$$A = 1, \ 3A + 3B = 6$$

This gives

$$A = 1, B = 1$$

This means that
$$a_n = 3^n(1+n)$$

Example 9. Consider the following recurrence relation

$$a_{n+2} = 2a_{n+1} - a_n, n \geq 0, a_0 = 1, a_1 = 3$$

Solution. Let $a_n = Cx^n$, C is a constant. The recurrence relation becomes

$$x^2 - 2x + 1 = 0$$

This means that $x = 1, 1$. Accordingly, we obtain

$$a_n = A(1)^n + Bn(1)^n = A + Bn$$

where A and B are constants that will be determined by using the initial values. Substituting $n = 0$ then $n = 1$ we find

$$A = 1, \ A + B = 3$$

This gives
$$A = 1, B = 2$$

This means that
$$a_n = 1 + 2n$$

10.1.3   Solving Recurrence Relation by Iteration

In recursion, the first term of a sequence or the first few terms of any sequence are defined. The given terms of a recurrence are called **initial values** or **initial conditions**. The other terms of a sequence can be completely determined by using the preceding term or terms and by using the given initial values.

It is clear that a recursion definition of a sequences consists of:
(i) one or more specified initial values;
(ii) a recursion relation that relates each subsequent term of the sequence to one or more of the preceding terms.

For the purpose of comparison between functions and recursion relations, we first consider the following explicit formula:

$$f(n) = 3n + 1$$

where we can easily evaluate $f(n)$ for $n = 1, 2, 3$, and so on. The formula gives an explicit rule that for every value of n from the domain, a corresponding value of $f(n)$ is easily determined. The value of $f(n)$ for each case depends on n only.

On the other hand, consider the recursion relation:

$$a_{n+1} = 3a_n + 1, \; a_1 = 2, \; n \geq 1$$

It is clear from this recursion relation that the initial value a_1 is defined, and each subsequent term of the sequence is determined only if the preceding term is known. To determine the second term a_2 for example, we substitute $n = 1$ in the recursion relation to obtain

$$a_2 \;\; = \;\; 3a_1 + 1 = 3 \times 2 + 1 = 7$$

obtained by using the given value of a_1. To determine the third term a_3, we substitute $n = 2$ in the relation to obtain

$$a_3 \;\; = \;\; 3a_2 + 1 = 3 \times 7 + 1 = 22$$

obtained by using the value that was determined before for a_2. We can proceed as before and set $n = 3$ to determine the fourth term a_4, hence we find

$$a_4 \;\; = \;\; 3a_3 + 1 = 3 \times 22 + 1 = 67$$

obtained by using the value that was determined before for a_3. Other terms can be determined in a recursion way as discussed before.

Example 10. Write the sequence defined by the recursion relation:

$$a_{n+1} = 2a_n + 3, \; n \geq 1$$

where

$$a_1 = 2$$

Solution.

For $n = 1$ we find

$$a_2 \;\; = \;\; 2a_1 + 3 = 2 \times 2 + 3 = 7$$

For $n = 2$ we find

$$a_3 \;\; = \;\; 2a_2 + 32 \times 7 + 3 = 17$$

For $n = 3$ we find

$$a_4 \; = \; 2a_3 + 3 = 2 \times 17 + 3 = 37$$

and so on. The sequence is given by

$$2, 7, 17, 37, 77, \cdots$$

Example 11. Write the sequence defined by the recursion relation:

$$a_{n+1} = 5a_n + (-1)^n, \; n \geq 1$$

where

$$a_1 = 3$$

Solution.

For $n = 1$ we find

$$a_2 \; = \; 5a_1 + (-1)^1 = 5 \times 3 - 1 = 14$$

For $n = 2$ we find

$$a_3 \; = \; 5a_2 + (-1)^2 = 5 \times 14 + 1 = 71$$

For $n = 3$ we find

$$a_4 \; = \; 5a_3 + (-1)^3 = 5 \times 71 - 1 = 354$$

and so on. The sequence is therefore

$$3, 14, 71, 354, \cdots$$

Example 12. Write the sequence defined by the recursion relation:

$$a_{n+2} = 3a_{n+1} - 2a_n + 2, \; n \geq 1$$

$$a_1 = 2, \; a_2 = 2$$

Solution. Notice here that two initial values are given because the recurrence relation is of second order.

For $n = 1$ we find

$$a_3 \; = \; 3a_2 - 2a_1 + 2 = 3 \times (2) - 2 \times (2) + 2 = 4$$

For $n = 2$ we find

$$a_4 \;=\; 3a_3 - 2a_2 + 2 = 3 \times (4) - 2 \times (2) + 2 = 10$$

For $n = 3$ we find

$$a_5 \;=\; 3a_4 - 2a_3 + 2 = 3 \times (10) - 2 \times (4) + 2 = 24$$

and so on. The sequence is therefore

$$2, 2, 4, 10, 24, \cdots$$

It is important to note that recursion is widely used in computer programs. The assignment statement:

$$x := x + 1$$

is used in computer programs that indicates that x is increased by 1. The statement has no meaning in mathematics, but in computer it has a useful meaning in programs. Usually, in computer we substitute the value of x in a previous step in the right hand side to evaluate the new value of x at the left hand side. The mathematics representation of this statement is

$$x_{n+1} = x_n + 1, \text{ for } n \geq 1$$

This is exactly a recursion statement when the starting value x_1 is given.

Example 13. Write the sequence of values of x defined by the assignment statement given in a computer program:

$$x = x + 2$$

given that the initial value of x is

$$x = 2$$

where this statement is executed 5 times.

Solution. The mathematics representation is given by the recursion relation:

$$x_{n+1} = x_n + 2, \text{ for } x_1 = 2, \; n \geq 1$$

For $n = 1$ we find

$$x_2 = x_1 + 2 = 2 + 2 = 4$$

For $n = 2$ we find

$$x_3 = x_2 + 2 = 4 + 2 = 6$$

and so on. We observe that x is increased by 2 at every time the loop is executed. Hence the values of x obtained are 4, 6, 8, 10 and 12 where the loop started at $x = 2$.

It is worth noting that the sequence obtained is an arithmetic sequence given by

$$2, 4, 6, 8, 10, \cdots$$

Exercises 10.1

In Exercises 1–4, construct a recurrence relation for each equation:

1. $a_n = 3^n + 1$

2. $a_n = 4(3^n) - 2$

3. $a_n = 3^n + 4^n$

4. $a_n = 2(3^n) + 7^n$

In Exercises 5–8, solve the first-order linear homogeneous recurrence relations:

5. $a_{n+1} = 6a_n, a_0 = 1, n \geq 0$

6. $a_{n+1} = 4a_n, a_0 = 3, n \geq 0$

7. $a_{n+1} = 3a_n, a_0 = -2, n \geq 0$

8. $a_{n+1} = -5a_n, a_0 = -5, n \geq 0$

In Exercises 9–16, solve the second-order linear homogeneous recurrence relations, where $n \geq 0$:

9. $a_{n+2} = 7a_{n+1} - 6a_n, a_0 = 7, a_1 = 32$

10. $a_{n+2} = 2a_{n+1} + 3a_n, a_0 = 2, a_1 = 2$

11. $a_{n+2} = 12a_{n+1} - 35a_n, a_0 = 3, a_1 = 17$

12. $a_{n+2} = 2a_{n+1} + 15a_n, a_0 = 2, a_1 = 2$

13. $a_{n+2} = 8a_{n+1} - 16a_n, a_0 = 1, a_1 = 8$

14. $a_{n+2} = 4a_{n+1} - 4a_n, a_0 = 3, a_1 = 16$

15. $a_{n+2} = 6a_{n+1} - 9a_n, a_0 = 2, a_1 = 9$

16. $a_{n+2} + 4a_{n+1} + 4a_n = 0, a_0 = 1, a_1 = -4$

In Exercises 17 – 28, write the first few terms of each of the recursion relations by iteration:

17. $a_{n+1} = 3a_n + 6$, where $a_1 = 4$, $n \geq 1$

18. $a_{n+1} = 3a_n + (-2)^n$, where $a_1 = -3$, $n \geq 1$

19. $a_{n+1} = 3na_n$, where $a_1 = -3$, $n \geq 1$

20. $a_{n+1} = (a_n)^2$, where $a_1 = -3$, $n \geq 1$

21. $a_{n+2} = a_{n+1} + 3a_n$, where $a_1 = 1, a_2 = 3$, $n \geq 1$

22. $a_{n+2} = a_{n+1} - 5a_n$, where $a_1 = 1, a_2 = 2$, $n \geq 1$

23. $a_{n+2} = a_{n+1} + 3a_n + (-2)^n$, where $a_1 = 1, a_2 = 3$, $n \geq 1$

24. $a_{n+2} = a_{n+1} + a_n$, where $a_1 = 2, a_2 = 3$, $n \geq 1$

25. $a_{n+2} = a_{n+1} \times a_n$, where $a_1 = 2, a_2 = 3$, $n \geq 1$

26. $a_{n+3} = a_{n+2} + a_{n+1} + a_n$, where $a_1 = 1, a_2 = 3, a_3 = 2$, $n \geq 1$

27. $a_{n+3} = a_{n+2} + a_{n+1} - a_n$, where $a_1 = 1, a_2 = 2, a_3 = 4$, $n \geq 1$

28. $a_{n+1} = (n+1)a_n$, where $a_1 = 1$, $n \geq 1$

10.2   Fibonacci Sequence

A useful and important example of the recursion relation is the well-known **Fibonacci sequence**. Fibonacci introduced the sequence of numbers

$$1, 1, 2, 3, 5, 8, 13, 21, 34, 55, 89, 144, 233, 377, 610, 987, 1597, 2584, 4181, \cdots$$

Fibonacci sequence was constructed as follows:
(i) the first two terms of the sequence are $F_1 = F_2 = 1$;
(ii) every subsequent term $F_k, k \geq 3$ is the sum of the preceding two terms.

Appendix C shows the first 30 Fibonacci numbers.

   In other words, Fibonacci, Leonardo of Pisa (1170 – 1250), introduced the recursion formula:

Initial values $F_1 = F_2 = 1,$

Recurrence relation $F_{n+2} = F_{n+1} + F_n, \, n \geq 1$

Example 1. Use the recursion formula for Fibonacci sequence to write the first few terms of the sequence as listed above:

Solution.

$$
\begin{aligned}
F_3 = F_2 + F_1 &= 1 + 1 = 2 \\
F_4 = F_3 + F_2 &= 2 + 1 = 3 \\
F_5 = F_4 + F_3 &= 3 + 2 = 5 \\
F_6 = F_5 + F_4 &= 5 + 3 = 8 \\
F_7 = F_6 + F_5 &= 8 + 5 = 13 \\
F_8 = F_7 + F_6 &= 13 + 8 = 21 \\
F_9 = F_8 + F_7 &= 21 + 13 = 34
\end{aligned}
$$

and so on.

Example 2. Given the two terms of Fibonacci sequence $F_{21} = 10946$ and $F_{22} = 17711$, find F_{23} and F_{20}.

Solution. Using the recursion formula for the Fibonacci sequence given before we set

$$
\underline{F_{20}} \qquad \underbrace{F_{21}}_{10946} \qquad \underbrace{F_{22}}_{17711} \qquad \underline{F_{23}}
$$

Accordingly, we get

$$
F_{23} = F_{22} + F_{21} = 17711 + 10946 = 28657
$$

and

$$
F_{20} = F_{22} - F_{21} = 17711 - 10946 = 6765
$$

10.2.1 Properties of Fibonacci Sequence

It is interesting to note here that Fibonacci and many mathematicians have established several interesting properties and identities related to Fibonacci sequence. These identities and properties are usually introduced in a number theory course. However, in this text, we will discuss some of these properties and identities:

1. Fibonacci Prime Numbers: Many Fibonacci numbers are prime numbers. Examples of Fibonacci prime numbers are:

$$2, 3, 5, 13, 89, 233, 1597, 28657, 514229$$

2. Prime Factorization of F_n: All composite Fibonacci numbers can be expressed as the product of primes, such as $8 = 2^3, 21 = 3 \cdot 7$. However, it was found by Zerger that the product $F_6 F_7 F_8 F_9$ is equal to the product of the first seven prime numbers. This means that

$$F_6 F_7 F_8 F_9 = 510510 = 2 \cdot 3 \cdot 5 \cdot 7 \cdot 11 \cdot 13 \cdot 17$$

3. In Fibonacci sequence, there must be at least four Fibonacci numbers and at most five Fibonacci numbers that have the same number of digits. For example, each number of $F_{17} - F_{20}$ has four digits, whereas $F_{21} - F_{25}$ has five digits for each Fibonacci number.

4. Any two consecutive Fibonacci numbers are relatively prime. This means that any two consecutive Fibonacci numbers do not have a common factors. For example, $F_8 = 21 = 3 \cdot 7$, and $F_9 = 34 = 2 \cdot 17$. This is applicable to all Fibonacci numbers and can bee seen from Appendix C.

5. The sum of any ten consecutive Fibonacci numbers is divisible by 11. For example, the sum of the Fibonacci numbers F_8 to F_{17} is $4147 = 11 \cdot 377$. Also, the sum of F_{20} to F_{29} is $1335323 = 11 \cdot 121393$.

6. The sum of first n Fibonacci numbers can be found by using the formula

$$F_1 + F_2 + F_3 + F_4 + \cdots + F_n = F_{n+2} - 1$$

This property will be proved later in Section 10.2.3. For example:

$$F_1 + F_2 + F_3 + F_4 + \cdots + F_{18} = F_{20} - 1 = 6765 - 1 = 6764$$

7. Petals on Flowers: It was found that on many plants, the number of petals on flowers is a Fibonacci number. Assuming all petals of a flower exist, we find the following number of petals on flowers:
(i) 3 petals: lily, iris.
(ii) 5 petals: buttercup, wild rose, larkspur, columbine, pinks.
(iii) 8 petals: delphiniums.
(iv) 13 petals: ragwort, corn marigold, cineraria.
(v) 21 petals: aster, black-eyed susan, chicory.
(vi) 34 petals: plantain, pyrethrum.
(vii) 55 and 89 petals: michaelmas daisies, the asteraceae family.

10.2.2   The Golden Ratio:

The ratio $\dfrac{F_{n+1}}{F_n} = 1.61803398875...$ as n approaches a large number. The ratio between any two consecutive Fibonacci numbers as n is large is called the **golden number**. This property is not easily seen for small n, but as n gets large, the ratio becomes clear. For example, if we selected $F_{16} = 987$ and $F_{15} = 610$, then we find

$$\frac{F_{16}}{F_{15}} = \frac{987}{610} \approx 1.618$$

We can also prove that the ratio $\dfrac{F_n}{F_{n+1}} \approx 0.618$ as n gets large. In other words we can easily show that

(i) $\lim_{n\to\infty} \dfrac{F_{n+1}}{F_n} \approx 1.618$

(ii) $\lim_{n\to\infty} \dfrac{F_{n+2}}{F_n} \approx 2.618$

(iii) $\lim_{n\to\infty} \dfrac{F_n}{F_{n+1}} \approx 0.618$

    The golden ratio is used in many applications. In what follows we discuss two examples to show the use of this concept.

Example 3. Use the golden ratio and related ratios to approximate the following expression

$$4\frac{F_{101}}{F_{100}} - 2\frac{F_{91}}{F_{92}} + 3\frac{F_{112}}{F_{111}} - 5\frac{F_{200}}{F_{201}}$$

Solution. Using the golden ratio and its reciprocal we obtain

$$4 \times 1.618 - 2 \times 0.618 + 3 \times 1.618 - 5 \times 0.618 \approx 7$$

Example 4. Use the golden ratio to approximate the following expression

$$\left(\frac{F_{71}}{F_{70}} - 2\frac{F_{82}}{F_{83}} + \frac{F_{91}}{F_{90}}\right) \times \left(5\frac{F_{41}}{F_{40}} - 6\frac{F_{78}}{F_{79}} + \frac{F_{66}}{F_{64}}\right)$$

Solution. Using the golden ratio and its reciprocal we obtain

$$(1.618 - 2 \times 0.618 + 1.618) \times (5 \times 1.618 - 6 \times 0.618 + 2.618) \approx 14$$

Powers of the Golden Ratio

We set the golden ratio

$$\frac{F_{n+1}}{F_n} = \alpha$$

and hence the reciprocal of the golden ratio is given by

$$\frac{F_n}{F_{n+1}} = \frac{1}{\alpha}$$

It is clear that the golden ration α and its reciprocal $\frac{1}{\alpha}$ differ by one. In other words

$$\alpha = 1 + \frac{1}{\alpha}$$

This in turn gives

$$\alpha^2 = \alpha + 1$$

From this fact we can determine the following identities for the powers of α:

$$\begin{aligned}
\alpha^3 &= \alpha \cdot \alpha^2 = \alpha(\alpha + 1) = \alpha^2 + \alpha = 2\alpha + 1 \\
\alpha^4 &= \alpha \cdot \alpha^3 = \alpha(2\alpha + 1) = 2\alpha^2 + \alpha = 3\alpha + 2 \\
\alpha^5 &= \alpha \cdot \alpha^4 = \alpha(3\alpha + 2) = 3\alpha^2 + 2\alpha = 5\alpha + 3 \\
\alpha^6 &= \alpha \cdot \alpha^5 = \alpha(5\alpha + 3) = 5\alpha^2 + 3\alpha = 8\alpha + 5
\end{aligned}$$

and so on. Combining the aforementioned results we find

$$\begin{aligned}
\alpha &= 1\alpha + 0 & \alpha^7 &= 13\alpha + 8 \\
\alpha^2 &= 1\alpha + 1 & \alpha^8 &= 21\alpha + 13 \\
\alpha^3 &= 2\alpha + 1 & \alpha^9 &= 34\alpha + 21 \\
\alpha^4 &= 3\alpha + 2 & \alpha^{10} &= 55\alpha + 34 \\
\alpha^5 &= 5\alpha + 3 & \alpha^{11} &= 89\alpha + 55 \\
\alpha^6 &= 8\alpha + 5 & \alpha^{12} &= 144\alpha + 89
\end{aligned}$$

$$\vdots \qquad \vdots$$

We can easily observe that the coefficients of the powers of α at the right side of each identity form the Fibonacci sequence. Moreover, the constants the right side of each identity form the Fibonacci equation as well. We can also prove the following generalized formula

$$\alpha^n = F_n\alpha + F_{n-1}, F_0 = 0, n \geq 1$$

This formula can be proved by mathematical induction.

10.2.3   Identities of Fibonacci Sequence

Several identities that relate Fibonacci numbers were formally derived and proved by many methods. In this section, we present some of these identities with proofs. However, other identities are left to the reader as exercises.

1. **Sum of Fibonacci Numbers:** The following identity

$$F_1 + F_2 + F_3 + F_4 + \cdots + F_n = F_{n+2} - 1$$

can be easily derived as follows:

$$
\begin{aligned}
F_1 &= F_3 - F_2 \\
F_2 &= F_4 - F_3 \\
F_3 &= F_5 - F_4 \\
&\vdots \\
F_{n-1} &= F_{n+1} - F_n \\
F_n &= F_{n+2} - F_{n+1}
\end{aligned}
$$

Adding these equations term by term gives

$$F_1 + F_2 + F_3 + F_4 + \cdots + F_n = F_{n+2} - F_2 = F_{n+2} - 1$$

Notice that this identity can also be proved by using mathematical induction (see Exercises 18). This identity means that if want to find the sum

$$F_1 + F_2 + F_3 + \cdots + F_{12}$$

We substitute $n = 12$ in the last identity given above to find that

$$F_1 + F_2 + F_3 + F_4 + \cdots + F_{12} = F_{14} - 1 = 376$$

2. **Sum of Fibonacci Numbers with Odd Subscripts:** The following identity

$$F_1 + F_3 + F_5 + F_7 + \cdots + F_{2n-1} = F_{2n}$$

will be derived as follows:

$$
\begin{aligned}
F_1 &= F_2 \\
F_3 &= F_4 - F_2 \\
F_5 &= F_6 - F_4 \\
&\vdots \\
F_{2n-3} &= F_{2n-2} - F_{2n-4} \\
F_{2n-1} &= F_{2n} - F_{2n-2}
\end{aligned}
$$

Adding these equations term by term gives

$$F_1 + F_3 + F_5 + F_7 + \cdots + F_{2n-1} = F_{2n}$$

This identity can also be proved by using mathematical induction (see Exercises 19). Consequently, if we want to find the sum

$$F_1 + F_3 + F_5 + \cdots + F_{13}$$

we substitute $2n - 1 = 13, n = 7$ in the last identity given before to find that

$$F_1 + F_3 + F_5 + F_7 + \cdots + F_{13} = F_{14} = 377$$

3. Sum of Fibonacci Numbers with Even Subscripts: The following identity

$$F_2 + F_4 + F_6 + F_8 + \cdots + F_{2n} = F_{2n+1} - 1$$

is left as an exercise (see Exercises 20).

4. Sum of Squares of Fibonacci Numbers: Lucas developed the following identity that gives the sum of squares of Fibonacci numbers. The identity is given by

$$F_1^2 + F_2^2 + F_3^2 + F_4^2 + \cdots + F_n^2 = F_n F_{n+1}$$

This identity will be proved by mathematical induction.
For $n = 1$, LHS $= F_1^2 = 1$, RHS$= F_1 F_2 = 1$, therefore the identity is true for $n = 1$.
Assume that the identity is true for $n = k$; this means that we assume that

$$F_1^2 + F_2^2 + \cdots + F_k^2 = F_k F_{k+1}$$

Now we should show that the identity is true for $n = k + 1$; i.e. we should show that

$$F_1^2 + F_2^2 + F_3^2 + F_4^2 + \cdots + F_k^2 + F_{k+1}^2 = F_{k+1} F_{k+2}$$

We notice that

$$
\begin{aligned}
(F_1^2 + F_2^2 + \cdots + F_k^2) + F_{k+1}^2 &= F_k F_{k+1} + F_{k+1}^2, \text{ by induction hypothesis} \\
&= F_{k+1}(F_k + F_{k+1}), \text{ by factoring} \\
&= F_{k+1} F_{k+2}
\end{aligned}
$$

We point out here that this identity can also be proved by using the equations

$$\begin{aligned} F_1^2 &= F_1 F_2 \\ F_2^2 &= F_2 F_3 - F_1 F_2 \end{aligned}$$

and so on. This approach of proof will be left as an exercise (see Exercise 25).

5. **Cassini's Formula:** The following identity

$$F_n^2 - F_{n-1} F_{n+1} = (-1)^{n+1}$$

is true for $n \geq 2$.

This identity will be proved by mathematical induction.

For $n = 2$, LHS $= F_2^2 - F_1 F_3 = -1$, RHS $= (-1)^3 = -1$, therefore the identity is true for $n = 1$.

Assume that the identity is true for $n = k$; this means that we assume that

$$F_k^2 - F_{k-1} F_{k+1} = (-1)^{k+1}$$

It remains to show that the identity is true for $n = k + 1$; this means that we should show that

$$F_{k+1}^2 - F_k F_{k+2} = (-1)^{k+2}$$

We notice that

$$\begin{aligned} F_{k+1}^2 - F_k F_{k+2} &= F_{k+1}^2 - (F_{k+1} - F_{k-1})(F_k + F_{k+1}), \\ &= -(F_{k+1} F_k - F_k F_{k-1} - F_{k-1} F_{k+1}), \\ &= -\left(F_{k+1} F_k - F_k F_{k-1} - F_k^2 + (-1)^{k+1} \right), \\ &= -(F_{k+1} F_k - F_k(F_{k-1} + F_k)) + (-1)^{k+2}, \\ &= -(F_{k+1} F_k - F_{k+1} F_k) + (-1)^{k+2}, \\ &= (-1)^{k+2} \end{aligned}$$

The identity is thus true for $n \geq 2$.

6. **Binet's Formula** In 1843, the French mathematician *Binet* introduced a formula for the Fibonacci number F_n given by

$$F_n = \frac{\alpha^n - \beta^n}{\alpha - \beta}, n \geq 1$$

or equivalently

$$F_n = \frac{\alpha^n - \beta^n}{\sqrt{5}}, n \geq 1$$

where

$$\alpha = \frac{1 + \sqrt{5}}{2}, \beta = \frac{1 - \sqrt{5}}{2}$$

The formula can be proved by using the concept of solving the recurrence relation introduced in the previous section. The Fibonacci recurrence relation

$$F_{n+2} = F_{n+1} + F_n, \text{ with initial values } F_1 = F_2 = 1, n \geq 1$$

The substitution $F_n = Cx^n, C$ is a constant, carries this relation into the equation

$$x^2 - x - 1 = 0$$

By solving this equation we find a positive root α and a negative root β given by

$$\alpha = \frac{1 + \sqrt{5}}{2}, \beta = \frac{1 - \sqrt{5}}{2}$$

This means that

$$F_n = A\alpha^n + B\beta^n$$

where A and B are constants that will be determined by using the initial values $F_1 = F_2 = 1$. Substituting $n = 1$ then $n = 2$ into the last equation we get

$$
\begin{aligned}
A\alpha + B\beta &= 1 \\
A\alpha^2 + B\beta^2 &= 1
\end{aligned}
$$

Solving this system of equations, and noting that $\alpha - 1 = -\beta, \beta - 1 = -\alpha$, we find

$$A = \frac{1}{\sqrt{5}} \approx 1.618, B = \frac{1}{\sqrt{5}} \approx 0.618$$

Exercises 10.2

1. If $F_{20} = 6765$, $F_{21} = 10946$, Find F_{19}, F_{22}

2. If $F_{25} = 75025$, $F_{26} = 121393$, Find F_{24}, F_{27}

3. If $F_{29} = 514229$, $F_{30} = 832040$, Find F_{28}, F_{31}

4. If $F_{27} = 196418$, $F_{25} = 75025$, Find F_{26}, F_{28}

5. If $F_{31} = 1346269$, $F_{29} = 514229$, Find F_{30}, F_{33}

In Exercises 6–10, find an approximation to the expressions by using the golden ratio:

6. $3\dfrac{F_{127}}{F_{126}} - \dfrac{F_{271}}{F_{272}} + 2\dfrac{F_{99}}{F_{98}} - 4\dfrac{F_{39}}{F_{40}}$

7. $6\dfrac{F_{27}}{F_{26}} + 4\dfrac{F_{271}}{F_{270}} - 8\dfrac{F_{97}}{F_{98}} - 2\dfrac{F_{139}}{F_{140}}$

8. $\left(3\dfrac{F_{27}}{F_{26}} + 4\dfrac{F_{271}}{F_{270}} - 7\dfrac{F_{97}}{F_{98}}\right) \times \left(\dfrac{F_{139}}{F_{138}} + \dfrac{F_{61}}{F_{60}} - 2\dfrac{F_{40}}{F_{41}}\right)$

9. $\left(6\dfrac{F_{22}}{F_{21}} + 3\dfrac{F_{24}}{F_{22}} - 9\dfrac{F_{97}}{F_{98}}\right) \div \left(\dfrac{F_{31}}{F_{30}} - 3\dfrac{F_{27}}{F_{28}} + 2\dfrac{F_{40}}{F_{39}}\right)$

10. $\left(8\dfrac{F_{202}}{F_{200}} + 2\dfrac{F_{24}}{F_{23}} - 7\dfrac{F_{97}}{F_{98}} - 3\dfrac{F_{34}}{F_{35}}\right) \div \left(2\dfrac{F_{71}}{F_{70}} - 3\dfrac{F_{27}}{F_{28}} + \dfrac{F_{41}}{F_{40}}\right)$

11. Construct a Fibonacci type sequence where the first term is the even prime number, and the third term is the third odd prime number. Show that the golden ratio is valid here.

12. Construct a Fibonacci type sequence by selecting 12 as the third term of the sequence, where the first two numbers of the sequence are prime numbers. Show that the ratio of any two consecutive numbers gives the golden ratio as n gets large.

13. Construct a Fibonacci type sequence by selecting 34 as the third number of the sequence, where the first two numbers are the first two perfect numbers. Show that the ratio of any two consecutive numbers gives the golden ratio as n gets large.

14. Determine which of the following is a Fibonacci type sequence. If so, write few more terms of the sequence:

(a) $2, 4, 6, 10, \cdots$

(b) $2, 6, 10, 14, 18, \cdots$

(c) $2, 5, 7, 12, 19, \cdots$

(d) $1, 2, 4, 8, 16, \cdots$

15. In Fibonacci sequence, show that $F_1 + F_2 + F_3 + \cdots + F_8 = F_{10} - 1$

16. In Fibonacci sequence, show that $F_1 + F_3 + F_5 + F_7 = F_8$

17. In Fibonacci sequence, show that $F_7^2 + F_8^2 = F_{15}$

In Exercises 18–24, use mathematical induction to prove each statement for $n \geq 1$:

18. $F_1 + F_2 + F_3 + \cdots + F_n = F_{n+2} - 1$

19. $F_1 + F_3 + F_5 + \cdots + F_{2n-1} = F_{2n}$

20. $F_2 + F_4 + F_6 + \cdots + F_{2n} = F_{2n+1} - 1$

21. $F_n = \frac{\alpha^n - \beta^n}{\sqrt{5}}$

22. $F_{n+1}^2 - F_n^2 = F_{n-1}F_{n+2}$

23. $F_1 - F_2 + F_3 - F_4 + \cdots + (-1)^{n+1}F_n = (-1)^{n+1}F_{n-1} + 1$

24. $F_n < 2^n$

25. Use the equations

$$F_1^2 = F_1 F_2, \ F_2^2 = F_2 F_3 - F_1 F_2, \cdots$$

to show that

$$F_1^2 + F_2^2 + F_3^2 + \cdots + F_n^2 = F_n F_{n+1}$$

26. Use the equation

$$\frac{F_n}{F_{n-1}F_{n+1}} = \frac{F_{n+1} - F_{n-1}}{F_{n-1}F_{n+1}} = \frac{1}{F_{n-1}} - \frac{1}{F_{n+1}}$$

to show that $\sum_{n=2}^{\infty} \frac{F_n}{F_{n-1}F_{n+1}} = 2$

27. Prove by mathematical induction

$$\alpha^n = F_n \alpha + F_{n-1}, F_0 = 0, n \geq 1$$

28. Prove by mathematical induction

$$\alpha^n = \alpha^{n-1} + \alpha^{n-2}, n \geq 2$$

29. Prove that

$$\sum_{n=1}^{\infty} \frac{1}{\alpha^n} = \alpha$$

30. Prove that

$$\sum_{n=1}^{\infty} \frac{1}{\alpha^{2n}} = \frac{1}{\alpha}$$

10.3   Lucas Sequence

Another useful and important example of the recurrence relation is the well-known **Lucas sequence** established by using the sense of Fibonacci sequence. Lucas introduced the sequence of numbers

$$1, 3, 4, 7, 11, 18, 29, 47, 76, 123, 199, 322, 521, 843, 1364, 2207, 3571, \cdots$$

Lucas sequence was established as follows:
(i) the first two terms of the sequence are $L_1 = 1$ and $L_2 = 3$;
(ii) every subsequent term $L_k, k \geq 3$ is the sum of the preceding two terms.
    In other words, Lucas introduced the recursion formula:

Initial values $L_1 = 1, L_2 = 3,$

Recurrence relation $L_{n+2} = L_{n+1} + L_n, n \geq 1$

Appendix C shows the first 30 Lucas numbers.

Example 1. Use the recursion formula for Lucas sequence to write the first few terms of the sequence as listed above:

Solution.

$$\begin{aligned}
L_3 = L_2 + L_1 &= 3 + 1 = 4 \\
L_4 = L_3 + L_2 &= 4 + 3 = 7 \\
L_5 = L_4 + L_3 &= 7 + 4 = 11 \\
L_6 = L_5 + L_4 &= 11 + 7 = 18 \\
L_7 = L_6 + L_5 &= 18 + 11 = 29 \\
L_8 = L_7 + L_6 &= 29 + 18 = 47 \\
L_9 = L_8 + L_7 &= 47 + 29 = 76
\end{aligned}$$

and so on.

Example 2. Given the two terms of Lucas sequence numbers

$$L_{21} = 24476, L_{22} = 39603$$

Find L_{20} and L_{24}

Solution. Using the recursion formula for the Lucas sequence given before we set

$$\underline{L_{20}} \qquad \underbrace{L_{21}}_{24476} \qquad \underbrace{L_{22}}_{39603} \qquad \underline{L_{23}}$$

Accordingly, we get

$$L_{23} = L_{22} + L_{21} = 24476 + 39603 = 64079$$

and

$$L_{20} = L_{22} - L_{21} = 39603 - 24476 = 15127$$

10.3.1 Properties of Lucas Sequence

Lucas and many others have established several interesting properties and identities related to Lucas sequence. These identities and properties are usually introduced in a number theory course. However, in this text, we will discuss some of these properties and identities:

1. **Lucas Prime Numbers:** Many Lucas numbers are prime numbers. Examples of Fibonacci prime numbers are:

$$3, 7, 11, 29, 47, 199, 521, 2207, 3571, 9349, \cdots$$

2. **Prime Factorization of L_n:** All composite Lucas numbers, other than L_1 can be expressed as the product of primes, such as $4 = 2^2, 18 = 3^2 \cdot 2, 76 = 2^3 \cdot 19$.

3. **The Golden Ratio:** The ratio $\dfrac{L_{n+1}}{L_n} = 1.61803398875\cdots$. The ratio between two consecutive Lucas numbers is called the **golden number**. This property is not easily seen for small n, but as n gets large, the ratio becomes clear. For example, if we selected $L_{22} = 39603$ and $L_{23} = 64079$, then we find

$$\frac{L_{23}}{L_{22}} = \frac{64079}{39603} \approx 1.618$$

We can also prove that the reciprocal ratio $\dfrac{L_n}{L_{n+1}} \approx 0.618$ as n approaches infinity. We can easily show that the golden ratio and other related ratios as follows:

(i) $\lim_{n\to\infty} \dfrac{L_{n+1}}{L_n} \approx 1.618$

(ii) $\lim_{n\to\infty} \dfrac{L_{n+2}}{L_n} \approx 2.618$

(iii) $\lim_{n\to\infty} \dfrac{L_n}{L_{n+1}} \approx 0.618$

Example 3. Use the golden ratio to approximate the following expression

$$4\frac{L_{101}}{L_{100}} - 2\frac{L_{91}}{L_{92}} + 3\frac{L_{112}}{L_{111}} - 5\frac{L_{200}}{L_{201}}$$

Solution. Using the golden ratio and other related ratios we obtain

$$4 \times 1.618 - 2 \times 0.618 - 3 \times 1.618 - 5 \times 0.618 \approx 7$$

Example 4. Use the golden ratio to approximate the following expression

$$\left(2\frac{L_{71}}{L_{70}} - 7\frac{L_{82}}{L_{83}} + 5\frac{L_{91}}{L_{90}}\right) - \left(5\frac{L_{41}}{L_{40}} - 9\frac{L_{78}}{L_{79}} + 4\frac{L_{66}}{L_{64}}\right)$$

Solution. Using the golden ratio and other ratios we obtain

$$(2 \times 1.618 - 7 \times 0.618 + 5 \times 1.618)$$
$$- \quad (5 \times 1.618 - 9 \times 0.618 + 4 \times 2.618) \approx -6$$

10.3.2   Identities of Lucas Sequence

Several identities that relate Lucas numbers were formally derived and proved by different approaches. In this section, we present some of these identities with proofs. However, other identities are left to the reader as exercises.

1. **Sum of Lucas Numbers:** The following identity

$$L_1 + L_2 + L_3 + L_4 + \cdots + L_n = L_{n+2} - 3$$

can be easily derived as follows:

$$
\begin{aligned}
L_1 &= L_3 - L_2 \\
L_2 &= L_4 - L_3 \\
L_3 &= L_5 - L_4 \\
L_4 &= L_6 - L_5 \\
&\vdots \\
L_{n-1} &= L_{n+1} - L_n \\
L_n &= L_{n+2} - L_{n+1}
\end{aligned}
$$

Adding these equations term by term gives

$$L_1 + L_2 + L_3 + L_4 + \cdots + L_n = L_{n+2} - L_2 = L_{n+2} - 3$$

Notice that this identity can also be proved by using mathematical induction. This identity means that if want to find the sum

$$L_1 + L_2 + L_3 + \cdots + L_{12}$$

we substitute $n = 12$ in the last identity to find that

$$L_1 + L_2 + L_3 + L_4 + \cdots + L_{12} = L_{14} - 3 = 843 - 3 = 840$$

2. Sum of Lucas Numbers with Odd Subscripts: The following identity

$$L_1 + L_3 + L_5 + L_7 + \cdots + L_{2n-1} = L_{2n} - 2$$

will be derived as follows:

$$
\begin{aligned}
L_1 &= L_2 - 2 \\
L_3 &= L_4 - L_2 \\
L_5 &= L_6 - L_4 \\
L_7 &= L_8 - L_6 \\
&\vdots \\
L_{2n-3} &= L_{2n-2} - L_{2n-4} \\
L_{2n-1} &= L_{2n} - L_{2n-2}
\end{aligned}
$$

Adding these equations term by term gives

$$L_1 + L_3 + L_5 + L_7 + \cdots + L_{2n-1} = L_{2n} - 2$$

This identity can also be proved by using mathematical induction. Consequently, if want to find the sum

$$L_1 + L_3 + L_5 + \cdots + L_{13}$$

we substitute $2n - 1 = 13, n = 7$ in the last identity to find that

$$L_1 + L_3 + L_5 + L_7 + \cdots + L_{13} = L_{14} - 2 = 841$$

3. **Sum of Lucas Numbers with Even Subscripts:** The following identity

$$L_2 + L_4 + L_6 + L_8 + \cdots + L_{2n} = L_{2n+1} - 1$$

is left as an exercise.

4. **Sum of Squares of Lucas Numbers:** The following identity that gives the sum of squares of Lucas numbers

$$L_1^2 + L_2^2 + L_3^2 + L_4^2 + \cdots + L_n^2 = L_n L_{n+1} - 2$$

will be proved by mathematical induction.
For $n = 1$, LHS $= L_1^2 = 1$, RHS$=L_1 L_2 = 1 \cdot 3 - 2 = 1$, therefore the identity is true for $n = 1$
Assume that the identity is true for $n = k$; this means that we assume that

$$L_1^2 + L_2^2 + L_3^2 + L_4^2 + \cdots + L_k^2 = L_k L_{k+1} - 2$$

It remains to show that the identity is true for $n = k + 1$; this means that we should show that

$$L_1^2 + L_2^2 + L_3^2 + L_4^2 + \cdots + L_k^2 + L_{k+1}^2 = L_{k+1} L_{k+2} - 2$$

We notice that

$$
\begin{aligned}
\text{LHS} = (L_1^2 + L_2^2 + \cdots + L_k^2) + L_{k+1}^2 &= L_k L_{k+1} + L_{k+1}^2 - 2, \\
&= L_{k+1}(L_k + L_{k+1}) - 2, \\
&= L_{k+1} L_{k+2} - 2,
\end{aligned}
$$

5. **Binet's Formula:** In 1843, the French mathematician *Binet* introduced a formula for the Lucas number L_n given by

$$L_n = \alpha^n + \beta^n, n \geq 1$$

where

$$\alpha = \frac{1 + \sqrt{5}}{2}, \beta = \frac{1 - \sqrt{5}}{2}$$

The formula can be proved by using the concept of solving the recurrence relation introduced in the previous section. The Lucas recurrence relation

$$L_{n+2} = L_{n+1} + L_n, \text{ with initial values } L_1 = 1, L_2 = 3, n \geq 1$$

The substitution $L_n = Cx^n, C$ is a constant, carries this relation into the equation

$$x^2 - x - 1 = 0$$

By solving this equation we find a positive root α and a negative root β given by

$$\alpha = \frac{1 + \sqrt{5}}{2}, \ \beta = \frac{1 - \sqrt{5}}{2}$$

This means that

$$L_n = C\alpha^n + D\beta^n$$

where C and D are constants that will be determined by using the initial values $L_1 = 1, L_2 = 3$. Substituting $n = 1$ then $n = 2$ into the last equation we get

$$\begin{aligned} C\alpha + D\beta &= 1 \\ C\alpha^2 + D\beta^2 &= 3 \end{aligned}$$

Solving this system of equations we find

$$C = D = 1$$

Notice that

$$\alpha \approx 1.618, |\beta| \approx 0.618$$

Moreover, by using the equation $x^2 - x - 1 = 0$, we can easily observe that

$$\begin{aligned} \alpha^2 &= \alpha + 1 \approx 2.618 \\ \alpha^3 &= 2\alpha + 1 \\ \alpha^4 &= 3\alpha + 2 \end{aligned}$$

6. **Cassini Type Formula:** The following identity

$$L_n^2 - L_{n-1}L_{n+1} = 5(-1)^n$$

is true for all $n \geq 1$. The identity will be proved by using Binet's formula. We first note that

$$L_n = \alpha^n + \beta^n, \ \alpha\beta = -1, \ \alpha^2 + \beta^2 = 3$$

Accordingly, we find

$$
\begin{aligned}
L_n^2 - L_{n-1}L_{n+1} &= (\alpha^n + \beta^n)^2 - (\alpha^{n-1} + \beta^{n-1})(\alpha^{n+1} + \beta^{n+1}) \\
&= 2(\alpha\beta)^n - (\alpha\beta)^n(\frac{\beta}{\alpha} + \frac{\alpha}{\beta}) \\
&= 2(-1)^n - (-1)^n(-3) \\
&= 5(-1)^n
\end{aligned}
$$

The proof for this identity by mathematical induction will be left as an exercise (see Exercise 9).

Exercises 10.3

1. If $L_{15} = 1364$, $L_{16} = 2207$, find L_{14}, L_{17}

2. If $L_{23} = 64079$, $L_{24} = 103682$, find L_{22}, L_{25}

3. If $L_{26} = 271443$, $L_{28} = 710647$, find L_{27}, L_{29}

4. If $L_{31} = 3010349$, $L_{33} = 7881196$, find L_{32}, L_{30}

In Exercises 5–8, use the golden ratio and its reciprocals to approximate the following expressions:

5. $7\dfrac{L_{32}}{L_{31}} + 3\dfrac{L_{42}}{L_{41}} - 8\dfrac{L_{21}}{L_{22}} - 2\dfrac{L_{30}}{L_{31}}$

6. $4\dfrac{L_{42}}{L_{40}} + 3\dfrac{L_{48}}{L_{47}} - 5\dfrac{L_{52}}{L_{53}} - 2\dfrac{L_{60}}{L_{61}}$

7. $\left(\dfrac{L_{32}}{L_{30}} - \dfrac{L_{62}}{L_{63}}\right) \times \left(\dfrac{L_{72}}{L_{71}} - \dfrac{L_{58}}{L_{59}}\right)$

8. $\left(6\dfrac{L_{32}}{L_{30}} + 3\dfrac{L_{24}}{L_{22}} - 9\dfrac{L_{60}}{L_{61}}\right) \div \left(\dfrac{L_{82}}{L_{81}} - 3\dfrac{L_{27}}{L_{28}} + 2\dfrac{L_{32}}{L_{31}}\right)$

9. Use mathematical induction to show that

$$L_n^2 - L_{n-1}L_{n+1} = 5(-1)^n$$

is true for $n \geq 1$.

10. Use mathematical induction to show that

$$5F_n = L_{n+2} - L_{n-2}, n \geq 3$$

11. Use mathematical induction to show that

$$F_{n+1} + F_{n-1} = L_n, n \geq 2$$

12. Use Exercise 11 to show that

$$F_{n+2} - F_{n-2} = L_n$$

13. Use Binet formula to show that the identity of Exercise 10

$$L_{n+2} - L_{n-2} = 5F_n$$

is true for $n \geq 3$

14. Use Binet formula to show that

$$F_{2n} = F_n L_n$$

15. $\lim_{n \to \infty} \dfrac{F_n}{L_n}$

16. $\lim_{n \to \infty} \dfrac{L_n^2}{F_n^2}$

Evaluate Exercises 17–20 in terms of α and β

17. $\dfrac{(\sqrt{5}F_n + L_n)^2 + (\sqrt{5}F_n - L_n)^2}{4}$

18. $\sqrt{5}L_n F_n$

19. $L_n^2 - 5F_n^2$

20. $\dfrac{5F_n^2 + L_n^2}{2}$

21. If $F_r = 2F_{101} + F_{100}$ Find r

22. If $L_r = 2L_{150} + L_{149}$ Find r

23. If $F_r = 3F_{12} + 2F_{11}$ Find r

24. If $L_r = 2C0L_7 + 2C1L_8 + 2C2L_9$ Find r

25. If $F_r = 2P0F_{10} + 2P1F_{11} + 2C0F_{12}$ Find r

26. If $L_r = 2C0L_7 + 3C1L_8 + 3P1F_9 + 3P0F_{10}$ Find r

27. If $F_r = 4P0F_{10} + 3P1F_{11} + 3C1F_{12} + 4C0F_{13}$ Find r

28. If $F_r = 5F_{117} + 3F_{116} + 2C0F_{12}$ Find r

Chapter 11

Matrices

11.1   Matrices

A **matrix** is a rectangular table of elements written in brackets. The elements of a matrix are called **entries**. The following is an example of a matrix:

$$A = \begin{bmatrix} 30 & 33 & 42 & 41 \\ 28 & 35 & 40 & 51 \end{bmatrix}$$

The matrix shown above has two rows and four columns. The dimension of this matrix is 2×4 (read 2 by 4) which indicates the number of rows by the number of columns. Matrices (plural of a matrix) are usually denoted by capital letters and mostly subscripted by its dimensions.

A matrix $B_{3\times4}$ is a matrix that contains 12 entries arranged in 3 rows and 4 columns. The matrix

$$C = \begin{bmatrix} 3 & 4 \\ -1 & 0 \\ 2 & 6 \end{bmatrix}$$

has the dimension 3×2 suitably written as $C_{3\times2}$. A matrix that consists of one row is called a **row** matrix such as

$$F_{1\times4} = \begin{bmatrix} 1 & 2 & 6 & 4 \end{bmatrix}$$

A matrix that consists of one column is called a **column** matrix such as

$$G_{3\times 1} = \begin{bmatrix} 1 \\ 3 \\ -4 \end{bmatrix}$$

A matrix where the number of rows is equal to the number of columns is called a **square** matrix. The following examples are 2×2 and 3×3 square matrices:

$$A_{2\times 2} = \begin{bmatrix} 1 & 2 \\ -2 & 3 \end{bmatrix}, \quad B_{3\times 3} = \begin{bmatrix} 1 & 2 & -1 \\ 0 & 4 & 1 \\ 3 & 2 & -1 \end{bmatrix}$$

The entries (elements) of a matrix are denoted by subscripted small letters as a_{ij}, such a notation refers to the entry located at the ith row and the jth column. In the matrix $A_{2\times 2}$ given above, we note that a_{21} indicates the entry located at the position where the second row and the first column of the matrix A pass through. Hence we write

$$a_{21} = -2$$

Similarly, in the matrix $B_{3\times 3}$ given above, we find

$$b_{23} = 1, \; b_{32} = 2, \; b_{33} = -1$$

It is important to note here that square matrices, such as the matrices $A_{2\times 2}$ and $B_{3\times 3}$, each has a main diagonal and a back diagonal. The elements on the main diagonal of $B_{3\times 3}$ are b_{11}, b_{22}, and b_{33} given by 1, 4, and -1 respectively. The elements on the back diagonal are b_{13}, b_{22}, and b_{31} given by -1, 4, and 3 respectively. The definitions related to a matrix and its elements will be illustrated by the following examples.

Example 1. Given the matrix

$$C = \begin{bmatrix} 1 & 6 & -3 \\ 2 & 4 & 0 \end{bmatrix}$$

Answer the following:

(a) Find the dimension of C

(b) Find c_{12}, c_{21}, and c_{23}

(c) If $c_{i2} = 4$, find i

Solution.

(a) The matrix has 2 rows and 3 columns, hence the dimension of C is 2×3. We therefore write $C_{2\times3}$

(b) $c_{12} = 6$, $c_{21} = 2$, and $c_{23} = 0$

(c) The entry 4 is the entry located at the second row and the second column, denoted by c_{22}, hence $i = 2$

Example 2. Construct a square matrix $A_{3\times3}$ that contains 1, 2, and 3 on the main diagonal, the entries 3, 2, 4 on the back diagonal, and 1 elsewhere.

Solution. The matrix $A_{3\times3}$ is given by

$$A_{3\times3} = \begin{bmatrix} 1 & 1 & 3 \\ 1 & 2 & 1 \\ 4 & 1 & 3 \end{bmatrix}$$

Example 3. Construct a square matrix $B_{4\times2}$ with entries given by the formula:

$$b_{ij} = i \times j$$

Solution. The dimension of the matrix is 4×2. This means that $1 \leq i \leq 4$ and $1 \leq j \leq 2$. Moreover, it is clear that each entry is determined by multiplying the two subscripts of that entry as defined by $i \times j$. For example, the entry $b_{32} = 3 \times 2 = 6$. Consequently, we find

$$b_{11} = 1 \times 1 = 1, \; b_{12} = 1 \times 2 = 2$$

$$b_{21} = 2 \times 1 = 2, \; b_{22} = 2 \times 2 = 4$$

$$b_{31} = 3 \times 1 = 3, \; b_{32} = 3 \times 2 = 6$$

$$b_{41} = 4 \times 1 = 4, \; b_{42} = 4 \times 2 = 8$$

Accordingly, the matrix $B_{4\times2}$ is given by

$$B_{4\times2} = \begin{bmatrix} 1 & 2 \\ 2 & 4 \\ 3 & 6 \\ 4 & 8 \end{bmatrix}$$

Definitions

1. A square matrix $A =$ is called **upper triangular** if all entries below the main diagonal are zeros, that is $a_{mn} = 0$ for $m > n$.

2. A square matrix $B =$ is called **lower triangular** if all entries above the main diagonal are zeros, that is $b_{mn} = 0$ for $m < n$.

Example 4. Construct an upper triangular matrix $A = [a_{mn}]$, $1 \le m \le 3$, $1 \le n \le 3$, with entries $a_{mn} = m + n$ for $m \le n$, and 0 otherwise

Solution. It is clear that each entry is determined by adding the two subscripts of that entry as defined by $m + n$ for $m \le n$ and 0 otherwise, hence we find

$$a_{11} = 1 + 1 = 2,\ a_{12} = 1 + 2 = 3,\ a_{13} = 1 + 3 = 4$$
$$a_{21} = 0, \qquad a_{22} = 2 + 2 = 4,\ a_{23} = 2 + 3 = 5$$
$$a_{31} = 0, \qquad a_{32} = 0, \qquad a_{33} = 3 + 3 = 6$$

Accordingly, we get

$$C_{3\times3} = \begin{bmatrix} 2 & 3 & 4 \\ 0 & 4 & 5 \\ 0 & 0 & 6 \end{bmatrix}$$

Exercises 11.1

1. Given the matrix

$$A = \begin{bmatrix} 1 & 2 & 1 & 0 \\ -3 & -4 & 1 & 2 \\ 4 & 3 & 1 & 6 \end{bmatrix}$$

(a) Find the dimension of A

(b) Find a_{23}, a_{32}, a_{34}, a_{14}

(c) If $a_{2j} = 2$, find j

2. Construct the matrix $B = [b_{ij}]$, $1 \leq i \leq 3$, $1 \leq j \leq 2$, with entries b_{ij} defined by $b_{ij} = i + j + 1$. Find the dimension of this matrix.

3. Construct the matrix $C = [c_{ij}]$, $1 \leq i \leq 2$, $1 \leq j \leq 3$, with entries c_{ij} defined by $c_{ij} = i^2 + j^2 - 1$. Find the dimension of this matrix.

4. Construct a matrix $B_{2\times2}$ by interchanging the rows and the columns of the matrix $A_{2\times2}$, where

$$A = \begin{bmatrix} 2 & 3 \\ 4 & 5 \end{bmatrix}$$

5. Construct the matrix $D = [d_{ij}]$, $1 \leq i \leq 3$, $1 \leq j \leq 3$, with entries d_{ij} defined by $d_{ij} = i \times j$. Check if the entries are similar on both sides of the main diagonal or not.

6. Construct the matrix $D = [d_{ij}]$, $1 \leq i \leq 3$, $1 \leq j \leq 3$, with entries $d_{ij} = d_{ji} = 2$, for $i \neq j$, and $d_{ij} = d_{ji} = 3$, for $i = j$. Check if the entries are similar on both sides of the main diagonal or not.

7. Construct a square matrix 3×3 where 3, 4, and 5 are entries of the main diagonal, and 2 elsewhere.

8. Construct a square matrix 3×3 where entries of the main diagonal are 1 for each and 0 elsewhere.

9. Construct a matrix $A_{1\times3}$, called a row vector, that consists of one row, whose entries are defined by $a_{ij} = i - j$, $i = 1$, $1 \leq j \leq 3$

10. Construct a matrix $B_{3\times1}$, called a column vector, that consists of one column, whose entries are defined by $b_{ij} = 3i + j$, $j = 1$, $1 \leq i \leq 3$

11. Construct a square matrix 3×3 where entries of the main diagonal and all entries above the main diagonal are 1 for each entry and 0 otherwise

12. Construct a square matrix 3×3 where entries of the main diagonal and all entries below the main diagonal are 1 for each entry and 0 otherwise

13. A square matrix A is called **upper triangular** if all entries below

the main diagonal are zeros; i.e $a_{mn} = 0$ for $m > n$. Construct an upper triangular matrix $A = [a_{mn}]$, $1 \le m \le 3$, $1 \le n \le 3$, with entries $a_{mn} = mn$ for $m \le n$, and 0 otherwise

14. A square matrix B is called **lower triangular** if all entries above the main diagonal are zeros; i.e $b_{mn} = 0$ for $m < n$. Construct a lower triangular matrix $B = [b_{mn}]$, $1 \le m \le 3$, $1 \le n \le 3$, with entries $a_{mn} = mn$ for $m \ge n$, and 0 otherwise

11.2   Operations on Matrices

In this section we will discuss algebraic operations that can be performed on matrices. The algebraic operations are:

I. Matrix equations: Two matrices A and B are equal if :

(a) A and B have the same dimension $m \times n$ for example.

(b) Corresponding entries of A and B are equal.
    The equality operation is explained by the following examples.

Example 1. Find x, y, u, and v if

$$\begin{bmatrix} x+1 & y-3 \\ u-1 & v+1 \end{bmatrix} = \begin{bmatrix} 2 & -1 \\ 2 & 5 \end{bmatrix}$$

Solution. It is clear that the two matrices have the same dimension 2×2. The equality of the two matrices indicates that the corresponding entries are equal. Therefore, we set

$$x + 1 = 2, \; y - 3 = -1$$

$$u - 1 = 2, \; v + 1 = 5$$

This gives
$$x = 1, \; y = 2, \; u = 3, \; v = 4$$

Example 2. Find x, y, u, and v if

$$\begin{bmatrix} x^2-1 & y^2+1 & u+1 \\ x^2+1 & y^2-1 & u+v \end{bmatrix} = \begin{bmatrix} 3 & 5 & 4 \\ 5 & 3 & 7 \end{bmatrix}$$

Solution. It is clear that the two matrices have the same dimension 2×3. The equality of the two matrices indicates that the corresponding entries are equal. Therefore, we set

$$x^2 - 1 = 3, \ x^2 + 1 = 5$$

$$y^2 + 1 = 5, \ y^2 - 1 = 3$$

$$u + 1 = 4, \ u + v = 7$$

This gives

$$x = \pm 2, \ y = \pm 2, \ u = 3, \ v = 4$$

Example 3. Find $a, b, c,$ and d if

$$\begin{bmatrix} a + b & c + d \\ c - d & a - b \end{bmatrix} = \begin{bmatrix} 3 & 7 \\ -1 & -1 \end{bmatrix}$$

Solution. It is clear that the two matrices have the same dimension 2×2. The equality of the two matrices indicates that the corresponding entries are equal. Equating the corresponding entries gives the following systems of equations:

$$\begin{aligned} a + b &= 3 \\ a - b &= -1 \end{aligned}$$

and

$$\begin{aligned} c + d &= 7 \\ c - d &= -1 \end{aligned}$$

Solving the first system gives

$$a = 1, \ b = 2$$

and by solving the second system we obtain

$$c = 3, \ d = 4$$

II. Addition of matrices: Two matrices A and B can be added only if they have the same dimension $m \times n$. The resulting matrix will have the same dimension and the entries can be obtained by adding the corresponding entries of A and B. This operation will be explained by the following illustrative examples.

Example 4. Given the matrices

$$A_{2\times3} = \begin{bmatrix} 1 & 2 & 4 \\ -1 & 0 & 3 \end{bmatrix}, \quad B_{2\times3} = \begin{bmatrix} 2 & 3 & -4 \\ 1 & 2 & 5 \end{bmatrix}$$

Find $A + B$

Solution.

$$A + B = \begin{bmatrix} 1 & 2 & 4 \\ -1 & 0 & 3 \end{bmatrix} + \begin{bmatrix} 2 & 3 & -4 \\ 1 & 2 & 5 \end{bmatrix}$$

$$= \begin{bmatrix} 1+2 & 2+3 & 4+(-4) \\ -1+1 & 0+2 & 3+5 \end{bmatrix} = \begin{bmatrix} 3 & 5 & 0 \\ 0 & 2 & 8 \end{bmatrix}$$

Example 5. Given the matrices

$$A_{2\times2} = \begin{bmatrix} 30 & 28 \\ 25 & 22 \end{bmatrix}, \quad B_{2\times2} = \begin{bmatrix} 25 & 26 \\ 21 & 20 \end{bmatrix}$$

Find $A + B$

Solution.

$$A + B = \begin{bmatrix} 30 & 28 \\ 25 & 22 \end{bmatrix} + \begin{bmatrix} 25 & 26 \\ 21 & 20 \end{bmatrix}$$

$$= \begin{bmatrix} 30+25 & 28+26 \\ 25+21 & 22+20 \end{bmatrix} = \begin{bmatrix} 55 & 54 \\ 46 & 42 \end{bmatrix}$$

III. Subtraction of matrices: The subtraction of the matrices A and B can be performed if both have the same dimension. The resulting matrix will have the same dimension and the entries can be obtained by subtracting the corresponding entries of A and B. This operation will be explained by the following illustrative examples.

Example 6. Given the matrices

$$A_{2\times2} = \begin{bmatrix} 5 & 6 \\ 7 & -2 \end{bmatrix}, B_{2\times2} = \begin{bmatrix} -2 & 4 \\ 1 & 3 \end{bmatrix}$$

Find $A - B$

Solution.

$$A - B = \begin{bmatrix} 5 & 6 \\ 7 & -2 \end{bmatrix} - \begin{bmatrix} -2 & 4 \\ 1 & 3 \end{bmatrix}$$

$$= \begin{bmatrix} 5-(-2) & 6-4 \\ 7-1 & -2-3 \end{bmatrix} = \begin{bmatrix} 7 & 2 \\ 6 & -5 \end{bmatrix}$$

Example 7. Given the matrices

$$C_{2\times2} = \begin{bmatrix} 30 & 28 \\ 25 & -22 \end{bmatrix}, D_{2\times2} = \begin{bmatrix} 14 & 15 \\ 13 & -12 \end{bmatrix}$$

Find $C - D$

Solution.

$$C - D = \begin{bmatrix} 30 & 28 \\ 25 & -22 \end{bmatrix} - \begin{bmatrix} 14 & 15 \\ 13 & -12 \end{bmatrix}$$

$$= \begin{bmatrix} 30-14 & 28-15 \\ 25-13 & -22-(-12) \end{bmatrix} = \begin{bmatrix} 16 & 13 \\ 12 & -10 \end{bmatrix}$$

IV. Multiplying a matrix by a scalar: A matrix A may be multiplied by a real number, called a scalar. The resulting matrix is a matrix of the same dimension as A, and obtained by multiplying each entry of the matrix A by that real number. This can be explained by the following examples.

Example 8. Given the matrix A defined by

$$A_{2\times3} = \begin{bmatrix} 1 & 2 & -1 \\ 3 & 4 & 2 \end{bmatrix}$$

Find $3A$

Solution.

$$3A = \begin{bmatrix} 3(1) & 3(2) & 3(-1) \\ 3(3) & 3(4) & 3(2) \end{bmatrix} = \begin{bmatrix} 3 & 6 & -3 \\ 9 & 12 & 6 \end{bmatrix}$$

Example 9. Given the matrices A and B defined by

$$A_{2\times2} = \begin{bmatrix} 1 & 2 \\ -2 & 1 \end{bmatrix}, \; B_{2\times2} = \begin{bmatrix} 2 & 4 \\ 6 & -8 \end{bmatrix}$$

Find $2A - \frac{1}{2}B$

Solution.

$$2A - \frac{1}{2}B = 2\begin{bmatrix} 1 & 2 \\ -2 & 1 \end{bmatrix} - \frac{1}{2}\begin{bmatrix} 2 & 4 \\ 6 & -8 \end{bmatrix}$$

$$= \begin{bmatrix} 2 & 4 \\ -4 & 2 \end{bmatrix} - \begin{bmatrix} 1 & 2 \\ 3 & -4 \end{bmatrix} = \begin{bmatrix} 1 & 2 \\ -7 & 6 \end{bmatrix}$$

In the following example, we combine all operations that we discussed so far.

Example 10. Find x, y, u, and v if

$$2\begin{bmatrix} x+1 & y-1 \\ u+2 & v-2 \end{bmatrix} = 2\begin{bmatrix} 4 & 1 \\ 3 & 3 \end{bmatrix} - \frac{1}{3}\begin{bmatrix} 6 & -6 \\ -18 & 0 \end{bmatrix}$$

Solution.

$$\begin{bmatrix} 2x+2 & 2y-2 \\ 2u+4 & 2v-4 \end{bmatrix} = \begin{bmatrix} 8 & 2 \\ 6 & 6 \end{bmatrix} - \begin{bmatrix} 2 & -2 \\ -6 & 0 \end{bmatrix} = \begin{bmatrix} 6 & 4 \\ 12 & 6 \end{bmatrix}$$

Equating the corresponding entries from both sides we find

$$2x+2 = 6, \; 2y-2 = 4$$

$$2u+4 = 12, \; 2v-4 = 6$$

This gives

$$x = 2, \; y = 3, \; u = 4, \; v = 5$$

Example 11. Find $x, y, u,$ and v if

$$\begin{bmatrix} x+y & xy \\ u-v & uv \end{bmatrix} = 5\begin{bmatrix} 2 & 2 \\ 1 & 0 \end{bmatrix} - 6\begin{bmatrix} 1 & 1 \\ 1 & -2 \end{bmatrix}$$

Solution.

$$\begin{bmatrix} x+y & xy \\ u-v & uv \end{bmatrix} = \begin{bmatrix} 10 & 10 \\ 5 & 0 \end{bmatrix} - \begin{bmatrix} 6 & 6 \\ 6 & -12 \end{bmatrix} = \begin{bmatrix} 4 & 4 \\ -1 & 12 \end{bmatrix}$$

Equating the corresponding entries from both sides we find

$$x+y = 4, \; xy = 4,$$

$$u-v = -1, \; uv = 12$$

We first solve the system involving x and y. From the first equation, we set

$$y = 4 - x$$

Substituting in the second equation, we find

$$\begin{aligned} x(4-x) &= 4 \\ x^2 - 4x + 4 &= 0 \\ x &= 2 \\ y &= 2 \end{aligned}$$

We then solve the system involving u and v. From the first equation of this system, we set

$$v = u + 1$$

Substituting in the second equation, we find

$$
\begin{aligned}
u(u+1) &= 12 \\
u^2 + u - 12 &= 0 \\
u &= 3, -4 \\
v &= 4, -3
\end{aligned}
$$

V. Transpose of a matrix: Given a matrix $A_{m \times n} = [a_{ij}]$, where the subscripts i and j are such that $1 \leq i \leq m, 1 \leq j \leq n$, the transpose of A, denoted by A^T, is obtained from A by interchanging the rows and the columns of A. This means that A^T is obtained by writing the rows of A as columns for A^T, or by writing the columns of A as rows for A^T. Based on this interchanging of rows and columns of A, the dimension of the resulting matrix A^T is $n \times m$. The transpose concept will be explained by discussing the following example.

Example 12. Find A^T if A is given by

$$A = \begin{bmatrix} 1 & 2 & 3 \\ 6 & 5 & 4 \end{bmatrix}$$

Solution. We write the first row of A as the first column of A^T, and the second row of A as the second column of A^T, hence we obtain

$$A = \begin{bmatrix} 1 & 6 \\ 2 & 5 \\ 3 & 4 \end{bmatrix}$$

Notice that the dimension of A^T is 3×2, recalling that the matrix A has the dimension 2×3.

As stated before, if A has dimension $m \times n$, then the transpose matrix A^T has dimension $n \times m$. As a result, the transpose of a square matrix $A_{m \times m}$ has the same dimension $m \times m$.

Exercises 11.2

1. Given the matrices

$$A = \begin{bmatrix} 1 & 2 \\ 3 & -1 \end{bmatrix}, B = \begin{bmatrix} -1 & 1 \\ 3 & 2 \end{bmatrix}, C = \begin{bmatrix} 3 & 1 \\ 2 & 0 \end{bmatrix}$$

Find
(a) $A + B$ (b) $B - C$ (c) $2A + 3B$ (d) $A - B - C$

2. Given the matrices

$$A = \begin{bmatrix} 1 & 2 & -1 \\ 1 & 3 & -2 \end{bmatrix}, B = \begin{bmatrix} 1 & 3 & -2 \\ -1 & 2 & 4 \end{bmatrix}, C = \begin{bmatrix} 0 & 4 & 1 \\ 2 & 1 & 5 \end{bmatrix}$$

Find
(a) $A + B$ (b) $C - B$ (c) $2A - 3B$ (d) $A + B - C$

3. Given the matrices

$$A = \begin{bmatrix} 1 & 2 \\ 2 & 3 \\ 3 & 4 \end{bmatrix}, B = \begin{bmatrix} 3 & 6 \\ 0 & -3 \\ 6 & -6 \end{bmatrix}$$

Find
(a) $2A + \frac{1}{3}B$ (b) $A - \frac{1}{3}B$ (c) $A^T + B^T$ (d) $A^T - B^T$

4. Given the matrices

$$A = \begin{bmatrix} 1 & 2 \\ -2 & 1 \end{bmatrix}, B = \begin{bmatrix} 3 & 4 \\ -5 & 6 \end{bmatrix}$$

Find
(a) $A + B$ (b) $A - B$ (c) $A^T + B^T$ (d) $A^T - B^T$

5. Use Exercise 4 to show that

(a) $(A^T)^T = A$ (b) $(A + B)^T = A^T + B^T$

(c) $(A - B)^T = A^T - B^T$ (d) $(3A)^T = 3(A)^T$

6. Find x, y, u and v if

$$\begin{bmatrix} x+2 & u+1 \\ y-2 & v-9 \end{bmatrix} = \begin{bmatrix} 4 & 4 \\ -4 & 4 \end{bmatrix}$$

7. Find x, y, u and v if

$$\begin{bmatrix} x+1 & 5 & u-6 \\ 2 & v+5 & 1 \end{bmatrix} = \begin{bmatrix} 4 & y+1 & -1 \\ x-1 & 0 & 1 \end{bmatrix}$$

8. Find x, y, u and v if

$$\begin{bmatrix} 2x+1 & 2u+3 \\ 3y-6 & 2v+8 \end{bmatrix} = \begin{bmatrix} 5 & -1 \\ 6 & 0 \end{bmatrix}$$

9. Find a, b, c and d if

$$\begin{bmatrix} 2a+b & 2c+d \\ c-2d & a-2b \end{bmatrix} = \begin{bmatrix} 16 & 22 \\ -9 & -7 \end{bmatrix}$$

10. Find a, b, c and d if

$$\begin{bmatrix} a+b & c-d \\ c+d & a-b \end{bmatrix} = 3\begin{bmatrix} 3 & -1 \\ 3 & 1 \end{bmatrix} + 2\begin{bmatrix} 1 & 1 \\ 3 & -2 \end{bmatrix}$$

11. Find a, b, c and d if

$$\begin{bmatrix} a+b & c+d \\ cd & ab \end{bmatrix} = \begin{bmatrix} 15 & 16 \\ 30 & 20 \end{bmatrix} - \begin{bmatrix} 4 & 1 \\ -26 & -10 \end{bmatrix}$$

12. Find a, b, c and d if

$$\begin{bmatrix} a+b & a^2+b^2 \\ c+d & c^2+d^2 \end{bmatrix} = \begin{bmatrix} 3 & 5 \\ 5 & 13 \end{bmatrix}$$

13. Find x, y, u and v if

$$\begin{bmatrix} 3x-1 & 3v-5 \\ 2y+1 & 3u+2 \end{bmatrix} = \begin{bmatrix} 9 & 8 \\ 12 & 11 \end{bmatrix}^T - \begin{bmatrix} -2 & -5 \\ 2 & 0 \end{bmatrix}^T$$

14. Find x, y, u and v if

$$\begin{bmatrix} x+y & u+v \\ uv-9 & x^2-y \end{bmatrix} = \begin{bmatrix} 6 & 8 \\ 3 & 8 \end{bmatrix}^T + \frac{1}{3}\begin{bmatrix} 9 & 9 \\ 18 & 9 \end{bmatrix}^T$$

15. Find x, y, u and v if

$$\begin{bmatrix} x+2y & uv \\ u+2v & xy \end{bmatrix} = 2\begin{bmatrix} 2 & 4 \\ 7 & 1 \end{bmatrix}^T - \begin{bmatrix} -1 & -3 \\ 2 & 0 \end{bmatrix}^T$$

16. Find x, y, u and v if

$$\begin{bmatrix} 2x+3y & 30 \\ 2u+3v & 6 \end{bmatrix} = 3\begin{bmatrix} 4 & 8 \\ 0 & 0 \end{bmatrix}^T + \begin{bmatrix} 1 & 4 \\ uv & xy \end{bmatrix}^T$$

17. Find x, y, u and v if

$$\begin{bmatrix} 1 & 2 & x+y \\ 0 & 4 & y+v \\ 0 & 0 & 2v+x \end{bmatrix} + \begin{bmatrix} 1 & 0 & 0 \\ u-x & 4 & 0 \\ 3 & 6 & 9 \end{bmatrix} = \begin{bmatrix} 2 & 2 & 3 \\ 2 & 8 & 6 \\ 3 & 6 & 18 \end{bmatrix}^T$$

18. Find x, y, u and v if

$$\begin{bmatrix} x^2-y^2 & u^2-v^2 & x-u \\ x+y & u+v & y-v \end{bmatrix} = \begin{bmatrix} 5 & 5 & 5 \\ 5 & 5 & 5 \end{bmatrix} + \begin{bmatrix} 2 & 2 \\ -2 & -2 \\ -3 & -3 \end{bmatrix}^T$$

11.3   Matrix Multiplication

We have discussed in the previous section how we can multiply a matrix by a scalar. In this section we will study the operation of multiplying two matrices A and B. It is important to note that a necessary condition is required to perform the matrix multiplication. To multiply any two matrices A and B, the number of columns of A must be equal to the number of rows of B. In other words, if given $A_{m \times n}$ and $B_{k \times r}$, with dimensions as shown by the subscripts, then n must be equal to k for the product AB to be defined. If $n \neq k$, then the multiplication operation cannot be carried out, and the multiplication is not defined in this case.

In addition, the dimension of the resulting matrix will be $m \times r$. Given $n = k$, the multiplication process is expressed by

$$A_{m \times n} B_{k \times r} = C_{m \times r}$$

Consequently, the matrix $C_{m \times r}$ has the following representation

$$C = \begin{bmatrix} c_{11} & c_{12} & c_{13} & \cdots & c_{1r} \\ c_{21} & c_{22} & c_{23} & \cdots & c_{2r} \\ c_{31} & c_{32} & c_{33} & \cdots & c_{3r} \\ \vdots & \vdots & \vdots & \vdots & \vdots \\ c_{m1} & c_{m2} & c_{m3} & \cdots & c_{mr} \end{bmatrix}$$

To determine the entries $c_{ij}, 1 \leq i \leq m, 1 \leq j \leq r$ of the matrix C, we use the definition of these entries given by:

c_{ij} is obtained by multiplying entries of the ith row of A by the entries of the jth column of B and by adding the resulting products. This means, for example, that c_{12} is obtained by multiplying entries of the first row of A by entries of the second column of B and by adding the resulting products. In a similar manner, c_{21} is obtained by multiplying entries of the second row of A by entries of the first column of B and by adding the resulting products.

The matrix multiplication, the condition needed to perform the multiplication and other related operations will be explained by the following illustrative examples.

Example 1. Check if each of the following matrix multiplication can be performed or not. Find the dimension of the resulting matrix for the multiplications that are defined.

(a) $A_{3\times2}\,B_{2\times5}$

(b) $A_{3\times2}\,A_{3\times2}$

(c) $A_{3\times2}\,(A_{3\times2})^T$

(d) $C_{3\times5}\,D_{5\times3}$

(e) $D_{5\times3}\,C_{3\times5}$

(f) $K_{5\times3}(G_{5\times3})^T$

Solution.

(a) Number of columns of A is equal to number of rows of B, hence

$$A_{3\times2}\,B_{2\times5} = F_{3\times5}$$

(b) The multiplication here is not defined.

(c) Since $A_{3\times2}\,(A_{3\times2})^T = A_{3\times2}\,A_{2\times3}^T$, the multiplication is defined, hence

$$A_{3\times2}\,A_{2\times3}^T = G_{3\times3}$$

(d) The multiplication $C_{3\times5}\,D_{5\times3}$ is defined, hence

$$C_{3\times5}\,D_{5\times3} = H_{3\times3}$$

(e) The multiplication $D_{5\times3}\,C_{3\times5}$ is defined, hence

$$D_{5\times3}\,C_{3\times5} = K_{5\times5}$$

(f) $K_{5\times3}$, G has dimension 5×3, hence, $G_{3\times5}^T$, hence

$$K_{5\times3}G_{3\times5}^T = H_{5\times5}$$

Example 2. Given the matrices

$$A = \begin{bmatrix} 1 & 2 \\ 3 & 4 \end{bmatrix},\ B = \begin{bmatrix} 5 & 6 \\ 7 & 8 \end{bmatrix}.$$

Find AB and BA

Solution.

(a) $AB = A_{2\times 2} B_{2\times 2} = C_{2\times 2} = \begin{bmatrix} c_{11} & c_{12} \\ c_{21} & c_{22} \end{bmatrix}$

where the entries are determined by

$$c_{11} = \begin{bmatrix} 1 & 2 \end{bmatrix} \begin{bmatrix} 5 \\ 7 \end{bmatrix} = 1(5) + 2(7) = 19$$

$$c_{12} = \begin{bmatrix} 1 & 2 \end{bmatrix} \begin{bmatrix} 6 \\ 8 \end{bmatrix} = 1(6) + 2(8) = 22$$

$$c_{21} = \begin{bmatrix} 3 & 4 \end{bmatrix} \begin{bmatrix} 5 \\ 7 \end{bmatrix} = 3(5) + 4(7) = 43$$

$$c_{22} = \begin{bmatrix} 3 & 4 \end{bmatrix} \begin{bmatrix} 6 \\ 8 \end{bmatrix} = 3(6) + 4(8) = 50$$

Thus we have

$$C = \begin{bmatrix} 19 & 22 \\ 43 & 50 \end{bmatrix}$$

(b) $B = B_{2\times 2} A_{2\times 2} = D_{2\times 2} = \begin{bmatrix} d_{11} & d_{12} \\ d_{21} & d_{22} \end{bmatrix}$

where the entries are determined by

$$d_{11} = \begin{bmatrix} 5 & 6 \end{bmatrix} \begin{bmatrix} 1 \\ 3 \end{bmatrix} = 5(1) + 6(3) = 23$$

$$d_{12} = \begin{bmatrix} 5 & 6 \end{bmatrix} \begin{bmatrix} 2 \\ 4 \end{bmatrix} = 5(2) + 6(4) = 34$$

$$d_{21} = \begin{bmatrix} 7 & 8 \end{bmatrix} \begin{bmatrix} 1 \\ 3 \end{bmatrix} = 7(1) + 8(3) = 31$$

$$d_{22} = \begin{bmatrix} 7 & 8 \end{bmatrix} \begin{bmatrix} 2 \\ 4 \end{bmatrix} = 7(2) + 8(4) = 46$$

Thus we have

$$D = \begin{bmatrix} 23 & 34 \\ 31 & 46 \end{bmatrix}$$

An important conclusion can be made here in that the matrix multiplication is in general not commutative. In other words, $AB \neq BA$. Sometimes, we can find specific matrices A and B where the products AB and BA are defined and $AB = BA$ as well (see exercise 14). However, this will not affect the general rule that $AB \neq BA$.

Further, it is possible sometimes to determine the product AB, whereas the product BA does not even exist. This can be easily seen by considering the matrices $A_{3 \times 2}$ and $B_{2 \times 5}$. It is clear in this example that the product AB is defined, whereas the product BA is not defined.

Recall that in algebra, the product of any two variables is commutative, where $xy = yx$. In matrices, $AB \neq BA$ in general.

Example 3. Given the matrices

$$A = \begin{bmatrix} 1 & 2 & 3 \\ 0 & 1 & 4 \end{bmatrix}, \, B = \begin{bmatrix} 1 & 2 \\ 2 & 3 \\ 3 & 4 \end{bmatrix}$$

Find AB and BA if exist.

Solution.

(a) $AB = A_{2 \times 3} B_{3 \times 2} = C_{2 \times 2} = \begin{bmatrix} c_{11} & c_{12} \\ c_{21} & c_{22} \end{bmatrix}$

where the entries are determined by

$$c_{11} = \begin{bmatrix} 1 & 2 & 3 \end{bmatrix} \begin{bmatrix} 1 \\ 2 \\ 3 \end{bmatrix} = 1(1) + 2(2) + 3(3) = 14$$

$$c_{12} = \begin{bmatrix} 1 & 2 & 3 \end{bmatrix} \begin{bmatrix} 2 \\ 3 \\ 4 \end{bmatrix} = 1(2) + 2(3) + 3(4) = 20$$

$$c_{21} = \begin{bmatrix} 0 & 1 & 4 \end{bmatrix} \begin{bmatrix} 1 \\ 2 \\ 3 \end{bmatrix} = 0(1) + 1(2) + 4(3) = 14$$

$$c_{22} = \begin{bmatrix} 0 & 1 & 4 \end{bmatrix} \begin{bmatrix} 2 \\ 3 \\ 4 \end{bmatrix} = 0(2) + 1(3) + 4(4) = 19$$

Thus we have

$$C = \begin{bmatrix} 14 & 20 \\ 14 & 19 \end{bmatrix}$$

(b) $BA = B_{3\times2} A_{2\times3} = D_{3\times3} = \begin{bmatrix} d_{11} & d_{12} & d_{13} \\ d_{21} & d_{22} & d_{23} \\ d_{31} & d_{32} & d_{33} \end{bmatrix}$

where the entries of the matrix D are determined as follows

$$d_{11} = 1(1) + 2(0) = 1, \ d_{12} = 1(2) + 2(1) = 4, \ d_{13} = 1(3) + 2(4) = 11$$

$$d_{21} = 2(1) + 3(0) = 2, \ d_{22} = 2(2) + 3(1) = 7, \ d_{23} = 2(3) + 3(4) = 18$$

$$d_{31} = 3(1) + 4(0) = 3, \ d_{32} = 3(2) + 4(1) = 10, \ d_{33} = 3(3) + 4(4) = 25$$

The entries d_{ij} are obtained in a similar manner to that used above in determining the entries of the matrix C. Thus we have

$$D = \begin{bmatrix} 1 & 4 & 11 \\ 2 & 7 & 18 \\ 3 & 10 & 25 \end{bmatrix}$$

It is clear from this example that both products AB and BA exist, but $AB \neq BA$ as stated above. Besides, notice that the product AB produces a 2×2 matrix, whereas the product BA gives a 3×3 matrix.

Examples 2 and 3 that we discussed above confirm the fact that, although the products AB and BA exist, but $AB \neq BA$ in general.

Example 4. Find x, y, u and v if

$$\begin{bmatrix} 1 & x \\ 3 & y \end{bmatrix} \begin{bmatrix} 4 & -1 \\ -2 & 3 \end{bmatrix}^T = \begin{bmatrix} 6 & 0 \\ 4 & -2 \end{bmatrix} - 4 \begin{bmatrix} 1 & u-5 \\ -1 & v-7 \end{bmatrix}$$

Solution. The left side of the equation contains a transpose and a multiplication processes. The right side contains a subtraction and a multiplication by a scalar. We next apply the matrix multiplication at the left hand side and the subtraction of two matrices at the right hand side to obtain

$$\begin{bmatrix} 1 & x \\ 3 & y \end{bmatrix} \begin{bmatrix} 4 & -2 \\ -1 & 3 \end{bmatrix} = \begin{bmatrix} 6 & 0 \\ 4 & -2 \end{bmatrix} - \begin{bmatrix} 4 & 4u-20 \\ -4 & 4v-28 \end{bmatrix}$$

Applying the matrix multiplication and the matrix subtraction gives

$$\begin{bmatrix} 4-x & -2+3x \\ 12-y & -6+3y \end{bmatrix} = \begin{bmatrix} 2 & 20-4u \\ 8 & 26-4v \end{bmatrix}$$

Equating corresponding entries in both sides, the variables x, y, u, and v can be easily obtained. Therefore, we set

$$4 - x = 2$$
$$x = 2$$

$$
\begin{aligned}
-2 + 3x &= 20 - 4u \\
u &= 4 \\
12 - y &= 8 \\
y &= 4 \\
-6 + 3y &= 26 - 4v \\
v &= 5
\end{aligned}
$$

11.3.1 Properties of Matrix Multiplication

If the matrices A, B and C are of appropriate dimensions, then

(i) $AB \neq BA$ in general.

(ii) $A(BC) = (AB)C$.

(iii) $A(B + C) = AB + AC$.

(iv) $(A + B)C = AC + BC$.

(v) $A^2 = A \cdot A, A^3 = A \cdot A \cdot A, \quad A$ is a square matrix.

(vi) $A^n = \underbrace{A \cdot A \cdot A \cdots A}_{n\,\text{factors}}, \quad A$ is a square matrix.

(vii) $(AB)^T = B^T A^T$.

(viii) $(A^T)^T = A$.

(ix) $(kA)^T = kA^T$, where k is a scalar.

(x) $A_{m \times m} I_m = I_m A_{m \times m}$, where I_m is the identity matrix

$$
I_m = \begin{bmatrix}
1 & 0 & 0 & \cdots & 0 \\
0 & 1 & 0 & \cdots & 0 \\
0 & 0 & 1 & \cdots & 0 \\
& & \vdots & & \\
0 & 0 & 0 & \cdots & 1
\end{bmatrix}
$$

The identity matrix I_m is the square matrix that all entries of the main diagonal are 1 for each and 0 otherwise.

Example 5. Find x, y, u and v given that

$$AA^T = \begin{bmatrix} xy & 2u - 3v \\ y - x & uv \end{bmatrix}$$

where

$$A = \begin{bmatrix} 1 & 1 & 0 \\ 0 & 1 & 1 \end{bmatrix}$$

Solution. It is clear that

$$A^T = \begin{bmatrix} 1 & 0 \\ 1 & 1 \\ 0 & 1 \end{bmatrix}$$

Therefore we find

$$AA^T = \begin{bmatrix} 2 & 1 \\ 1 & 2 \end{bmatrix} = \begin{bmatrix} xy & 2u - 3v \\ y - x & uv \end{bmatrix}$$

Equating the corresponding entries from both sides we find

$$y - x = 1, \; xy = 2$$

and

$$2u - 3v = 1, \; uv = 2$$

Solving these two systems we find

$$x = 1, -2, \; y = 2, -1, \; u = 2, -\frac{3}{2}, \; v = 1, -\frac{4}{3}$$

Example 6. Find x, y, u and v given that

$$A^T A = \begin{bmatrix} x & y & 0 \\ x & u & x \\ 0 & y & v \end{bmatrix}$$

where A is the matrix defined in Example 5.

Solution. Both matrices A and A^T are defined above, therefore we find

$$A^T A = \begin{bmatrix} 1 & 1 & 0 \\ 1 & 2 & 1 \\ 0 & 1 & 1 \end{bmatrix} = \begin{bmatrix} x & y & 0 \\ x & u & x \\ 0 & y & v \end{bmatrix}$$

Equating the corresponding entries from both sides we find

$$x = 1, \, y = 1, \, u = 2, \, , \, v = 1$$

Example 7. Given the matrices

$$A = \begin{bmatrix} 3 & 4 \\ 6 & 2 \end{bmatrix}, \, B = \begin{bmatrix} 5x & y \\ 6 & 4 \end{bmatrix}$$

Find x and y if $AB = BA$

Solution. Evaluating AB and BA and equating the results we obtain

$$\begin{bmatrix} 60x + 96 & 12y + 64 \\ 120x + 48 & 24y + 32 \end{bmatrix} = \begin{bmatrix} 60x + 24y & 80x + 8y \\ 168 & 128 \end{bmatrix}$$

Notice that we have two unknowns only x and y. Hence, we equate two corresponding entries from both sides to obtain

$$60x + 96 = 60x + 24y, \, 120x + 48 = 168$$

This in turn gives
$$x = 1, \, y = 4$$

Exercises 11.3

1. Check if each of the following multiplication can be performed or not. Find the dimension of the resulting matrix for the products that are defined.

(a) $A_{2 \times 5} \, B_{5 \times 3}$

(b)$C_{3\times4}\, D_{4\times6}$

(c) $D_{4\times6}\, C_{3\times4}$

(d) $A_{2\times5}\, A_{2\times5}$

(e) $A_{2\times5}\, (A_{2\times5})^T$

In Exercises 2 – 5, find AB and BA

2. $A = \begin{bmatrix} 4 & -2 \\ 3 & 1 \end{bmatrix}$, $B \begin{bmatrix} 1 & 2 \\ -3 & -2 \end{bmatrix}$

3. $A = \begin{bmatrix} 3 & 2 & 1 \\ 0 & 2 & 3 \end{bmatrix}$, $B \begin{bmatrix} 2 & 3 \\ 3 & 2 \\ 1 & -1 \end{bmatrix}$

4. $A = \begin{bmatrix} 5 & 3 \\ 2 & 3 \end{bmatrix}$, $B \begin{bmatrix} 2 & -3 \\ -3 & 5 \end{bmatrix}$

5. $A = \begin{bmatrix} 1 & 2 & 1 \\ 0 & 1 & -1 \\ 2 & 0 & 1 \end{bmatrix}$, $B \begin{bmatrix} 1 & 0 & -1 \\ 1 & 0 & 1 \\ 1 & 2 & 0 \end{bmatrix}$

6. Find AA^T if

$$A = \begin{bmatrix} 1 & -1 & 0 \\ -1 & 2 & -1 \end{bmatrix}$$

7. Find $A^T A$ if

$$A = \begin{bmatrix} 1 & 2 \\ 2 & 1 \\ 3 & -1 \end{bmatrix}$$

8. Find AB and BA for the following matrices:

$$A = \begin{bmatrix} 1 & 2 & 3 \end{bmatrix}, B = \begin{bmatrix} 3 \\ 4 \\ 5 \end{bmatrix}$$

9. Find AB and BA if given

$$A^T = \begin{bmatrix} 1 & -1 \\ -2 & 3 \end{bmatrix}, B^T = \begin{bmatrix} 2 & 3 \\ 3 & 5 \end{bmatrix}$$

Hint: Recall that $(A^T)^T = A$

10. Find $(AB)C$ and $A(BC)$ for the following matrices:

$$A = \begin{bmatrix} 1 & 2 \\ 3 & -1 \end{bmatrix}, B = \begin{bmatrix} 2 & 1 \\ 3 & 2 \end{bmatrix}, C = \begin{bmatrix} -1 & 2 \\ 1 & -2 \end{bmatrix}$$

11. Find $x, y, u,$ and v if

$$\begin{bmatrix} 1 & x \\ 2 & u \end{bmatrix} \begin{bmatrix} y & -1 \\ v & 1 \end{bmatrix} = \begin{bmatrix} 1 & 0 \\ 0 & 1 \end{bmatrix}$$

12. Find $x, y, u,$ and v if

$$\begin{bmatrix} x-1 & y-3 \\ u+1 & v-1 \end{bmatrix}^T = \begin{bmatrix} 1 & 2 & -1 \\ -1 & 1 & 0 \end{bmatrix} \begin{bmatrix} 1 & 2 & -1 \\ -1 & 1 & 0 \end{bmatrix}^T$$

13. Find $a, b, c,$ and d if

$$\begin{bmatrix} a-2 & b-3 \\ c-4 & d-5 \end{bmatrix}^T = \begin{bmatrix} 7 & 5 \\ 4 & 3 \end{bmatrix} \begin{bmatrix} 3 & -5 \\ -4 & 7 \end{bmatrix}$$

14. Given the matrices

$$A = \begin{bmatrix} 2 & 3 \\ 4 & 1 \end{bmatrix}, B = \begin{bmatrix} x & 3 \\ y & 2 \end{bmatrix}$$

Find x and y if $AB = BA$

15. Find $x, y, u,$ and v if

$$\begin{bmatrix} 2x + 13 & 6y - 5 \\ 3u - 6 & 4v - 17 \end{bmatrix} - \begin{bmatrix} 1 & -2 \\ 0 & -2 \end{bmatrix} = \begin{bmatrix} 1 & 3 \\ 3 & -4 \end{bmatrix}^T \begin{bmatrix} 5 & 3 \\ 3 & 2 \end{bmatrix}$$

16. Find $x, y, u,$ and v if

$$\begin{bmatrix} x + 4 & 6y + 2 \\ 3u - 7 & 2v + 1 \end{bmatrix} + 6 \begin{bmatrix} 1 & -2 \\ 4 & 3 \end{bmatrix} = \begin{bmatrix} 1 & -2 \\ 3 & -4 \end{bmatrix}^T \begin{bmatrix} 5 & 2 \\ 2 & 1 \end{bmatrix}$$

17. Find $x, y, u,$ and v if

$$\begin{bmatrix} x + 2y & 2x - y \\ 2u + v & u - 2v \end{bmatrix} = \begin{bmatrix} 5 & 16 \\ 0 & -7 \end{bmatrix}^T \begin{bmatrix} 1 & 0 \\ 0 & 1 \end{bmatrix}$$

18. Find $x, y, u,$ and v if

$$\begin{bmatrix} 2x + 14 & 6y - 6 \\ 3u + 3 & 4v - 16 \end{bmatrix} - \frac{1}{2} \begin{bmatrix} 2 & -2 \\ 0 & -4 \end{bmatrix} = \begin{bmatrix} 10 & -3 \\ -3 & 1 \end{bmatrix} \begin{bmatrix} 2 & -3 \\ 1 & 5 \end{bmatrix}^T$$

19. Find $x, y, u,$ and v if

$$\begin{bmatrix} 6y & u + v \\ u^2 & x^2 + y \end{bmatrix} = \begin{bmatrix} 1 & 2 \\ 3 & 4 \end{bmatrix} \begin{bmatrix} 4 & 1 \\ 3 & 2 \end{bmatrix}^T$$

20. Find $x, y, u,$ and v if

$$\begin{bmatrix} x^2 + y & 2u - v \\ u^2 + v & y - 2x \end{bmatrix} = \begin{bmatrix} -1 & -3 \\ 2 & 0 \end{bmatrix}^T \begin{bmatrix} -2 & 1 \\ 5 & -1 \end{bmatrix}$$

21. Given the matrices

$$A = \begin{bmatrix} 3 & 4 \\ 6 & 2 \end{bmatrix}, B = \begin{bmatrix} x & y \\ 3 & 2 \end{bmatrix}$$

Find x and y if $AB = BA$

22. Given the matrices

$$A = \begin{bmatrix} 4 & 6 \\ 3 & 3 \end{bmatrix}, B = \begin{bmatrix} x & 6 \\ y & 2 \end{bmatrix}$$

Find x and y if $AB = BA$

23. Given the matrices

$$A = \begin{bmatrix} 3 & -2 \\ x & y \end{bmatrix}$$

Find x and y if $AA^T = A^T A$

22. Given the matrices

$$A = \begin{bmatrix} 2 & -1 & 0 \\ -1 & 0 & 2 \end{bmatrix}$$

Find $x, y, u,$ and v if

$$AA^T = \begin{bmatrix} x+y & 2u-2v \\ 2x-y & u+v \end{bmatrix}$$

23. Find $x, y, u,$ and v if

$$A^T A = \begin{bmatrix} x+y & -v & -2 \\ 2x-y & 1 & 2u-3v \\ 2x-y & 0 & u+x \end{bmatrix}$$

where A is the matrix defined in Exercise 22

24. Given

$$A = \begin{bmatrix} 1 & 2 \\ 3 & 1 \end{bmatrix}$$

Find $A^2 - A - 6I_2$

25. Given

$$A = \begin{bmatrix} \cos x & \sin x \\ -\sin x & \cos x \end{bmatrix}$$

Show that $AA^T = A^T A = I_2$

26. Given

$$A = \begin{bmatrix} 1 & 0 & 0 \\ 0 & \cos x & \sin x \\ 0 & -\sin x & \cos x \end{bmatrix}$$

Show that $AA^T = A^T A = I_3$

27. Given

$$A = \begin{bmatrix} 1 & 0 & 0 \\ 0 & \frac{1}{\sqrt{a}} & -\frac{1}{\sqrt{a}} \\ 0 & -\frac{1}{\sqrt{a}} & -\frac{1}{\sqrt{a}} \end{bmatrix}$$

Show that $AA^T = A^T A = I_3$ only if $a = 2$

28. Given

$$A = \begin{bmatrix} 2 & 3 & 4 \\ 4 & 3 & 2 \end{bmatrix}$$

Find x, y, u and v if

$$AA^T = \begin{bmatrix} x+y & x-y \\ u-v & u+v \end{bmatrix} + 2 \begin{bmatrix} 2 & 16 \\ 18 & -16 \end{bmatrix}$$

29. Given

$$A = \begin{bmatrix} 1 & -1 & 1 \\ -1 & 1 & 1 \end{bmatrix}$$

Find x, y, u and v if

$$6AA^T = \begin{bmatrix} 2x & 3y - 4 \\ 2u + 1 & v + 1 \end{bmatrix} + \begin{bmatrix} 8 & -20 \\ -25 & 7 \end{bmatrix}$$

30. Given

$$A = \begin{bmatrix} 2 & -2 & 1 \\ 1 & -2 & 2 \end{bmatrix}$$

Find x, y, u and v if

$$AA^T = \begin{bmatrix} x + y & uv \\ xy & u + v \end{bmatrix}$$

11.4   Solving Equations by Matrices

In this section we will study solving systems of equations by using matrices. The method that will be used is called the **Gauss-Jordan reduction method**. Our study will be focused on system of two equations with two unknowns x and y. Systems of more than two equations with more than two unknowns may be solved by this method in a like manner.

However, another method that employs the inverse of a matrix, denoted by A^{-1}, will be introduced in the coming chapter.

Gauss-Jordan Reduction Method

We can write the system of equations in the form of matrix products as follows

$$AX = B$$

where A is the matrix of the coefficients, X is the matrix of the unknowns, and B is the matrix of the constants at the right side. For simplicity reasons, we use a system of two equations defined by

$$a_{11}x + a_{12}y = b_{11}$$
$$a_{21}x + a_{22}y = b_{21}$$

The matrices A, X and B are given by

$$A = \begin{bmatrix} a_{11} & a_{12} \\ a_{21} & a_{22} \end{bmatrix}, \quad X = \begin{bmatrix} x \\ y \end{bmatrix}, \quad B = \begin{bmatrix} b_{11} \\ b_{21} \end{bmatrix}$$

We first adjoin the matrix B to the matrix A to form the **augmented matrix** defined by

$$\begin{bmatrix} a_{11} & a_{12} & \vdots & b_{11} \\ a_{21} & a_{22} & \vdots & b_{21} \end{bmatrix}$$

The augmented matrix should be transformed to the form

$$\begin{bmatrix} I_2 & \vdots & C \end{bmatrix}$$

or equivalently

$$\begin{bmatrix} 1 & 0 & \vdots & c_1 \\ 0 & 1 & \vdots & c_2 \end{bmatrix}$$

where c_1 and c_2 are constants. The transformation can be performed by using the **elementary row operations** that can be explained as follows:

(i) The first nonzero entry of the first row should be 1. This can be achieved by dividing the first row by the first nonzero entry to make it 1, or by interchanging two rows.

(ii) The first entry of the second row should be converted to 0. This can be done by multiplying the first row by a constant and subtracting all entries of the first row from the corresponding entries of the second row.

(iii) The second entry of the second row should become 1 by dividing the second row by that entry.

(iv) The second row should be used to make the second entry of the first row 0.

(v) Having obtained the reduced form as shown above, the solution of the system is therefore $x = c_1, y = c_2$.

We point out that the Gauss-Jordan method can be used for any system of more than two equations.

Example 1. Consider the system

$$x + y = 5$$
$$2x + 3y = 14$$

Solution. This system can be written in the form of matrix product as follows:

$$AX = B$$

where A is the matrix of the coefficients, X is the matrix of the unknowns, and B is the matrix of the constants at the right side. The matrices A, X and B are given by

$$A = \begin{bmatrix} 1 & 1 \\ 2 & 3 \end{bmatrix}, \quad X = \begin{bmatrix} x \\ y \end{bmatrix}, \quad B = \begin{bmatrix} 5 \\ 14 \end{bmatrix}$$

We first adjoin the matrix B to the matrix A to form the **augmented matrix** defined by

$$\begin{bmatrix} 1 & 1 & \vdots & 5 \\ 2 & 3 & \vdots & 14 \end{bmatrix}$$

The augmented matrix should be transformed to the form

$$\begin{bmatrix} 1 & 0 & \vdots & c_1 \\ 0 & 1 & \vdots & c_2 \end{bmatrix}$$

where c_1 and c_2 are constants. Using the Gauss-Jordan reduction method, we observe that the first entry of the first row in the augmented matrix is 1. Multiplying the first row by -2 and adding all resulting entries of the first row to the second we find

$$\begin{bmatrix} 1 & 1 & \vdots & 5 \\ 0 & 1 & \vdots & 4 \end{bmatrix}$$

We next subtract the first row from the second to get

$$\begin{bmatrix} 1 & 0 & \vdots & 1 \\ 0 & 1 & \vdots & 4 \end{bmatrix}$$

This gives $x = 1, y = 4$.

Example 2. Consider the system

$$\begin{aligned} 2x + y &= 9 \\ 3x - y &= 26 \end{aligned}$$

Solution. We first adjoin the matrix B to the matrix A to form the **augmented matrix** to find

$$\begin{bmatrix} 2 & 1 & \vdots & 9 \\ 3 & -1 & \vdots & 26 \end{bmatrix}$$

Dividing the first row by 2 we find

$$\begin{bmatrix} 1 & \frac{1}{2} & \vdots & \frac{9}{2} \\ 3 & -1 & \vdots & -26 \end{bmatrix}$$

Multiplying the first row by -3 and adding all resulting entries of the first row to the second we find

$$\begin{bmatrix} 1 & \frac{1}{2} & \vdots & \frac{9}{2} \\ 0 & -\frac{5}{2} & \vdots & \frac{25}{2} \end{bmatrix}$$

Dividing the second row by $-\frac{5}{2}$ we find

$$\begin{bmatrix} 1 & \frac{1}{2} & \vdots & \frac{9}{2} \\ 0 & 1 & \vdots & -5 \end{bmatrix}$$

We finally multiply the second row by $-\frac{1}{2}$ and add all resulting entries to the first row to get

$$\begin{bmatrix} 1 & 0 & \vdots & 7 \\ 0 & 1 & \vdots & -5 \end{bmatrix}$$

This gives $x = 7, y = -5$.

It is interesting to point out that other systems of more than two equations can be solved in a similar way. This will be left to a linear algebra course. However, solving systems of equations will be solved also by using the inverse of a matrix that will be presented in Chapter 12.

Exercises 11.4

Solve the following systems of equations by using the Gauss-Jordan reduction method:

1. $x + 5y = 16$
    $3x - 4y = -9$

2. $3x + y = 9$
    $4x - y = 5$

3. $5x - 2y = 7$
    $3x + 5y = 29$

4. $2x + y = 7$
    $3x - y = 28$

5. $x + y = 17$
    $2x - y = 7$

6. $2x + y = 35$
    $3x - y = 20$

Use the gauss-Jordan reduction method to solve the systems of equations in Exercises 7–20:

7. $x + y + z = 6$
    $3x - y + z = 4$
    $2x + y - z = 1$

8. $x + y + z = 6$
    $2x - y + z = 2$
    $x + y - 5z = 0$

9. $x + y + z = 6$
    $x + y - 5z = 0$
    $x - y + z = 2$

10. $x + y + z = 1$
    $x + y - z = 3$
    $x - y + z = -1$

11. $x + y + z = 8$
    $x - y + z = 2$
    $3x + y - 5z = 2$

12. $x + y + z = 12$
    $x + y - z = 2$
    $x - y + z = 4$

13. $x - 3y + 3z = -4$
    $2x + 3y - z = 15$
    $4x - 3y - z = 19$

14. $4x + 2y - 2z = 10$
    $2x + 8y + 4z = 32$
    $30x + 12y - 4z = 24$

15. $4x - 2y + 3z = 1$
    $x + 3y - 4z = -7$
    $3x + y + 2z = 5$

16. $x + y = 3$
    $y + z = 5$
    $x + z = 4$

17. $x - 3z = -5$
    $2x - y + 2z = 16$
    $7x - 3y - 5z = 19$

18. $x - y + z = 10$
    $3x + y + 2z = 34$
    $-5x + 2y - z = -14$

19. $x - y + z = -1$
    $4x + y - 2z = 5$
    $3x + 3y + 4z = 16$

20. $6x - y + 3z = -9$
    $x + y - z = 4$
    $3x - y + 4z = -5$

21. $x + y + z = 6$
     $-x + 2y + 2z = 9$
     $2x - y + 3z = 9$

22. $x + 2y = 5$
     $y + 2z = 8$
     $x + 2z = 7$

Chapter 12

Determinants

12.1 Determinants

In this chapter we will study the concept of **determinants**. It is important to note first that determinants concept is related only to square matrices. The determinant of a square matrix A is denoted by $\det(A)$, or by $|A|$. Recall that a matrix A is a rectangular arrangement of numbers or letters written in rows and columns. However, $\det(A)$ is a real number or a scalar that will be determined. To evaluate the determinant of any square matrix, we consider the following 2×2 and 3×3 matrices:

1. We first discuss the determinant of a 2×2 matrix of the form

$$A = \begin{bmatrix} a & b \\ c & d \end{bmatrix}$$

where a, b, c, and d are real numbers. The determinant of A is defined by

$$\det(A) = \begin{vmatrix} a & b \\ c & d \end{vmatrix}$$
$$= ad - bc$$

obtained by multiplying the entries of the main diagonal and subtracting the product of the entries of the back diagonal. This rule is applicable only for 2×2 matrices.

The evaluation of a determinant of a 2×2 matrix is illustrated by the following examples.

Example 1. Evaluate $\det(A)$ where

$$A = \begin{bmatrix} 5 & 2 \\ -1 & 3 \end{bmatrix}$$

Solution.

$$\det(A) = \begin{vmatrix} 5 & 2 \\ -1 & 3 \end{vmatrix} = 5(3) - 2(-1) = 17$$

Example 2. Find x such that $\det(B) = 30$, where

$$B = \begin{bmatrix} x+1 & -2 \\ 3 & x-1 \end{bmatrix}$$

Solution.

$$\det(B) = \begin{vmatrix} x+1 & -2 \\ 3 & x-1 \end{vmatrix}$$
$$= (x+1)(x-1) - (-2)(3)$$
$$= x^2 - 1 + 6 = x^2 + 5$$

But we are given that $\det(B) = 30$, Therefore we set

$$x^2 + 5 = 30$$
$$(x+5)(x-5) = 0$$
$$x = \pm 5$$

Example 3. Find x such that $\det(C) = 0$, where

$$C = \begin{bmatrix} x & -2 \\ -1 & x-1 \end{bmatrix}$$

Solution.

$$\det(C) \;=\; \begin{vmatrix} x & -2 \\ -1 & x-1 \end{vmatrix} = 0$$

This gives

$$
\begin{aligned}
x^2 - x - 2 &= 0 \\
(x+1)(x-2) &= 0 \\
x &= -1, 2
\end{aligned}
$$

2. We now discuss the determinant of a 3×3 matrix of the form:

$$A = \begin{bmatrix} a_{11} & a_{12} & a_{13} \\ a_{21} & a_{22} & a_{23} \\ a_{31} & a_{32} & a_{33} \end{bmatrix}$$

The determinant of a 3×3 matrix can be evaluated by any of the following two methods:

(a) In the first method we decompose $\det(A)$ to 3 distinct 2×2 reduced determinants, called **minors**. Considering the entries of the first row, for example, we can express $\det(A)$ in the expression

$$
\begin{aligned}
|A| \;=&\; \begin{vmatrix} a_{11} & a_{12} & a_{13} \\ a_{21} & a_{22} & a_{23} \\ a_{31} & a_{32} & a_{33} \end{vmatrix} \\[2mm]
=&\; a_{11}\begin{vmatrix} a_{22} & a_{23} \\ a_{32} & a_{33} \end{vmatrix} - a_{12}\begin{vmatrix} a_{21} & a_{23} \\ a_{31} & a_{33} \end{vmatrix} + a_{13}\begin{vmatrix} a_{21} & a_{22} \\ a_{31} & a_{32} \end{vmatrix}
\end{aligned}
$$

where each reduced minor determinant is obtained by deleting the row and the column that pass through the entry that is a coefficient for that reduced determinant. For example, the minor associated with a_{11} is obtained by eliminating the first row and the first column that pass through the entry

a_{11}. Other minors can be obtained in the same manner. It is important to note that the coefficients a_{11}, a_{12}, and a_{13} of the first row are given alternating signs $+, -$, and $+$ respectively. It is now easy to evaluate the 2×2 determinants as discussed before.

It is important to note that this method can also be applied in a similar manner by using the entries of any row or any column. To use the entries of the second row a_{21}, a_{22}, and a_{23}, for example, we should also use the alternating signs $-, +$, and $-$ respectively. The pattern of the alternating signs is given by:

$$+ \quad - \quad +$$
$$- \quad + \quad -$$
$$+ \quad - \quad +$$

For simplicity reasons, we will evaluate the determinant in the coming example by using entries of the first row. This method of evaluation will be explained by the following example.

Example 4. Evaluate $\det(A)$ for the matrix

$$A = \begin{bmatrix} 1 & 2 & 3 \\ -1 & 1 & 2 \\ 3 & -2 & 4 \end{bmatrix}$$

Solution. Following the method discussed above we set

$$\det(A) = \begin{vmatrix} 1 & 2 & 3 \\ -1 & 1 & 2 \\ 3 & -2 & 4 \end{vmatrix}$$

$$= 1 \begin{vmatrix} 1 & 2 \\ -2 & 4 \end{vmatrix} - 2 \begin{vmatrix} -1 & 2 \\ 3 & 4 \end{vmatrix} + 3 \begin{vmatrix} -1 & 1 \\ 3 & -2 \end{vmatrix}$$

$$= 1(8) - 2(-10) + 3(-1)$$

$$= 25$$

(b) In the second method we rewrite the first two columns in addition to the columns of the determinant, hence we obtain

column 1   column 2   column 3   column 1   column 2

$$
\begin{array}{ccccc}
a_{11} & a_{12} & a_{13} & a_{11} & a_{12} \\
a_{21} & a_{22} & a_{23} & a_{21} & a_{22} \\
a_{31} & a_{32} & a_{33} & a_{31} & a_{32}
\end{array}
$$

It is easily observed that the new array has three lines from left to right, namely,

$$a_{11} \rightarrow a_{22} \rightarrow a_{33}$$

$$a_{12} \rightarrow a_{23} \rightarrow a_{31}$$

$$a_{13} \rightarrow a_{21} \rightarrow a_{32}$$

and three lines from right to left, namely,

$$a_{12} \rightarrow a_{21} \rightarrow a_{33}$$

$$a_{11} \rightarrow a_{23} \rightarrow a_{32}$$

$$a_{13} \rightarrow a_{22} \rightarrow a_{31}$$

Consequently, $\det(A)$ can be evaluated by adding the products of the entries of the three left to right lines, then by subtracting the products of the entries of the three right to left lines. In other words, $\det(A)$ can be expressed by

$$
\begin{aligned}
\det(A) \;=\; & (a_{11}a_{22}a_{33} + a_{12}a_{23}a_{31} + a_{13}a_{21}a_{32}) \\
& - (a_{12}a_{21}a_{33} + a_{11}a_{23}a_{32} + a_{13}a_{22}a_{31})
\end{aligned}
$$

In the following example, we will explain the evaluation of determinants by this method.

Example 5. Evaluate $\det(B)$ for the matrix

$$
B = \begin{bmatrix} 1 & 2 & 3 \\ 1 & 1 & 2 \\ 3 & -2 & 4 \end{bmatrix}
$$

Solution. Following the method discussed above, we rewrite the first two columns to obtain the array

column 1	column 2	column 3	column 1	column 2
1	2	3	1	2
1	1	2	1	1
3	-2	4	3	-2

Therefore, we find

$$
\begin{aligned}
|A| &= \{1(1)(4) + 2(2)(3) + 3(1)(-2)\} - \{2(1)(4) + 1(2)(-2) + 3(1)(3)\} \\
&= \{4 + 12 - 6\} - \{8 - 4 + 9\} = -3
\end{aligned}
$$

Two determinants of specific types of square matrices, called **upper triangular** and **lower triangular** matrices will be studied. The entries of these types of matrices are located **above** or **below** the main diagonal respectively, where all other entries are zeros. Examples of the upper and lower triangular matrices are given by

$$
A = \begin{bmatrix} 1 & 3 & 4 \\ 0 & 2 & 5 \\ 0 & 0 & 6 \end{bmatrix}
$$

and

$$
B = \begin{bmatrix} 1 & 0 & 0 & 0 \\ 2 & 4 & 0 & 0 \\ 1 & 2 & -3 & 0 \\ 4 & 3 & 1 & 2 \end{bmatrix}
$$

respectively.

It is important to note that these types of square matrices have interesting applications. In addition, the determinant of any matrix of this form can be easily obtained by multiplying the entries of the main diagonal only.

This can be easily proved by applying any of the two methods introduced before. For the matrices A and B given above, we have

$$\det(A) = (1)(2)(6) = 12$$
$$\det(B) = (1)(4)(-3)(2) = -24$$

In the following examples, we use this important technique for evaluating upper and lower triangular matrices.

Example 6. Find x if $\det(A) = 0$, where

$$A = \begin{bmatrix} x & 1 & 2 \\ 0 & x-1 & 3 \\ 0 & 0 & x+2 \end{bmatrix}$$

Solution. It is obvious that the matrix is an upper triangular matrix, hence by multiplying the entries of the main diagonal of the related determinant we find

$$\det(A) = x(x-1)(x+2)$$

Therefore, we set

$$x(x-1)(x+2) = 0$$
$$x = -2, 0, 1$$

Example 7. Find x if

$$\begin{vmatrix} x & 0 & 0 \\ 1 & x & 0 \\ 2 & 1 & 1 \end{vmatrix} = \begin{vmatrix} 1 & 1 & 1 \\ 0 & 1 & 2 \\ 0 & 0 & x+2 \end{vmatrix}$$

Solution. It is obvious that the determinant at the left hand side is a determinant of a lower triangular matrix and the determinant at the right hand side is a determinant of an upper triangular matrix, hence by multiplying the entries of the main diagonal of each determinant we find

$$x(x)(1) = 1(1)(x+2)$$
$$x^2 - x - 2 = 0$$
$$(x+1)(x-2) = 0$$
$$x = -1, 2$$

Example 8. Find x if

$$\begin{vmatrix} x & 0 & 0 \\ 0 & -1 & 0 \\ 0 & 1 & x \end{vmatrix} \begin{vmatrix} -1 & 1 & 2 \\ 0 & x & 3 \\ 0 & 0 & 1 \end{vmatrix} = \begin{vmatrix} 2x & 1 \\ 0 & x \end{vmatrix} + 3 \begin{vmatrix} 1 & 0 \\ 1 & x \end{vmatrix}$$

Solution. It is obvious that the left hand side contains determinants of a lower triangular matrix and an upper triangular matrix. The right hand side contains two determinants of 2×2 matrices. Hence, we set

$$\begin{aligned} \{x(-1)(x)\} \times \{(-1)(x)(1)\} &= (2x)(x) + 3(1)(x) \\ x^3 &= 2x^2 + 3x \\ x(x^2 - 2x - 3) &= 0 \\ x(x - 3)(x + 1) &= 0 \\ x &= -1, 0, 3 \end{aligned}$$

Exercises 12.1

1. Evaluate the determinant of each of the following 2×2 matrices:

(a) $A = \begin{bmatrix} 2 & -9 \\ 3 & 6 \end{bmatrix}$, (b) $B = \begin{bmatrix} -2 & 4 \\ -6 & 3 \end{bmatrix}$, (c) $C = \begin{bmatrix} 3 & -1 \\ -10 & -5 \end{bmatrix}$

2. Evaluate each of the following determinants:

(a) $A = \begin{vmatrix} 1 & 2 & -1 \\ 3 & 1 & 2 \\ 0 & -4 & 5 \end{vmatrix}$, (b) $B = \begin{vmatrix} -1 & -2 & 0 \\ 3 & 4 & 1 \\ 2 & 3 & 5 \end{vmatrix}$, (c) $C = \begin{vmatrix} 1 & 2 & 3 \\ -2 & 1 & 4 \\ -1 & 4 & 2 \end{vmatrix}$

3. Evaluate each of the following determinants:

(a) $A = \begin{vmatrix} 1 & 4 & 5 \\ 0 & -2 & 6 \\ 0 & 0 & -3 \end{vmatrix}$, (b) $B = \begin{vmatrix} -3 & 0 & 0 \\ 1 & -2 & 0 \\ 3 & 4 & -3 \end{vmatrix}$, (c) $C = \begin{vmatrix} 1 & 0 & 0 \\ 0 & 2 & 0 \\ 0 & 6 & 3 \end{vmatrix}$

4. Given the following matrices:

(i) $A = \begin{bmatrix} 1 & 2 \\ 3 & 1 \end{bmatrix}$, (ii) $B = \begin{bmatrix} -1 & -2 \\ 3 & 4 \end{bmatrix}$

Show for the square matrices A and B that $\det(A + B) \neq \det(A) + \det(B)$

5. Given that

$$\begin{vmatrix} x & u \\ v & y \end{vmatrix} = 4$$

Evaluate the determinants of the following matrices:

(a) $A = \begin{bmatrix} 2x & u \\ 2v & y \end{bmatrix}$, (b) $B = \begin{bmatrix} 2x & 2u \\ 2v & 2y \end{bmatrix}$, (c) $C = \begin{bmatrix} 2x & 2u \\ v & y \end{bmatrix}$

6. Find x if

$$\begin{vmatrix} x & 3 \\ 7 & x-4 \end{vmatrix} = 0$$

7. Find x if

$$\begin{vmatrix} x & x \\ 3 & x+1 \end{vmatrix} = -1$$

8. Find x if

$$\begin{vmatrix} x & x+1 \\ x-1 & x+3 \end{vmatrix} = 13$$

9. Find x if

$$\begin{vmatrix} 1 & 2 & 3 & 1 \\ 0 & x & 1 & 2 \\ 0 & 0 & x & 3 \\ 0 & 0 & 0 & 1 \end{vmatrix} = x+2$$

10. Find x if

$$\begin{vmatrix} 1 & -1 & 1 \\ -1 & x & 1 \\ 1 & -1 & -1 \end{vmatrix} = 0$$

11. Find x if

$$\begin{vmatrix} 1 & 2 & 3 \\ 0 & x & 1 \\ 0 & 0 & 2x \end{vmatrix} = 18$$

12. Find x if

$$\begin{vmatrix} 2 & 3 & -1 \\ 0 & x & 1 \\ 0 & 0 & 2x \end{vmatrix} + \begin{vmatrix} 1 & 0 & 0 \\ 2 & x & 0 \\ 3 & 1 & -1 \end{vmatrix} = 3$$

13. Find x if

$$\begin{vmatrix} 1 & 0 & 0 \\ 0 & x-1 & 0 \\ 0 & 0 & x+1 \end{vmatrix} \begin{vmatrix} 3 & 2 & 1 \\ 0 & 1 & 1 \\ 0 & 0 & 1 \end{vmatrix} = \begin{vmatrix} -3 & 1 \\ 0 & -3 \end{vmatrix}$$

14.

(a) Evaluate each of the following determinants:

(i) $\det(A) = \begin{vmatrix} 3 & 3 & -2 \\ 4 & 4 & 1 \\ 5 & 5 & 0 \end{vmatrix}$

(ii) $\det(B) = \begin{vmatrix} 1 & 2 & 3 \\ 1 & 2 & 3 \\ 4 & 5 & 6 \end{vmatrix}$

(b) State a conclusion from the answers you obtained above by investigating the first two columns of (i) and the first two rows of (ii)

(c) Use the conclusion you derived in (b) to evaluate the following 3×3 determinant

$$\begin{vmatrix} 1 & 1 & x \\ x & x & x^2 \\ x^2 & x^2 & 1 \end{vmatrix}$$

15. Find x if

$$\begin{vmatrix} x & 2 & 3 \\ 0 & x & 1 \\ 0 & 0 & x \end{vmatrix} \begin{vmatrix} 1 & 0 & 0 \\ 2 & x-1 & 0 \\ 3 & 4 & 1 \end{vmatrix} = \begin{vmatrix} x^3 & 2 \\ x^2 & x \end{vmatrix} - \begin{vmatrix} 3 & 2 \\ 0 & x \end{vmatrix}$$

16. Find x if

$$\begin{vmatrix} x & 2 & 3 \\ 0 & x & 1 \\ 0 & 0 & x \end{vmatrix} - \begin{vmatrix} 1 & 0 & 0 \\ 2 & x-1 & 0 \\ 3 & 4 & 1 \end{vmatrix} = \begin{vmatrix} x & -2 \\ x & 2x \end{vmatrix} + \begin{vmatrix} 1 & 2 \\ 0 & 1 \end{vmatrix}$$

17. Find $\det(A)$ if

$$A_{n \times n} = \begin{bmatrix} 1 & 0 & 0 & \cdots & 0 \\ 0 & 2 & 0 & \cdots & 0 \\ 0 & 0 & 3 & \cdots & 0 \\ & & \vdots & & \\ 0 & 0 & 0 & \cdots & n \end{bmatrix}$$

18. Find x if

$$\begin{vmatrix} x & 0 & 0 \\ 0 & -1 & 0 \\ 0 & 1 & x \end{vmatrix} \begin{vmatrix} -1 & 1 & 2 \\ 0 & x & 3 \\ 0 & 0 & 1 \end{vmatrix} = \begin{vmatrix} 2x & 1 \\ -1 & x \end{vmatrix} + \begin{vmatrix} 1 & 1 \\ 1 & -x \end{vmatrix}$$

19. Find x if

$$\begin{vmatrix} x & 2 & 3 \\ 0 & x & 1 \\ 0 & 0 & 3 \end{vmatrix} - \begin{vmatrix} 1 & 2 & 3 \\ 0 & x-1 & 1 \\ 0 & 0 & 1 \end{vmatrix} = \begin{vmatrix} x & -2 \\ x & 2x \end{vmatrix} + \begin{vmatrix} -1 & 2 \\ 0 & 1 \end{vmatrix}$$

20. Find x if

$$\begin{vmatrix} x & 1 & 1 \\ 0 & 1 & 1 \\ 0 & 0 & x \end{vmatrix} \div \begin{vmatrix} 2 & 1 & 3 \\ 0 & 1 & 2 \\ 0 & 0 & 3 \end{vmatrix} = \begin{vmatrix} x & 1 \\ 1 & 1 \end{vmatrix} + \begin{vmatrix} 1 & 2 \\ 0 & 1 \end{vmatrix}$$

12.2   Properties of Determinants

In this section, we will discuss a number of properties that are of interest in working with determinants. However, for simplicity reasons, some of these properties will be proved by using numerical examples only. To achieve our goal, we consider the square matrix A given by

$$A = \begin{bmatrix} 4 & 5 \\ 3 & 6 \end{bmatrix}$$

where

$$\det(A) = 9$$

The properties are:
1. If A is a square matrix and A^T is its transpose, then

$$\det(A^T) = \det(A)$$

This can be easily seen from

$$A^T = \begin{bmatrix} 4 & 3 \\ 5 & 6 \end{bmatrix}$$

that gives
$$\det(A^T) = 9$$

2. If a square matrix B is obtained by interchanging two rows (or two columns) of a square matrix A, then

$$\det(B) = -\det(A)$$

For example, interchange the two rows of A we get

$$B = \begin{bmatrix} 3 & 6 \\ 4 & 5 \end{bmatrix}$$

Therefore
$$\det(B) = -9$$

3. If two rows (or two columns) of a square matrix C are equal, then

$$\det(C) = 0$$

This can be seen by considering the matrix C given by

$$C = \begin{bmatrix} 3 & 6 \\ 3 & 6 \end{bmatrix}$$

Therefore
$$\det(C) = 0$$

4. If a row (or a column) of a square matrix is multiplied by a real number c, then the determinant of the resulting matrix B is given by

$$\det(B) = c \det(A)$$

Multiply the first row of A by 2 we find

$$B = \begin{bmatrix} 8 & 10 \\ 3 & 6 \end{bmatrix}$$

Therefore
$$\det(B) = 18 = 2\det(A)$$

5. If a square matrix is multiplied by a constant number k, then
$$\det(k(A_{n\times n})) = k^n \det(A)$$

6. If $A_{n\times n}$ and $B_{n\times n}$ are square matrices, then
$$\det(AB) = \det(A)\det(B)$$

7. In view of (6) we also set
$$\det((AB)^T) = \det(A^T)\det(B^T) = \det(A)\det(B)$$

8. If $A_{n\times n}$ and $B_{n\times n}$ are square matrices, then
$$\det(A+B) \neq \det(A) + \det(B)$$

The properties of determinants presented above will be explained by using the following illustrative example.

Example 1. Given the matrix

(a) $A = \begin{bmatrix} a_1 & a_2 & a_3 \\ b_1 & b_2 & b_3 \\ c_1 & c_2 & c_3 \end{bmatrix}$

such that
$$\det(A) = 3$$

Evaluate the determinants of the following matrices

(a) $B = \begin{bmatrix} b_1 & b_2 & b_3 \\ a_1 & a_2 & a_3 \\ c_1 & c_2 & c_3 \end{bmatrix}$,   (b) $C = \begin{bmatrix} 2a_1 & 2a_2 & 2a_3 \\ b_1 & b_2 & b_3 \\ c_1 & c_2 & c_3 \end{bmatrix}$

(c) $K = \begin{bmatrix} 2a_1 & 2a_2 & 2a_3 \\ 2b_1 & 2b_2 & 2b_3 \\ 2c_1 & 2c_2 & 2c_3 \end{bmatrix}$,   (d) $G = \begin{bmatrix} a_1 & b_1 & c_1 \\ a_2 & b_2 & c_2 \\ a_3 & b_3 & c_3 \end{bmatrix}$

(e) $H = \begin{bmatrix} 2a_1 & 2a_2 & 2a_3 \\ b_1 & b_2 & b_3 \\ 5c_1 & 5b_2 & 5c_3 \end{bmatrix}$,   (f) $M = \begin{bmatrix} 2a_1 & 8a_2 & 2a_3 \\ b_1 & 4b_2 & b_3 \\ c_1 & 4c_2 & c_3 \end{bmatrix}$

Solution.

(a) The matrix B is obtained by interchanging the first two rows, therefore

$$\det(B) = -3$$

(b) The matrix C is obtained by multiplying the first row of A by 2, therefore

$$\det(C) = 2 \times 3 = 6$$

(c) The matrix K is obtained by multiplying A by 2, therefore

$$\det(K) = 2^3 \times 3 = 24$$

(d) The matrix G is the transpose matrix of A , therefore

$$\det(G) = \det(A) = 3$$

(e) The matrix H is obtained by multiplying the first row A by 2 and the third row of (A) by 5, therefore

$$\det(H) = 2 \times 5 \times 3 = 30$$

(f) The matrix M is obtained by multiplying the first row A by 2 and the second column of (A) by 4, therefore

$$\det(M) = 2 \times 4 \times 3 = 24$$

Exercises 12.2

1. Given the matrix

$$A = \begin{bmatrix} a_1 & a_2 & a_3 \\ b_1 & b_2 & b_3 \\ c_1 & c_2 & c_3 \end{bmatrix}$$

such that

$$\det(A) = -5$$

Evaluate the determinants of the following matrices

(a) $B = \begin{bmatrix} b_1 & b_2 & b_3 \\ a_1 & a_2 & a_3 \\ c_1 & c_2 & c_3 \end{bmatrix}$,   (b) $C = \begin{bmatrix} 3a_1 & 3a_2 & 3a_3 \\ b_1 & b_2 & b_3 \\ c_1 & c_2 & c_3 \end{bmatrix}$

(c) $K = \begin{bmatrix} 3a_1 & 3a_2 & 3a_3 \\ 3b_1 & 3b_2 & 3b_3 \\ 3c_1 & 3c_2 & 3c_3 \end{bmatrix}$,   (d) $G = \begin{bmatrix} a_1 & b_1 & c_1 \\ a_2 & b_2 & c_2 \\ a_3 & b_3 & c_3 \end{bmatrix}$

(e) $H = \begin{bmatrix} 2a_1 & 2a_2 & 2a_3 \\ -3b_1 & -3b_2 & -3b_3 \\ 5c_1 & 5b_2 & 5c_3 \end{bmatrix}$,   (f) $M = \begin{bmatrix} 3a_1 & 15a_2 & -6a_3 \\ b_1 & 5b_2 & -2b_3 \\ c_1 & 5c_2 & -2c_3 \end{bmatrix}$

2. If $|A| = 3$, $A_{4\times4}$, find

(i) $|2A|$, (ii) $2|A|$, (iii) $|A^2|$, (iv) $|AA^T|$

3. If $|A| = 2$, $|B| = 3$, $A_{2\times2}$ and $B_{2\times2}$, find

(i) $|AB|$, (ii) $|BA|$, (iii) $|2A||2B|$, (iv) $2|AB|$

4. If $|A| \neq 0$, $A^2 = A$, find $|A|$

5. If $A_{3\times3}A_{3\times3}^T = I_3$, find $|A|$

6. If $|A| = 3$, $A_{4\times4}$, find $|2A| - 4|A^T| + |I_4|$

7. If $7|AA^T| = |2A| + |3A|$, $A_{3\times3}$, $|A| \neq 0$, find $|A|$

8. If $|2A||3A^T| = 35|AA^T| + |A| + 2|I_2|$, $A_{2\times2}$, find $|A|$

9. Given the matrix

$$A = \begin{bmatrix} f(x) & 1 & a \\ f(a) & 1 & a \\ f(b) & 1 & b \end{bmatrix}$$

find x if $\det(A) = 0$

10.   Given the matrix

$$A = \begin{bmatrix} f(x) & 1 & f(b) \\ a & 1 & a \\ b & 1 & b \end{bmatrix}$$

find x if $\det(A) = 0, b > a$

12.3   Inverse of a Matrix

To study the concept of the inverse of a matrix, it seems reasonable to review the **identity** matrices given by

$$I_2 = \begin{bmatrix} 1 & 0 \\ 0 & 1 \end{bmatrix}, I_3 = \begin{bmatrix} 1 & 0 & 0 \\ 0 & 1 & 0 \\ 0 & 0 & 1 \end{bmatrix}$$

The identity matrices defined above are diagonal matrices of order 2 and 3 respectively, where all entries of the main diagonal are 1, and 0 elsewhere. For square matrices $A_{n \times n}$, we can easily verify that

$$I_n A = A I_n = A$$

The role of the identity matrix is identical to the role of 1 when multiplied by any real number. We shall discuss first the inverse of matrices of dimension 2×2.

It is important to note that the inverse of a matrix $A_{n \times n}$ exists only if:

1. A is a square matrix. There is no inverse for matrices of dimension $m \times n, m \neq n$.

2. $\det(A) \neq 0$.

For a square matrix $A_{2 \times 2}$, the inverse of A, denoted by A^{-1} is a square matrix that satisfies

$$A A^{-1} = A^{-1} A = I_2$$

Given a matrix $A_{2\times2}$ defined by

$$A_{2\times2} = \begin{bmatrix} a & b \\ c & d \end{bmatrix}$$

It can be easily seen that the inverse of A is given by

$$A_{2\times2}^{-1} = \frac{1}{\det(A)} \begin{bmatrix} d & -b \\ -c & a \end{bmatrix}$$

so that the entries of the main diagonal switch locations, and the signs of the back diagonal are changed to opposite signs. The essential conditions for A^{-1} to exist can be observed from the need of the determinant of A. As discussed before, the determinant exists only for square matrices. In addition, $\det(A)$ appears in the denominator. This means that for A^{-1} to exist, then $\det(A) \neq 0$.

It is to be noted that the inverse of other matrices in general, of dimensions $m \times m$, $m \geq 3$, can be obtained by performing the the Gauss-Jordan reduction discussed in Chapter 11. In this case, we first adjoin A to the identity matrix I_m to obtain the $m \times 2m$ matrix given by

$$\begin{bmatrix} A & \vdots & I_m \end{bmatrix}$$

Using the Gauss-Jordan reduction method we convert this matrix to the form

$$\begin{bmatrix} I_m & \vdots & A^{-1} \end{bmatrix}$$

the m columns at the rightmost gives the inverse matrix A^{-1}. It is important to note that if we obtain zeros in the first m positions of a row, then the matrix A does not have an inverse. Furthermore, the inverse of a matrix can be obtained by using Maple or Mathematica software to reduce the size of calculations. The concept of the inverse of a matrix and the associated conditions will be illustrated by the following examples.

Example 1. Find A^{-1} for the matrix

$$A = \begin{bmatrix} 4 & -5 \\ -2 & 3 \end{bmatrix}$$

and show that $A A^{-1} = I_2$.

Solution. $\det(A) = 2$. Hence, by using the formula given above, we find

$$A^{-1} = \frac{1}{2} \begin{bmatrix} 3 & 5 \\ 2 & 4 \end{bmatrix} = \begin{bmatrix} \frac{3}{2} & \frac{5}{2} \\ 1 & 2 \end{bmatrix}$$

To show that the product $AA^{-1} = I_2$, we set

$$AA^{-1} = \begin{bmatrix} 4 & -5 \\ -2 & 3 \end{bmatrix} \begin{bmatrix} \frac{3}{2} & \frac{5}{2} \\ 1 & 2 \end{bmatrix} = \begin{bmatrix} 1 & 0 \\ 0 & 1 \end{bmatrix} = I_2$$

Example 2. Find A^{-1} for the matrix

$$A = \begin{bmatrix} 3 & 7 \\ 2 & 5 \end{bmatrix}$$

Solution. It is clear that $\det(A) = 1$. Hence, by using the formula given above, we find

$$A^{-1} = \frac{1}{1} \begin{bmatrix} 5 & -7 \\ -2 & 3 \end{bmatrix} = \begin{bmatrix} 5 & -7 \\ -2 & 3 \end{bmatrix}$$

Example 3. Find x such that A^{-1} exists for the matrix

$$A = \begin{bmatrix} x & 5 \\ 3 & x - 2 \end{bmatrix}$$

Solution. For A^{-1} to exist, then

$$\begin{aligned} \det(A) &\neq 0 \\ x(x - 2) - 15 &\neq 0 \\ x^2 - 2x - 15 &\neq 0 \\ (x - 5)(x + 3) &\neq 0 \\ x &\neq -3, 5 \end{aligned}$$

Example 4. Show that

$$\det(A^{-1}) = \frac{1}{\det(A)}$$

for the matrix

$$A = \begin{bmatrix} 7 & 8 \\ 3 & 4 \end{bmatrix}$$

Solution. Following the procedure discussed above we find

$$A^{-1} = \begin{bmatrix} 1 & -2 \\ -\frac{3}{4} & \frac{7}{4} \end{bmatrix}$$

Consequently, we find

$$\det(A) = 4$$

and

$$\det(A^{-1}) = \frac{1}{4}$$

Determinant of an inverse:

If A is a square matrix, and $\det(A) \neq 0$, then

$$\det(A^{-1}) = \frac{1}{\det(A)}$$

Example 5. Given $A_{4 \times 4}$ and $B_{4 \times 4}$ where $\det(A) = 4, \det(B) = 3$, find

(a) $3 \det(A) + 16 \det(A^{-1}) + \det(2A)$,   (b) $36 \det((AB)^{-1})$

Solution.

(a) $3 \det(A) + 16 \det(A^{-1}) + \det(2A) = 3 \times 4 + 16 \times \frac{1}{4} + 2^4 \times 4 = 80$

(b) $36 \det((AB)^{-1}) = 36 \times \dfrac{1}{\det(AB)} = 36 \times \dfrac{1}{\det(A)\det(B)} = 3$

Example 6. Given the matrix

$$A = \begin{bmatrix} 1 & -1 & 0 \\ -1 & 0 & 1 \end{bmatrix}$$

Find x, y, u, and v if

$$3(AA^T)^{-1} = \begin{bmatrix} 3y - 4x & y - x \\ u - v & 3u - 4v \end{bmatrix}$$

Solution. We first find AA^T

$$AA^T = \begin{bmatrix} 1 & -1 & 0 \\ -1 & 0 & 1 \end{bmatrix} \begin{bmatrix} 1 & -1 \\ -1 & 0 \\ 0 & 1 \end{bmatrix} = \begin{bmatrix} 2 & -1 \\ -1 & 2 \end{bmatrix}$$

Now we find

$$(AA^T)^{-1} = \frac{1}{3} \begin{bmatrix} 2 & 1 \\ 1 & 2 \end{bmatrix}$$

This means that

$$3(AA^T)^{-1} = \begin{bmatrix} 2 & 1 \\ 1 & 2 \end{bmatrix} = \begin{bmatrix} 3y - 4x & y - x \\ u - v & 3u - 4v \end{bmatrix}$$

Solving the systems of equations

$$\begin{aligned} 3y - 4x &= 2 \\ y - x &= 1 \end{aligned}$$

gives

$$x = 1, y = 2$$

However, solving the system

$$\begin{aligned} 3u - 4v &= 2 \\ u - v &= 1 \end{aligned}$$

gives

$$u = 2, v = 1$$

12.3.1   Solving Systems of equations

The concept of the inverse of a matrix can be used to solve a system of linear equations in two unknowns x and y given by

$$
\begin{aligned}
ax + by &= \alpha \\
cx + dy &= \beta
\end{aligned}
$$

The system can be written in a matrix multiplication form by

$$
\begin{bmatrix} a & b \\ c & d \end{bmatrix} \begin{bmatrix} x \\ y \end{bmatrix} = \begin{bmatrix} \alpha \\ \beta \end{bmatrix}
$$

Multiplying both sides by the inverse of the matrix of the coefficients gives

$$
\begin{bmatrix} a & b \\ c & d \end{bmatrix}^{-1} \begin{bmatrix} a & b \\ c & d \end{bmatrix} \begin{bmatrix} x \\ y \end{bmatrix} = \begin{bmatrix} a & b \\ c & d \end{bmatrix}^{-1} \begin{bmatrix} \alpha \\ \beta \end{bmatrix}
$$

which gives the formula

$$
\begin{bmatrix} x \\ y \end{bmatrix} = \begin{bmatrix} a & b \\ c & d \end{bmatrix}^{-1} \begin{bmatrix} \alpha \\ \beta \end{bmatrix}
$$

for determining x and y. In other words, the column matrix of the unknowns x and y is equal to the product of the inverse of the matrix of the coefficients by the column matrix of the constants. By equating the corresponding entries of both sides, x and y are immediately determined.

However, for an $n \times n$ system of equations of n unknowns, defined by

$$
\begin{aligned}
a_{11}x_1 + a_{12}x_2 + \cdots + a_{1n}x^n &= \alpha_1, \\
a_{21}x_1 + a_{22}x_2 + \cdots + a_{2n}x^n &= \alpha_2, \\
a_{31}x_1 + a_{32}x_2 + \cdots + a_{3n}x^n &= \alpha_3, \\
a_{41}x_1 + a_{42}x_2 + \cdots + a_{4n}x^n &= \alpha_4, \\
&\;\;\vdots \\
a_{n1}x_1 + a_{n2}x_2 + \cdots + a_{nn}x^n &= \alpha_n.
\end{aligned}
$$

We usually follow a similar method to find that

$$
\begin{bmatrix} x_1 \\ x_2 \\ \vdots \\ x_n \end{bmatrix} = \begin{bmatrix} a_{11} & a_{12} & \cdots & a_{1n} \\ a_{21} & a_{22} & \cdots & a_{2n} \\ & \vdots & & \\ a_{n1} & a_{n2} & \cdots & a_{nn} \end{bmatrix}^{-1} \begin{bmatrix} \alpha_1 \\ \alpha_2 \\ \vdots \\ \alpha_n \end{bmatrix}
$$

Example 7. Solve the following system of equations:

$$
\begin{aligned}
3x + 2y &= 7 \\
7x + 5y &= 17
\end{aligned}
$$

Solution. The matrix of the coefficients is given by

$$
A = \begin{bmatrix} 3 & 2 \\ 7 & 5 \end{bmatrix}
$$

Hence, A^{-1} is given by

$$
A^{-1} = \begin{bmatrix} 5 & -2 \\ -7 & 3 \end{bmatrix}
$$

Using the method introduced above we find

$$
\begin{bmatrix} x \\ y \end{bmatrix} = \begin{bmatrix} 5 & -2 \\ -7 & 3 \end{bmatrix} \begin{bmatrix} 7 \\ 17 \end{bmatrix} = \begin{bmatrix} 1 \\ 2 \end{bmatrix}
$$

By equating the corresponding entries of both sides, we find

$$
x = 1, \ y = 2
$$

Example 8. Solve the following system of equations:

$$
\begin{aligned}
2x + 5y + 4z &= 1 \\
x + 4y + 3z &= 1 \\
x - 3y - 2z &= 0
\end{aligned}
$$

given that

$$
\begin{bmatrix} 2 & 5 & 4 \\ 1 & 4 & 3 \\ 1 & -3 & -2 \end{bmatrix}^{-1} = \begin{bmatrix} -1 & 2 & 1 \\ -5 & 8 & 2 \\ 7 & -11 & -3 \end{bmatrix}
$$

Solution. Using the method introduced above we find

$$
\begin{bmatrix} x \\ y \\ z \end{bmatrix} = \begin{bmatrix} -1 & 2 & 1 \\ -5 & 8 & 2 \\ 7 & -11 & -3 \end{bmatrix} \begin{bmatrix} 1 \\ 1 \\ 0 \end{bmatrix} = \begin{bmatrix} 1 \\ 3 \\ -4 \end{bmatrix}
$$

By equating the corresponding entries of both sides, we find

$$
x = 1, \ y = 3, \ z = -4
$$

An alternative approach

It is interesting to note that systems of n equations of n variables, where the solutions of these variables exist, can be solved by using a concept called reduced row echelon form **rref**. This can be performed by using a calculator. We first construct the augmented matrix A, and then determine rref(A) by using the calculator. For Example 8, the augmented matrix is given by

$$
A = \begin{bmatrix} 2 & 5 & 4 & 1 \\ 1 & 4 & 3 & 1 \\ 1 & -3 & -2 & 0 \end{bmatrix} \tag{1}
$$

Using **rref** we obtain

$$
\text{rref}(A) = \begin{bmatrix} 1 & 0 & 0 & 1 \\ 0 & 1 & 0 & 3 \\ 0 & 0 & -2 & -4 \end{bmatrix} \tag{2}
$$

This means that

$$
x = 1, \ y = 3, \ z = -4
$$

Exercises 12.3

1. Find the inverse of each of the following matrices if exists:

(a) $A = \begin{bmatrix} 7 & 4 \\ 5 & 3 \end{bmatrix}$, (b) $B = \begin{bmatrix} 3 & 4 \\ 3 & 5 \end{bmatrix}$, (c) $C = \begin{bmatrix} 6 & 9 \\ 2 & 3 \end{bmatrix}$

2. Find $(AA^T)^{-1}$ for the matrix A given by

$$A = \begin{bmatrix} 1 & -1 & 0 \\ -1 & 2 & -1 \end{bmatrix}$$

3. Show that $(A^{-1})^{-1} = A$ for the matrix A given by

$$A = \begin{bmatrix} 5 & 3 \\ 3 & 2 \end{bmatrix}$$

4. Show that $(AB)^{-1} = B^{-1}A^{-1}$ for the matrices

$$A = \begin{bmatrix} 1 & -1 \\ -2 & 3 \end{bmatrix}, B = \begin{bmatrix} 2 & 3 \\ 3 & 5 \end{bmatrix}$$

5. Find x such that A^{-1} does not exist for the matrix

$$A = \begin{bmatrix} 1 & x \\ x-1 & 2 \end{bmatrix}$$

6. Find x such that A^{-1} does not exist for the matrix

$$A = \begin{bmatrix} x & x \\ 1 & x-1 \end{bmatrix}$$

7. Find x such that A^{-1} does not exist for the matrix

$$A = \begin{bmatrix} x^2 & 4 \\ x & x \end{bmatrix}$$

8. Find x such that A^{-1} does not exist for the matrix

$$A = \begin{bmatrix} 1 & 0 & 1 \\ 3 & x & 0 \\ 1 & -2 & x \end{bmatrix}$$

9. Find x such that A^{-1} does not exist for the matrix

$$A = \begin{bmatrix} x^2 & 0 & 3 \\ 5 & x & 2 \\ 3 & 0 & 1 \end{bmatrix}$$

10. Find x such that A^{-1} exists for the matrix

$$A = \begin{bmatrix} x^2 - 2x & x \\ 3 & 1 \end{bmatrix}$$

11. Solve the following system of equations:

$$\begin{aligned} 5x + 7y &= -2 \\ 11x + 5y &= 6 \end{aligned}$$

12. Solve the following system of equations:

$$\begin{aligned} 11x - 7y &= 29 \\ 9x + 11y &= 7 \end{aligned}$$

13. Solve the following system of equations:

$$\begin{aligned} 13x + 17y &= 73 \\ 15x - 11y &= 23 \end{aligned}$$

14. Solve the following system of equations:

$$\begin{aligned} 19x + 16y &= 67 \\ 22x - 13y &= -17 \end{aligned}$$

15. Find x, y, u, and v if

$$3(AA^T)^{-1} = \begin{bmatrix} 2x - 2 & 3y - 4 \\ 2u + 1 & v + 1 \end{bmatrix} + \begin{bmatrix} 2 & -3 \\ -6 & -2 \end{bmatrix}$$

where the matrix A is given by

$$A = \begin{bmatrix} 1 & 0 & 1 \\ 0 & 1 & 1 \end{bmatrix}$$

16. Find x, y, u, and v if

$$8(AA^T)^{-1} = \begin{bmatrix} 2x & 3y - 4 \\ 2u + 1 & v + 1 \end{bmatrix} + \begin{bmatrix} x & -1 \\ -6 & -y \end{bmatrix}$$

where the matrix A is given by

$$A = \begin{bmatrix} 1 & -1 & 1 \\ -1 & 1 & 1 \end{bmatrix}$$

17. Find x, y, u, and v if

$$5(AA^T)^{-1} = \begin{bmatrix} x - 1 & y - x \\ u - y & v - x \end{bmatrix}$$

where the matrix A is given by

$$A = \begin{bmatrix} 1 & 0 & -1 & 0 \\ 1 & -1 & 0 & 1 \end{bmatrix}$$

18. Given the matrix

$$A = \begin{bmatrix} 1 & 0 & 1 & -1 \\ -1 & 1 & 0 & 1 \end{bmatrix}$$

Find x, y, u, and v if

$$5(AA^T)^{-1} = \begin{bmatrix} v - x & v - u \\ 2y - v & 2y - 3x \end{bmatrix}$$

19. Given the matrix

$$A = \begin{bmatrix} 2 & 0 & 2 & -2 \\ -2 & 2 & 0 & 2 \end{bmatrix}$$

Find x, y, u, and v if

$$80(AA^T)^{-1} = \begin{bmatrix} x + u & 2v - 2y \\ v - u & 4x - u \end{bmatrix}$$

20. Find x, y, u, and v if

$$\begin{bmatrix} 2x + 14 & 6y - 6 \\ 2u + 3 & 4v - 14 \end{bmatrix} - \begin{bmatrix} 1 & -1 \\ 0 & -2 \end{bmatrix} = \begin{bmatrix} 1 & 3 \\ 3 & 10 \end{bmatrix}^{-1} - \begin{bmatrix} 2 & -3 \\ 1 & 5 \end{bmatrix}^T$$

21. Find x, y, u, and v if

$$A + A^T + A^{-1} = \begin{bmatrix} x + y & u + v \\ 2u - v & 2x - y \end{bmatrix}$$

where A is given by

$$A = \begin{bmatrix} 5 & 3 \\ 3 & 2 \end{bmatrix}$$

22. Evaluate the expression

$$\det(A) \det(A^{-1}) \det(A^T)$$

given that $\det(A) = 4$

23. Given the following system of equations:

$$3x + 6y = 21$$
$$kx + 12y = 41$$

For which values of k does the system have one solution. Find that solution for $k = 5$.

24. Solve the following system of equations:

$$2x + y = 5$$
$$-x + 2y + 2z = 15$$
$$y + z = 8$$

given that

$$\begin{bmatrix} 2 & 1 & 0 \\ -1 & 2 & 2 \\ 0 & 1 & 1 \end{bmatrix}^{-1} = \begin{bmatrix} 0 & -1 & 2 \\ 1 & 2 & -4 \\ -1 & -2 & 5 \end{bmatrix}$$

25. Solve the following system of equations:

$$-24x + 7y + z - 2u = -11$$
$$-10x + 3y - u = -5$$
$$-29x + 7y + 3z - 2u = -14$$
$$12x - 3y - z + u = 6$$

given that

$$\begin{bmatrix} -24 & 7 & 1 & -2 \\ -10 & 3 & 0 & -1 \\ -29 & 7 & 3 & -2 \\ 12 & -3 & -1 & 1 \end{bmatrix}^{-1} = \begin{bmatrix} 1 & -2 & -1 & -2 \\ 3 & -5 & -2 & -3 \\ 2 & -5 & -2 & -5 \\ -1 & 4 & 4 & 11 \end{bmatrix}$$

26. Solve the following system of equations:

$$\begin{aligned} x - y - z + u &= 1 \\ -y - z &= 4 \\ -x - z - 2u &= 2 \\ x + y + u &= 3 \end{aligned}$$

given that

$$\begin{bmatrix} 1 & -1 & -1 & 1 \\ 0 & -1 & -1 & 0 \\ -1 & 0 & -1 & -2 \\ 1 & 1 & 0 & 1 \end{bmatrix}^{-1} = \begin{bmatrix} 3 & -4 & 1 & -1 \\ -1 & 1 & 0 & 1 \\ 1 & -2 & 0 & -1 \\ -2 & 3 & -1 & 1 \end{bmatrix}$$

27. Use the concept of the augmented matrix, and a calculator command **rref** (reduced row echelon form) to solve the system of equations:

$$\begin{aligned} x + y + z + u - v &= 5 \\ x - y + z - u + v &= 3 \\ x + z + v &= 9 \\ y + z - u &= 1 \\ 2x - 3y + 4z - v &= 3 \end{aligned}$$

28. Use the concept of the augmented matrix, and a calculator command **rref** (reduced row echelon form) to solve the system of equations:

$$\begin{aligned} x - y - z - u - v &= 5 \\ x - y + z - u + v &= 3 \\ x + z + v &= 9 \\ y + z - u &= 1 \\ 2x - 3y + 4z - v &= 3 \end{aligned}$$

12.4 Fibonacci and Lucas Matrices

In this section we will study the link of Fibonacci and Lucas numbers to matrices. The matrices that we will present can be easily proved by using the method of mathematical induction presented before. In addition, these matrices can be used to derive some of the formulas that were discussed before such as the Cassini formula.

12.4.1 The Q-Matrix

To study the link between Fibonacci sequence and matrices we consider the *Q-Matrix* defined by

$$Q = \begin{bmatrix} 1 & 1 \\ 1 & 0 \end{bmatrix}$$

Using the multiplication of matrices, and noting that $F_0 = 0$, we can easily find that

$$Q = \begin{bmatrix} 1 & 1 \\ 1 & 0 \end{bmatrix} = \begin{bmatrix} F_2 & F_1 \\ F_1 & F_0 \end{bmatrix}$$

$$Q^2 = \begin{bmatrix} 2 & 1 \\ 1 & 1 \end{bmatrix} = \begin{bmatrix} F_3 & F_2 \\ F_2 & F_1 \end{bmatrix}$$

$$Q^3 = \begin{bmatrix} 3 & 2 \\ 2 & 1 \end{bmatrix} = \begin{bmatrix} F_4 & F_3 \\ F_3 & F_2 \end{bmatrix}$$

$$Q^4 = \begin{bmatrix} 5 & 3 \\ 3 & 2 \end{bmatrix} = \begin{bmatrix} F_5 & F_4 \\ F_4 & F_3 \end{bmatrix}$$

From this we set the following theorem
Theorem. For $n \geq 1$, we have

$$Q^n = \begin{bmatrix} F_{n+1} & F_n \\ F_n & F_{n-1} \end{bmatrix}$$

This theorem will be proved by the mathematical induction.

For $n = 1$

$$\text{LHS} = Q = \begin{bmatrix} F_2 & F_1 \\ F_1 & F_0 \end{bmatrix}$$

$$\text{RHS} = \begin{bmatrix} 1 & 1 \\ 1 & 0 \end{bmatrix}$$

Assume that it is true for $n = k$, i.e assume that

$$Q^k = \begin{bmatrix} F_{k+1} & F_k \\ F_k & F_{k-1} \end{bmatrix}$$

It remains to show that it is true for $n = k + 1$, i.e show that

$$Q^{k+1} = \begin{bmatrix} F_{k+2} & F_{k+1} \\ F_{k+1} & F_k \end{bmatrix}$$

$$Q^{k+1} = Q^k Q = \begin{bmatrix} F_{k+1} & F_k \\ F_k & F_{k-1} \end{bmatrix} \begin{bmatrix} 1 & 1 \\ 1 & 0 \end{bmatrix}$$

$$= \begin{bmatrix} F_{k+1} + F_k & F_{k+1} \\ F_k + F_{k-1} & F_k \end{bmatrix} = \begin{bmatrix} F_{k+2} & F_{k+1} \\ F_{k+1} & F_k \end{bmatrix}$$

Thus the identity is true for every $n \geq 1$.

One important conclusion can be made here is that

$$|Q^n| = |Q|^n = (-1)^n$$

This gives another proof to the Cassini formula that

$$\begin{vmatrix} F_{n+1} & F_n \\ F_n & F_{n-1} \end{vmatrix} = (-1)^n$$

12.4.2 The M-Matrix

Another link between Fibonacci sequence and matrices can be derived by using the *M-Matrix* defined by

$$M = \begin{bmatrix} 1 & 1 \\ 1 & 2 \end{bmatrix}$$

Using the multiplication of matrices we can find

$$M = \begin{bmatrix} 1 & 1 \\ 1 & 2 \end{bmatrix} = \begin{bmatrix} F_1 & F_2 \\ F_2 & F_3 \end{bmatrix}$$

$$M^2 = \begin{bmatrix} 2 & 3 \\ 3 & 5 \end{bmatrix} = \begin{bmatrix} F_3 & F_4 \\ F_4 & F_5 \end{bmatrix}$$

$$M^3 = \begin{bmatrix} 5 & 8 \\ 8 & 13 \end{bmatrix} = \begin{bmatrix} F_5 & F_6 \\ F_6 & F_7 \end{bmatrix}$$

$$M^4 = \begin{bmatrix} 13 & 21 \\ 21 & 34 \end{bmatrix} = \begin{bmatrix} F_7 & F_8 \\ F_8 & F_9 \end{bmatrix}$$

From this we set the following theorem
Theorem. For $n \geq 1$, we have

$$M^n = \begin{bmatrix} F_{2n-1} & F_{2n} \\ F_{2n} & F_{2n+1} \end{bmatrix}$$

This theorem will be proved by the mathematical induction.
For $n = 1$

$$\text{LHS} = M = \begin{bmatrix} F_1 & F_2 \\ F_2 & F_3 \end{bmatrix}$$

$$\text{RHS} = \begin{bmatrix} 1 & 1 \\ 1 & 2 \end{bmatrix}$$

Assume that it is true for $n = k$, i.e assume that

$$M^k = \begin{bmatrix} F_{2k-1} & F_{2k} \\ F_{2k} & F_{2k+1} \end{bmatrix}$$

It remains to show that it is true for $n = k+1$, i.e show that

$$M^{k+1} = \begin{bmatrix} F_{2k+1} & F_{2k+2} \\ F_{2k+2} & F_{2k+3} \end{bmatrix}$$

$$M^{k+1} = M^k M = \begin{bmatrix} F_{2k-1} & F_{2k} \\ F_{2k} & F_{2k+1} \end{bmatrix} \begin{bmatrix} 1 & 1 \\ 1 & 2 \end{bmatrix}$$

$$= \begin{bmatrix} F_{2k-1} + F_{2k} & F_{2k-1} + F_{2k} + F_{2k} \\ F_{2k=1} + F_{2k} & F_{2k} + F_{2k+1} + F_{2k+1} \end{bmatrix}$$

$$= \begin{bmatrix} F_{2k+1} & F_{2k+2} \\ F_{2k+2} & F_{2k+3} \end{bmatrix}$$

Thus the identity is true for every $n \geq 1$

One important conclusion can be made here is that

$$|M^n| = |M|^n = (1)^n = 1$$

This gives another proof to the formula

$$\begin{vmatrix} F_{2n-1} & F_{2n} \\ F_{2n} & F_{2n+1} \end{vmatrix} = 1$$

12.4.3   The R-Matrix

We finally consider the *R-Matrix* defined by

$$R = \begin{bmatrix} 1 & 2 \\ 2 & -1 \end{bmatrix}$$

Hogatt and Ruggles introduced a link that combines the R-matrix and the Q-matrix with Lucas numbers. We first notice that

$$RQ^2 = \begin{bmatrix} 4 & 3 \\ 3 & 1 \end{bmatrix} = \begin{bmatrix} L_3 & L_2 \\ L_2 & L_1 \end{bmatrix}$$

$$RQ^3 = \begin{bmatrix} 7 & 4 \\ 4 & 3 \end{bmatrix} = \begin{bmatrix} L_4 & L_3 \\ L_3 & L_2 \end{bmatrix}$$

$$RQ^4 = \begin{bmatrix} 11 & 7 \\ 7 & 4 \end{bmatrix} = \begin{bmatrix} L_5 & L_4 \\ L_4 & L_3 \end{bmatrix}$$

From this we set the following theorem

Theorem. For $n \geq 2$, we have

$$RQ^n = \begin{bmatrix} L_{n+1} & L_n \\ L_n & L_{n-1} \end{bmatrix}$$

This theorem will be proved by the mathematical induction.
It can be easily proved that it is true for $n = 2$
Assume that it is true for $n = k$, i.e assume that

$$RQ^k = \begin{bmatrix} L_{k+1} & L_k \\ L_k & L_{k-1} \end{bmatrix}$$

It remains to show that it is true for $n = k + 1$, i.e show that

$$RQ^{k+1} = \begin{bmatrix} L_{k+2} & L_{k+1} \\ L_{k+1} & L_k \end{bmatrix}$$

$$RQ^{k+1} = RQ^k Q = \begin{bmatrix} L_{k+1} & L_k \\ L_k & L_{k-1} \end{bmatrix} \begin{bmatrix} 1 & 1 \\ 1 & 0 \end{bmatrix}$$

$$= \begin{bmatrix} L_{k+1} + L_k & L_{k+1} \\ L_k + L_{k-1} & L_k \end{bmatrix} = \begin{bmatrix} L_{k+2} & L_{k+1} \\ L_{k+1} & L_k \end{bmatrix}$$

Thus the identity is true for every $n \geq 2$

One important conclusion can be made here is that

$$|RQ^n| = |R||Q|^n = (-5)(-1)^n = 5(-1)^{n+1}$$

This gives another proof to the Cassini-type formula proved before

$$\begin{vmatrix} L_{n+1} & L_n \\ L_n & L_{n-1} \end{vmatrix} = 5(-1)^{n+1}$$

Example 1.   (a) Use mathematical induction to show that

$$KM^n = \begin{bmatrix} L_{2n-3} & L_{2n-2} \\ F_{2n+2} & F_{2n+3} \end{bmatrix}$$

is true for $n \geq 2$, where M is the *M-matrix* defined before, and the matrix K is given by

$$K = \begin{bmatrix} 2 & -1 \\ 1 & 2 \end{bmatrix}$$

(b) Use part (a) to show that

$$\begin{vmatrix} L_{2n-3} & L_{2n-2} \\ F_{2n+2} & F_{2n+3} \end{vmatrix} = 5$$

Solution. It can be easily proved that it is true for $n = 2$
Assume that it is true for $n = k$, i.e assume that

$$KM^k = \begin{bmatrix} L_{2k-3} & L_{2k-2} \\ F_{2k+2} & F_{2k+3} \end{bmatrix}$$

It remains to show that it is true for $n = k + 1$, i.e show that

$$KM^{k+1} = \begin{bmatrix} L_{2k-1} & L_{2k} \\ F_{2k+4} & F_{2k+5} \end{bmatrix}$$

$$KM^{k+1} = KM^k M = \begin{bmatrix} L_{2k-3} & L_{2k-2} \\ F_{2k+2} & F_{2k+3} \end{bmatrix} \begin{bmatrix} 1 & 1 \\ 1 & 2 \end{bmatrix}$$

$$= \begin{bmatrix} L_{2k-3} + L_{2k-2} & L_{2k-3} + L_{2k-2} + L_{2k-2} \\ F_{2k+2} + F_{2k+3} & F_{2k+2}F_{2k+3} + F_{2k+3} \end{bmatrix}$$

$$= \begin{bmatrix} L_{2k-1} & L_{2k} \\ F_{2k+4} & F_{2k+5} \end{bmatrix}$$

Thus the identity is true for every $n \geq 2$

(b) It is clear that $|KM^n| = |K||M|^n = 5(1) = 5$

Example 2. (a) Use mathematical induction to show that

$$KQ^n = \begin{bmatrix} L_n & L_{n-1} \\ L_{n+1} & L_n \end{bmatrix}$$

is true for $n \geq 2$, where Q is the *Q-matrix* defined before, and the matrix K is given before by

$$K = \begin{bmatrix} 2 & -1 \\ 1 & 2 \end{bmatrix}$$

(b) Use part (a) to show that

$$\begin{vmatrix} L_{2n-3} & L_{2n-2} \\ F_{2n+2} & F_{2n+3} \end{vmatrix} = 5(-1)^n$$

Solution. It can be easily proved that it is true for $n = 2$

Assume that it is true for $n = k$, i.e assume that

$$KQ^k = \begin{bmatrix} L_k & L_{k-1} \\ L_{k+1} & L_k \end{bmatrix}$$

It remains to show that it is true for $n = k + 1$, i.e show that

$$KQ^{k+1} = \begin{bmatrix} L_{k+1} & L_k \\ L_{k+2} & L_{k+1} \end{bmatrix}$$

$$KQ^{k+1} = KM^k M = \begin{bmatrix} L_{k+1} & L_k \\ L_{k+2} & L_{k+1} \end{bmatrix} \begin{bmatrix} 1 & 1 \\ 1 & 0 \end{bmatrix}$$

$$= \begin{bmatrix} L_k + L_{k-1} & L_k \\ L_{k+1} + L_k & L_{k+1} \end{bmatrix} = \begin{bmatrix} L_{k+1} & L_k \\ L_{k+2} & L_{k+1} \end{bmatrix}$$

Thus the identity is true for every $n \geq 2$

(b) It is clear that $|KQ^n| = |K||Q|^n = 5(-1)^n$

Exercises 12.4

1. (a) Use mathematical induction to show that

$$RM^n = \begin{bmatrix} F_{2n+2} & F_{2n+3} \\ F_{2n-3} & F_{2n-2} \end{bmatrix}$$

is true for $n \geq 2$, where R and M are the *R-matrix* and the *M-matrix* defined before.

(b) Use part (a) to show that

$$\begin{vmatrix} F_{2n+2} & F_{2n+3} \\ F_{2n-3} & F_{2n-2} \end{vmatrix} = -5$$

2. (a) Use mathematical induction to show that

$$MQ^n = \begin{bmatrix} F_{n+2} & F_{n+1} \\ L_{n+1} & L_n \end{bmatrix}$$

is true for $n \geq 1$, where Q and M are the *Q-matrix* and the *R-matrix* defined before.

(b) Use part (a) to show that

$$
\begin{vmatrix}
F_{n+2} & F_{n+1} \\
L_{n+1} & L_n
\end{vmatrix}
= (-1)^{n+1}
$$

3. (a) Use mathematical induction to show that

$$
AQ^n =
\begin{bmatrix}
F_{n+2} & F_{n+1} \\
F_{n+3} & F_{n+2}
\end{bmatrix}
$$

is true for $n \geq 1$, where Q is the *Q-matrix* defined before and the matrix A is

$$
A =
\begin{bmatrix}
1 & 1 \\
2 & 1
\end{bmatrix}
$$

(b) Use part (a) to show that

$$
\begin{vmatrix}
F_{n+2} & F_{n+1} \\
F_{n+3} & F_{n+2}
\end{vmatrix}
= (-1)^{n+1}
$$

4. (a) Use mathematical induction to show that

$$
AM^n =
\begin{bmatrix}
F_{2n+1} & F_{2n+2} \\
L_{2n} & L_{2n+1}
\end{bmatrix}
$$

is true for $n \geq 1$, where M is the *M-matrix* defined before and the matrix A is the matrix defined in Exercises 3 and given by

$$
A =
\begin{bmatrix}
1 & 1 \\
2 & 1
\end{bmatrix}
$$

(b) Use part (a) to show that

$$\begin{vmatrix} F_{2n+1} & F_{2n+2} \\ L_{2n} & L_{2n+1} \end{vmatrix} = (-1)$$

5. (a) Use mathematical induction to show that

$$BQ^n = \begin{bmatrix} F_{n+4} & F_{n+3} \\ F_{n+2} & F_{n+1} \end{bmatrix}$$

is true for $n \geq 1$, where Q is the Q-matrix defined before and the matrix B is

$$B = \begin{bmatrix} 3 & 2 \\ 1 & 1 \end{bmatrix}$$

(b) Use part (a) to show that

$$\begin{vmatrix} F_{n+4} & F_{n+3} \\ F_{n+2} & F_{n+1} \end{vmatrix} = (-1)^n$$

Chapter 13

Relations and Trees

13.1 Relations and Digraphs

In Chapter 5, the sets and the subsets concepts were discussed. In this section, a new set will be constructed by using the elements of a set A, but the derived set is not a subset of A. The set to be established consists of ordered pairs (x, y), where $x \in A$, $y \in A$, and x and y satisfy a specified condition. In a set format, the new set, called a **binary relation** R, is defined by

$$R = \{(x, y) | x \in A, y \in A, x \text{ is related to } y\}$$

Consider a set A given by

$$A = \{1, 2, 3\}$$

It is suggested to form the set of ordered pairs (x, y), $x \in A$, $y \in A$ and the condition $x < y$. The set R of ordered pairs is thus given by

$$R = \{(1, 2), (1, 3), (2, 3)\}$$

However, if the condition $x < y$ is changed to $x \leq y$, then a distinct set $\bar{R}$ is constructed and defined by

$$\bar{R} = \{(1, 1), (1, 2), (1, 3), (2, 2), (2, 3), (3, 3)\}$$

It is useful to note that it is not always necessary that x and y belong to one set A, but it is also possible to construct the set of ordered pairs (x, y) where x and y belong to distinct sets A and B. This can be explained if we consider the sets

$$A = \{1, 2, 3\}$$

and

$$B = \{1, 2, 3, 4, 5, 6, 7, 8, 9\}$$

A set $\tilde{R}$ of ordered pairs (x, y), $x \in A$, $y \in B$, and the condition $y = 3x$ is thus given by

$$\tilde{R} = \{(1, 3), (2, 6), (3, 9)\}$$

which is equivalent to the set format

$$\tilde{R} = \{(x, y) | x \in A, y \in B, y = 3x\}$$

However, in this text, we will focus our study on the case where x and y belong to the same set A. The above discussed concept of the set of ordered pairs will be explained by the following examples.

Example 1. Given the set A where

$$A = \{0, 1, 2, 3\}$$

Write the binary relation of ordered pairs described by

$$R = \{(x, y) | x \in A, y \in A, x > y\}$$

Solution. The following set

$$R = \{(1, 0), (2, 0), (3, 0), (2, 1), (3, 1), (3, 2)\}$$

satisfies the given condition.

Example 2. Given the set A where

$$A = \{1, 2, 3, 4\}$$

Write the binary relation of ordered pairs described by

$$R = \{(x, y) | x \in A, y \in A, x = y\}$$

Solution. The following set

$$R = \{(1, 1), (2, 2), (3, 3), (4, 4)\}$$

satisfies the given condition.

13.1.1 Digraphs

It is well known that visual representations are frequently used in mathematics. Graphs, such as lines, parabolas or curves are used to display visual structure for functions. Moreover, visual representations are used to display the universal set and the related subsets in the Venn diagram that we discussed before. Graphs and Venn diagram are usually used to give an insight about the behavior of functions and the interrelation between sets.

It is then normal to consider a visual representation for binary relations. The visual representation of a binary relation is established and is called a **directed graph** or simply, a **digraph**. A digraph is constructed by representing each element of the set A by a circle, called **a node**, and the ordered pairs by **directed edges**, called arrows that start from the element x towards the element y.

Example 3. Consider the set A defined by

$$A = \{1, 2, 3\}$$

and the binary relation R defined on A given by

$$R = \{(1, 2), (1, 3), (2, 3)\}$$

constructed by using the condition $x < y$. Construct the digraph that represents the binary relation R.

Solution. We first represent each element of A by a circle, then each ordered pair of R by a directed edge from x to y. Consequently, we obtain the following digraph that represents the binary relation R given above.

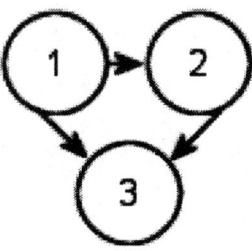

Example 4. Consider the set A defined by

$$A = \{1, 2, 3\}$$

and the binary relation R_1 defined on A given by

$$R_1 = \{(1,1), (1,2), (1,3), (2,2), (2,3), (3,3)\}$$

constructed by using the condition $x \leq y$. Construct the digraph that represents the binary relation R_1.

Solution. Proceeding as before we obtain

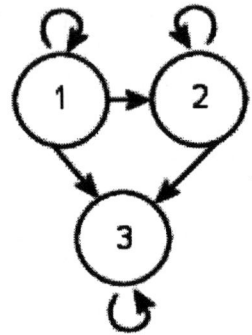

Example 5. Consider the set A defined by

$$A = \{1, 2, 3, 4\}$$

and the binary relation R_2 defined on A given by

$$R_2 = \{(1,1), (2,2), (3,3), (4,4)\}$$

constructed by using the condition $x = y$. Construct a proper digraph that represents R_2.

Solution.

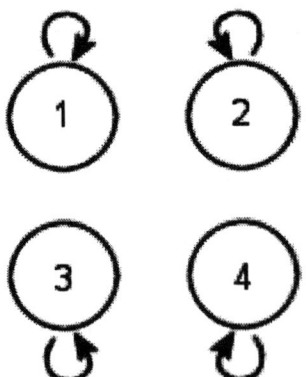

Proceeding as before, each element of A is represented by a circle. Each ordered pair of R is visualized by a directed edge from x to y, that is from each circle to itself. It is clear that there is no directed edge from any circle towards other circles. Consequently, we obtain the digraph shown above that represents the binary relation R_2 defined before.

Exercises 13.1

1. Consider the set A given by

$$A = \{1, 3, 5, 7\}$$

List the elements of the following relations:

(a) $R = \{(x, y) | x \in A, y \in A, x < y\}$
(b) $R = \{(x, y) | x \in A, y \in A, 3 | (x + y)\}$
(c) $R = \{(x, y) | x \in A, y \in A, x = y\}$
(d) $R = \{(x, y) | x \in A, y \in A, x > y\}$

2. Consider the set A given by

$$A = \{-1, 0, 1\}$$

Write the binary relation for each of the following descriptions:

(a) $R = \{(x, y) | x \in A, y \in A, y = |x|\}$
(b) $R = \{(x, y) | x \in A, y \in A, y = |x| - 1\}$
(c) $R = \{(x, y) | x \in A, y \in A, y = -x\}$
(d) $R = \{(x, y) | x \in A, y \in A, y = x\}$

3. Write the binary relation that represents the following digraph

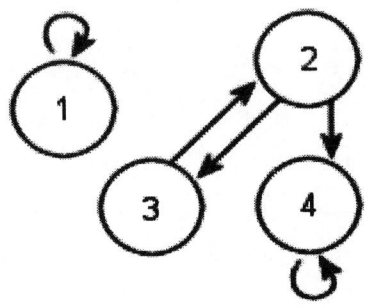

4. Write the binary relation that represents the following digraph

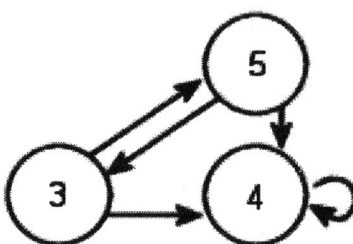

5. Write the binary relation that represents the following digraph

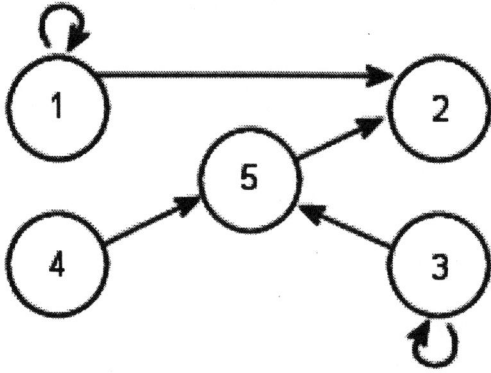

6. Write the binary relation that represents the following digraph

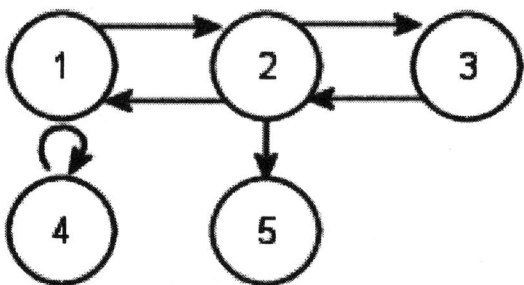

7. Construct a proper digraph for each of the binary relations given in Exercise 1

8. Construct a proper digraph for each of the binary relations given in Exercise 2

13.2 Trees

A **tree** is defined as a special type of graph mostly used in data structure and probability courses. A natural tree usually consists of a root, branches that differ in number from one tree to another, and leaves, that also differ in number, at the end of the branches. A tree diagram has been used before in working with probability concepts, where the root has been designated by the label "start", and the branches represent the output of the experiment performed.

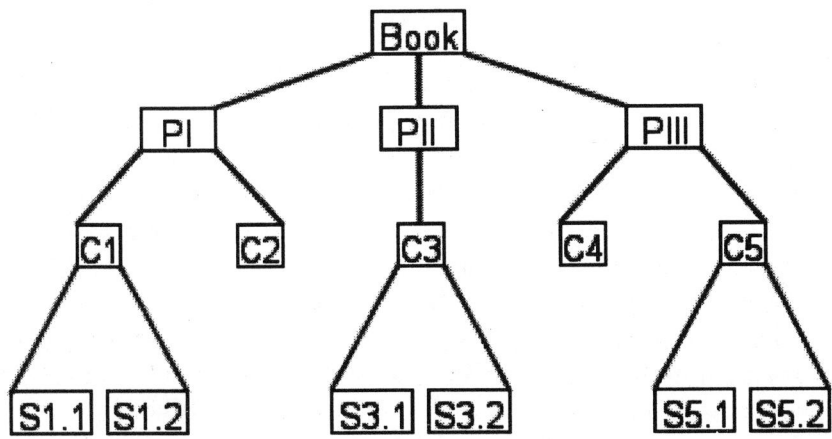

Trees are used to store information within a computer or a calculator. The hierarchical tree represents the classification of a certain book to parts, chapters, and sections. Usually the roots of a natural tree are drawn at the bottom. However, in graphs, trees are drawn with the root at the top with the branches of the tree extending downwards and the leaves at the lowest level. Observing the tree shown below, we introduce the following remarks:

1. The root of the tree at the top represents the book. The root is usually represented by a node. We call the root to be on level 0. The root represents the origin of the downward extending. The level 0 is identical to the origin in a coordinate plane.

2. Three distinct branches extend downwards from the root to the vertices PI, PII, and $PIII$. The vertices labeled PI, PII, and $PIII$ represent the three parts of the book, namely part 1, part 2, and part 3 respectively. Since the level of any vertex is defined as the distance from the root to the vertex, then vertices PI, PII, and $PIII$ are called to be on level 1.

3. Part 1 has two subbranches, namely, $C1$, and $C2$ that represent Chapter 1 and Chapter 2 respectively. Part II has one subbranch indicated by $C3$ that represents Chapter 3. Part III has two subbranches, namely, $C4$, and $C5$ that represent Chapter 4 and Chapter 5 respectively.

The vertices $C1, C2, C3, C4$, and $C5$ are said to be on level 2 from the root. This means that each vertex is 2 distances, or edges, from the root.

3. At level 3, the sections of each chapter are specified. Chapters 1, 3, and 5, each has two sections, whereas Chapters 2, and 4 have no sections. The sections of Chapter 1 are indicated by $S1.1$ and $S1.2$, the sections of Chapter 3 are indicated by $S3.1$ and $S3.2$, and the sections of Chapter 5 are indicated by $S5.1$ and $S5.2$.

Notice that the nodes at the end of the tree, where the tree does not extend, are called **leaves**. In the tree shown above, the nodes $S1.1, S1.2, C2, S3.1$, $S3.2, C4, S5.1$, and $S5.2$ are called the leaves of this tree.

It is worth noting that other trees, that represent several data information such as family trees and directories of a hard disc, are possible trees where branches differ in number and levels. However, in this text we will focus our study on **rooted binary trees only**. Binary trees are considered the most useful type of rooted trees. This type of trees has a typical node that represents the root, and only 0, 1, or 2 branches, or edges, are extended from the root node and from any vertex at other levels. Since three branches extended from the root of the tree shown above, therefore, the tree is not a binary tree. As stated before, the node with 0 edges is called a **leaf**. Moreover, the tree is called a binary tree because there are at most two directed edges, called left branch and right branch, that extend from the root.

To study how an arithmetic and algebraic expressions can be represented by **binary rooted trees**, we consider the following steps:

1. For the expression $A + B$, the main operator $+$ labels the root node, where A and B labels the nodes of the two branches that extend from the root. Hence, we obtain the following three-node tree

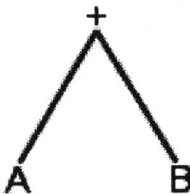

2. If $A = x + y$ and $B = u - v$, then two branches will extend from each

of the nodes A and B to represent the expression

$$A + B = (x + y) + (u - v)$$

Proceeding as before, we obtain the following tree

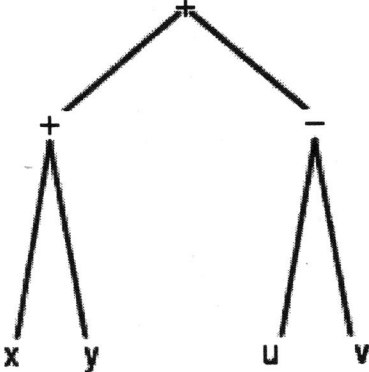

The structure of the binary trees will be explained by the following illustrative examples.

Example 1. Construct a binary tree for the arithmetic expression

$$(4 \times 6) + (3 \uparrow 2)$$

Solution. We note first that the $+$ operation is the main operator between the parentheses, therefore we use the $+$ to label the root node. Two branches extend from the root, the first is 4×6 that is represented by two subbranches that extend from the $\times$ operator.

A similar representation can be made for the second branch that represents $3 \uparrow 2$, where $\uparrow$ indicates exponent. The tree that represents this expression is as shown below.

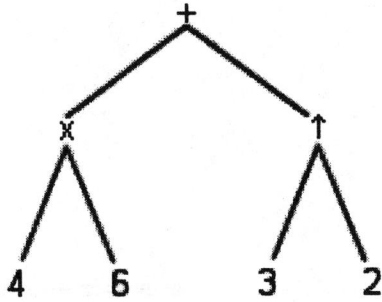

Example 2. Construct a binary tree for the arithmetic expression

$$(2 \times 4 + 3) - (6 \div 2)$$

Solution. We first set

$$A = (2 \times 4 + 3)$$

and

$$B = (6 \div 2)$$

The main operator is $-$, and this labels the root node as shown by the following graph

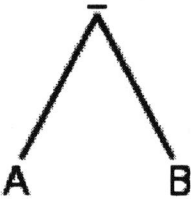

For $A = (2 \times 4) + 3$, it can be first represented by

For the expression 2×4, it can be represented by subbranches, hence A is represented by

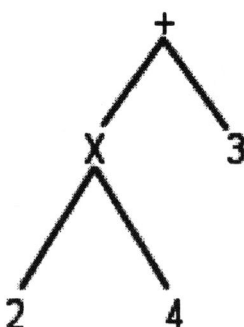

For the expression $B = 6 \div 2$, we find

Combining the above representations of A and B, we find

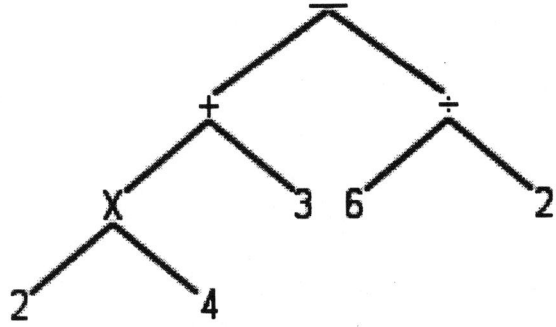

Example 3. Construct a binary tree for the arithmetic expression

$$((4 - x) \div 3) \uparrow ((x + y) \times 2)$$

Solution. It is clear that the operator $\uparrow$ between the two main parentheses is the main operator. We set

$$A = ((4 - x) \div 3)$$

and

$$B = ((x + y) \times 2)$$

Hence we have

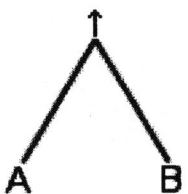

Following the previous example, A can be represented first by the following graph

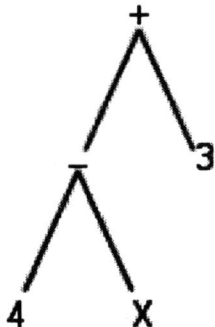

Similarly, for B we obtain the following:

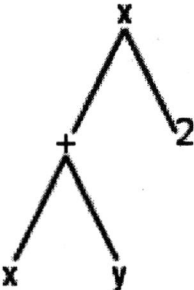

Combining all these parts we obtain the binary rooted tree

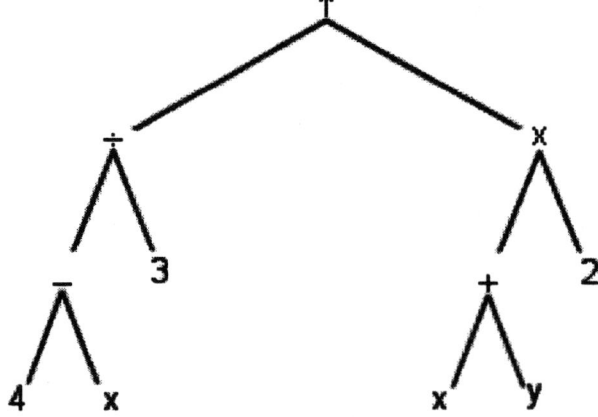

Expressing a binary tree by an algebraic expression

On the other hand, it is useful now to discuss how we can write the arithmetic or the algebraic expression if the binary rooted tree is given. We first notice the main operator at the node of the root. We then consider the branches of the tree. Each branch is read from left to rigth and from bottom to upper level.

Expressing a binary tree by an algebraic expression or by arithmetic expression can be explained by considering the binary tree

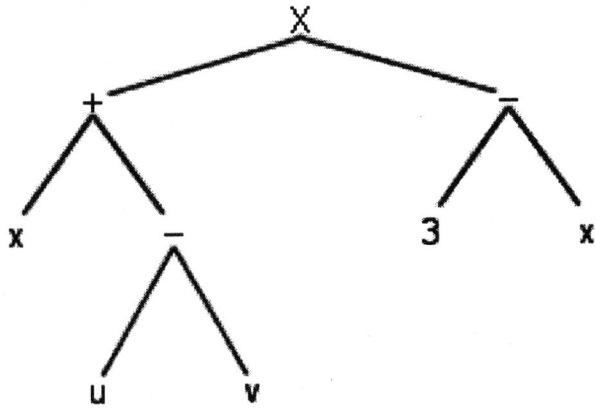

The main operator is × represented by the node of the root. This tree is given by two branches A and B, where A is given by the left part and B is given by the right part. The left part A can be read from left to right and from bottom to upper level. Accordingly, we obtain

$$A = x + (u - v)$$

In a similar manner, the right part B can also be read from left to right and from bottom to top. Hence, we obtain

$$B = 3 - x$$

Using the operator × to combine the two parts we obtain the equivalent expression given by

$$\text{Expression} = (x + (u - v)) \times (3 - x)$$

Another example that will translate the binary rooted tree to an equivalent algebraic expression is given by the following example.

Example 4. Write the algebraic expression for the following binary tree.

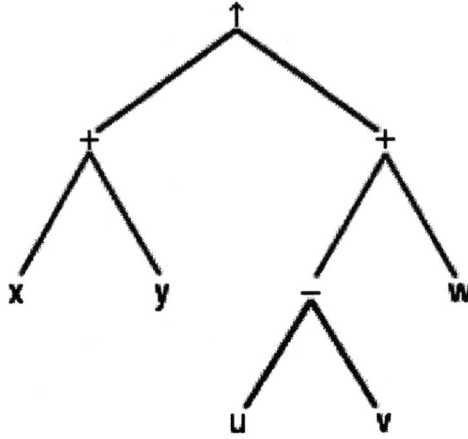

Solution. The main operator is ↑ represented by the node of the root. This tree is given by two branches A and B, where A is given by the left part and B is given by the right part. The left part A can be read from left to right and from bottom to upper level. Accordingly, we obtain

$$A = (x + y)$$

In a similar manner, the right part B can also be read from left to right and from bottom to top. Hence, we obtain

$$B = ((u - v) + w)$$

Using the operator × to combine the two parts we obtain the equivalent expression given by

$$\text{Expression} = (x + y) \uparrow ((u - v) + w)$$

Exercises 13.2

In Exercises 1 – 8, graph the rooted tree that represents each of the following expressions:

1. $(a + b) \times (c + d)$

2. $((6 + 5) \div 3) + (6^2)$

3. $(a \div b) \times (c + d)$

4. $(x - y) \uparrow (a + (b \div c))$

5. $(6 + (a \div b)) \times (x - (y \times z))$

6. $(6 - (a \div b)) \times (y \times x)$

7. $((a \div b) \times c) \uparrow ((a - (b \div c))$

8. $(4 \times (a \div b)) + (a \times b)$

9. Write the expression represented by the following binary tree

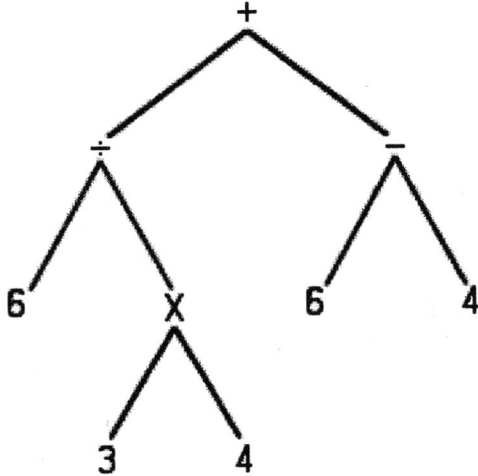

10. Write the expression represented by the following binary tree

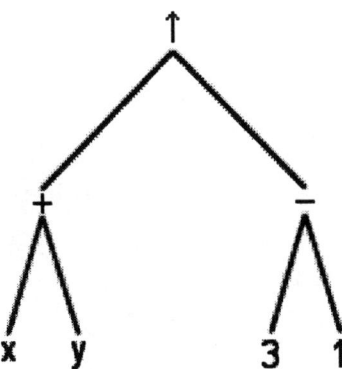

11. Write the expression represented by the following binary tree

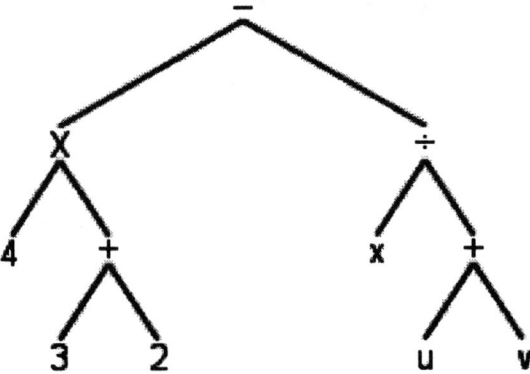

12. Write the expression represented by the following binary tree

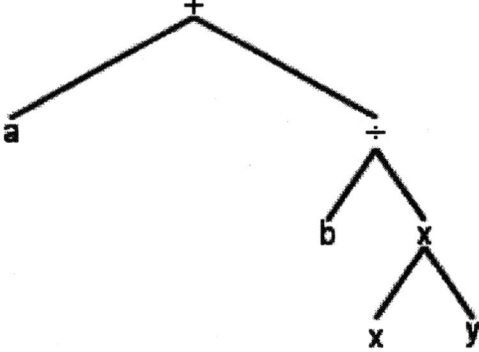

13. Write the expression represented by the following binary tree

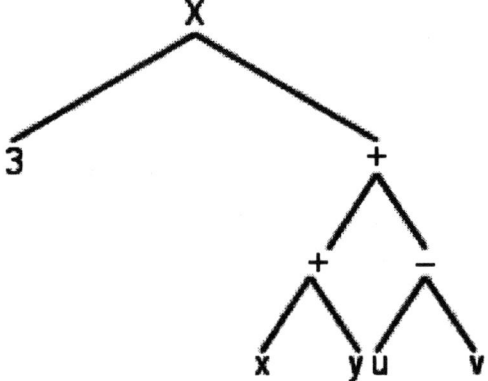

14. Write the expression represented by the following binary tree

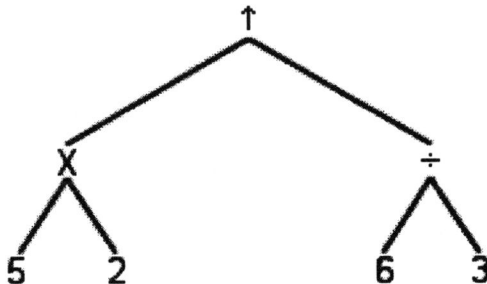

15. Write the expression represented by the following binary tree

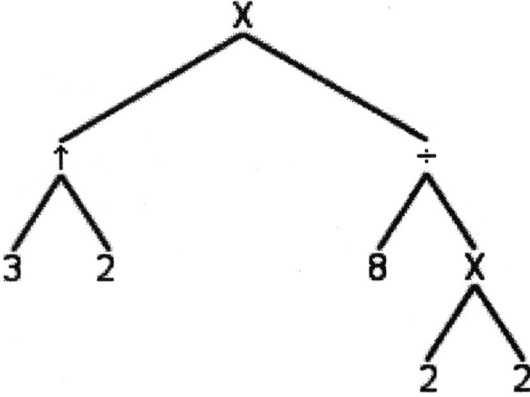

16. Write the expression represented by the following binary tree

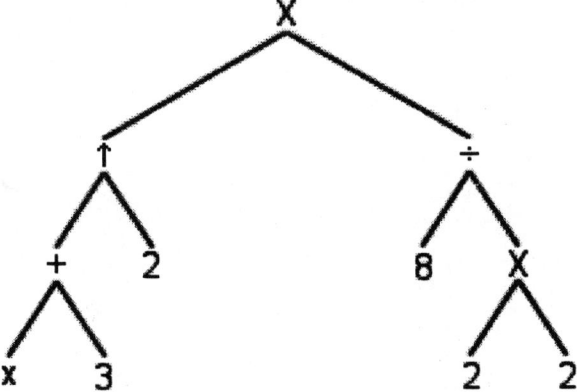

13.3   Tree Searching

In Section 10.2, we discussed how we can represent an algebraic or arithmetic expression by a rooted binary tree. We have also seen that the binary operators such as $+, -, \times$, etc., are represented by nodes or vertices of the tree. The variables $x, y, z, \cdots$, or the numbers $1, 2, 3, \cdots$ are represented by the leaves at the end of the branches of a tree.

Computers usually store the information represented by a rooted tree by traversing that tree from node to node. This can be done in two ways:

1.   Preorder: We first define what is known by **Polish notation**. In this notation, we usually represent any binary expression such as $(a + b)$ by the representation $(+ab)$, where the binary operator $+$ that acts on the variables a and b precedes the two variables. Similarly, the expression $(x \uparrow y)$ is expressed in Polish notation by $(\uparrow xy)$, and $(6 \div 2)$ is expressed as $(\div 62)$. The binary operators $\uparrow$ and $\div$ precede each pairs of variables or numbers that act on. However, the information $\times 43$ in Polish notation is equivalent to 4×3. In the same way, we can calculate the following expression given in Polish notation by $(+ \div 624)$. We start the calculation work by inserting the inner operator $\div$ between the two numbers that follow this operator, hence we obtain

$+ \div 624 = +(6 \div 2)4 = +34$

We then insert the operator $+$ between the two numbers that follow this operator, hence we find

$+34 = 3 + 4 = 7$

Based on the ideas discussed about the Polish notation, the computer views any tree by using the preorder, or Polish notation. In this method, the root and each node is visited once. The root is visited first, then every node is visited from up to down and from left to right. This is illustrated by considering the following tree.

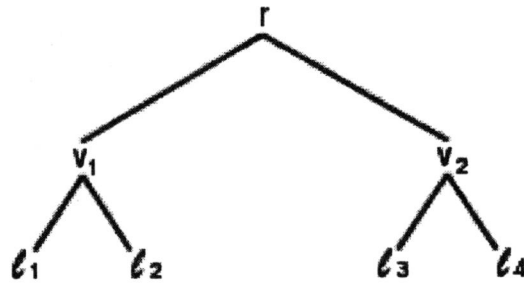

As mentioned above, we first start by visiting the root r. Then moving from the top to the bottom, the vertex v_1 is visited next. Proceeding downwards, the leaves l_1 and l_2 are visited respectively from left to right.

Next, the vertex v_2 is visited to proceed from there to visit the leaves l_3 and l_4. Accordingly, the preorder output of the nodes of the tree is given by

$$rv_1\ell_1\ell_2v_2\ell_3\ell_4$$

The preorder listing of the vertices of trees can be explained more by examining the following examples.

Example 1. Write the preorder listing of the vertices of the Following tree.

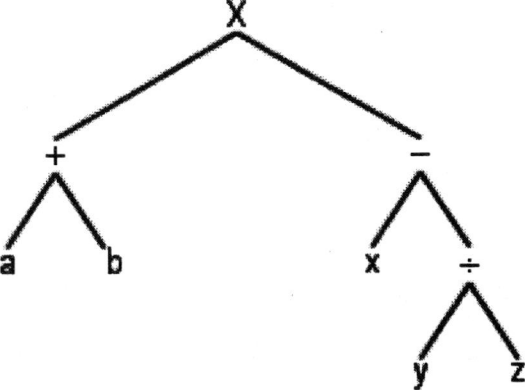

Solution. The vertex $\times$ is visited first followed by visiting the $+$ operator then down to the leaves a and b respectively. The operator $-$ is visited next, moving to the leaf x. Finally, the node $\div$ is visited to move to the leaves y and z respectively. Accordingly, the preorder listing is defined by

$$\times + ab - x \div yz$$

It is useful to note that the algebraic expression that is equivalent to this preorder listing can be obtained as follows. We first start by inserting the operator $\div$ between the variables y and z. We then insert the operator $-$ between the two quantities that follow this operator. We continue in this manner and apply the following steps:

$$\times + ab - x \div yz \quad = \quad \times + ab - x(y \div z)$$
$$= \quad \times + ab - x(\frac{y}{z})$$

$$= \ \times + ab(x - \frac{y}{z})$$

$$= \ \times(a+b)(x - \frac{y}{z})$$

$$= \ (a+b) \times (x - \frac{y}{z})$$

Example 2. Repeat Example 1 for the binary tree given by

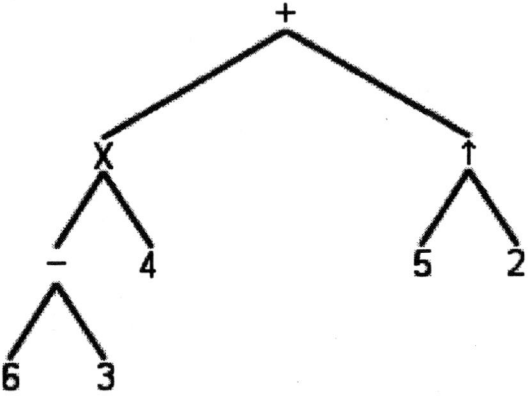

Solution. Following the discussion made above, the preorder listing is

$$+ \times -634 \uparrow 52$$

To determine the equivalent expression by using the preorder listing, we proceed as discussed in the previous example by using the operator $\uparrow$ first, and proceed as before. Therefore, we find

$$
\begin{aligned}
+ \times -634 \uparrow 52 \ &= \ + \times -634(5 \uparrow 2) \\
&= \ + \times -634(5^2) \\
&= \ + \times -634(25) \\
&= \ + \times (6-3)4(25) \\
&= \ + \times (3)4(25) \\
&= \ +(3 \times 4)(25) \\
&= \ +(12)(25) \\
&= \ (12 + 25) = 37
\end{aligned}
$$

The same result can be obtained by using the tree and the conversion to an equivalent expression.

2. Postorder: In this notation we use what is called the reverse Polish notation. In this notation the binary operator such as $+, -, \times$, etc., follows the two variables or numbers that acts on. For example, the expression $a + b$ is expressed by $ab+$. Also, the expression, $6 - 4$ is expressed by $64-$.

On the other hand, the postorder notation $xy+$ represents the expression $x + y$. In addition, the postorder notation $123 + \div$ represents the expression $1 \div (2 + 3)$. In other words, the operator follows the variables.

Unlike the preorder notation, the root in the postorder methods is visited at last. We start first by visiting the leaves, beginning with the leftmost one before visiting the vertex of these leaves. We then continue to visit the leaves of the other vertex before visiting that vertex. Finally, we end by visiting the root at the top of the tree.

The postorder traversal is illustrated by discussing the following binary tree.

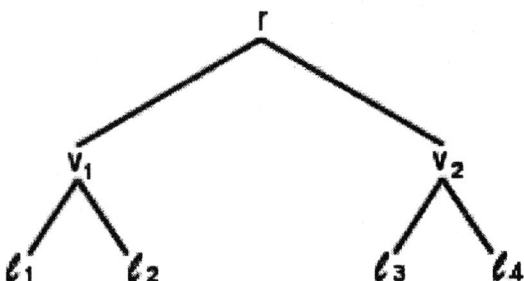

In writing the postorder listing of this tree, the leaves ℓ_1 and ℓ_2 respectively are visited first, then we proceed from here to visit the vertex v_1 of these two leaves. In a similar manner, we visit the leaves ℓ_3 and ℓ_4 respectively. We then proceed from here to visit the vertex v_2 of these two leaves. Finally, we visit the root r.

Based on this, the postorder listing of this tree is shown by

$$\ell_1 \ell_2 v_1 \ell_3 \ell_4 v_2 r$$

As indicated before, the leaves of the binary tree are visited first from left to right, then the vertices are visited, and the root at the top is visited at last.

Example 3. Write the postorder listing of vertices of the binary tree given before in Example 1.

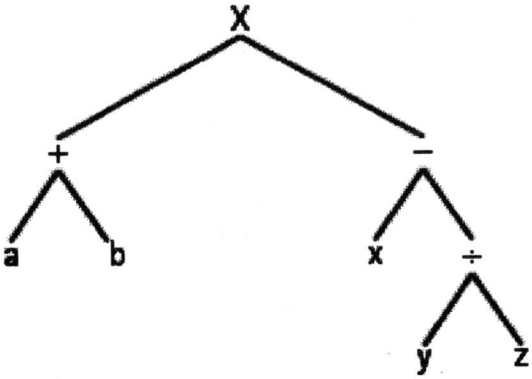

Solution. The postorder listing is given by

$$ab + xyz \div -\times$$

Proceeding as discussed before we get

$$
\begin{aligned}
ab + xyz \div -\times \;&=\; (a+b)xyz \div -\times \\
&=\; (a+b)x(y \div z) - \times \\
&=\; (a+b)x(\tfrac{y}{z}) - \times \\
&=\; (a+b)(x - \tfrac{y}{z})\times = (a+b) \times (x - \tfrac{y}{z})
\end{aligned}
$$

Example 4. Write the post order listing of the binary tree of Example 2 and given by

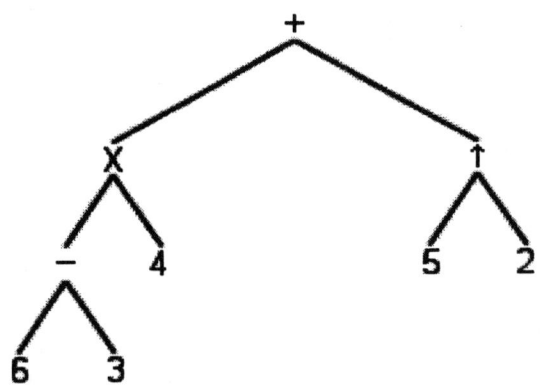

Solution. The postorder is given by

$$63 - 4 \times 52 \uparrow +$$

The equivalent arithmetic expression can be found by listing each operator to act on the preceding two numbers or resulting quantities. Hence, we have

$$
\begin{aligned}
63 - 4 \times 52 \uparrow + \ &= \ 63 - 4 \times 52 \uparrow + \\
&= \ (6 - 3)4 \times 52 \uparrow + \\
&= \ (3)4 \times 52 \uparrow + \\
&= \ (3 \times 4)52 \uparrow + \\
&= \ (12)52 \uparrow + \\
&= \ (12)(5 \uparrow 2) + \\
&= \ (12)(25) + \\
&= \ (12) + (25) \\
&= \ 37
\end{aligned}
$$

Exercises 13.3

In Exercises 1 – 14, write the algebraic expression or the arithmetic value of each of the following expressions written in Polish notation.

1. $\uparrow xy + -ab$

2. $+ \div ab - xy$

3. $+ \uparrow 23 \div 62$

4. $\uparrow +34 + -342$

In Exercises 5 – 8, write the algebraic expression or the arithmetic value of each of the following expressions written in reverse Polish notation.

5. $xy \times xy + -$

6. $ab \times ab + c - \div$

7. $42 \div 2 + 43 + 2 \uparrow \times$

8. $81 + 3 - 82 \div 2 + \div$

In Exercises 9 – 12, graph the rooted tree for each algebraic expression and use the tree to write the expression in Polish notation.

9. $(x - y) \div (x + y)$

10. $(a + b) \uparrow ((a - b) + c)$

11. $((6 \div 2) + 1) \times ((4 + 3) \uparrow 2)$

12. $((12 - 3) \div 9) + ((3 + 1) \uparrow (4 - 2))$

In Exercises 13 – 16, write the expression in reverse Polish notation.

13. $(x - y) \div (x + y)$

14. $(a + b)^{a - b + c}$

15. $((6 \div 2) + 1)(4 + 3)^2$

16. $((1 - 3)^2) \times (6 + 4)$

In Exercises 17 – 20, evaluate the preorder expression if $a = 2$, $b = 4$, and $c = 5$.

17. $+ \uparrow ab - ca$

18. $\times \div ba + a - cb$

19. $\uparrow + \times abca$

20. $+ \times -cbab$

In Exercises 21 – 24, evaluate the post order expressions if $A = 3$, $B = 4$, and $C = 2$.

21. $AB - AC \times +$

22. $BA \times BC \div +$

23. $CB - A \times B +$

24. $AB - C + CBA \times \times +$

Appendix A

Prime Numbers

A.1 Prime Numbers

2	3	5	7	11	13	17	19	23	29
31	37	41	43	47	53	59	61	67	71
73	79	83	89	97	101	103	107	109	113
127	131	137	139	149	151	157	163	167	173
179	181	191	193	197	199	211	223	227	229
233	239	241	251	257	263	269	271	277	281
283	293	307	311	313	317	331	337	347	349
353	359	367	373	379	383	389	397	401	409
419	421	431	433	439	443	449	457	461	463
467	479	487	491	499	503	509	521	523	541
547	557	563	569	571	577	587	593	599	601
607	613	617	619	631	641	643	647	653	659
661	673	677	683	691	701	709	719	727	733
739	743	751	757	761	769	773	787	797	809

811	821	823	827	829	839	853	857	859	863
877	881	883	887	907	911	919	929	937	941
947	953	967	971	977	983	991	997	1009	1013
1019	1021	1031	1033	1039	1049	1051	1061	1063	1069
1087	1091	1093	1097	1103	1109	1117	1123	1129	1151
1153	1163	1171	1181	1187	1193	1201	1213	1217	1223
1229	1231	1237	1249	1259	1277	1279	1283	1289	1291
1297	1301	1303	1307	1319	1321	1327	1361	1367	1373
1381	1399	1409	1423	1427	1429	1433	1439	1447	1451
1453	1459	1471	1481	1483	1487	1489	1493	1499	1511
1523	1531	1543	1549	1553	1559	1567	1571	1579	1583
1597	1601	1607	1609	1613	1619	1621	1627	1637	1657
1663	1667	1669	1693	1697	1699	1709	1721	1723	1733
1741	1747	1753	1759	1777	1783	1787	1789	1801	1811
1823	1831	1847	1861	1867	1871	1873	1877	1879	1889
1901	1907	1913	1931	1933	1949	1951	1973	1979	1987

A.2   Mersenne Prime Numbers

#	p	Digits in M_p	Discovery Date	Discoverer
1	2	1	ancient	ancient
2	3	1	ancient	ancient
3	5	2	ancient	ancient
4	7	3	ancient	ancient
5	13	4	1456	anonymous
6	17	6	1588	Cataldi
7	19	6	1588	Cataldi
8	31	10	1772	Euler
9	61	19	1883	Pervushin
10	89	27	1911	Powers
11	107	33	1914	Powers
12	127	39	1876	Lucas
13	521	157	1952	Robinson
14	607	183	1952	Robinson
15	1,279	386	1952	Robinson
16	2,203	664	1952	Robinson
17	2,281	687	1952	Robinson
18	3,217	969	1957	Riesel
19	4,253	1,281	1961	Hurwitz
20	4,423	1,332	1961	Hurwitz

#	p	Digits in M_p	Discovery Date	Discoverer
21	9,689	2,917	1963	Gillies
22	9,941	2,993	1963	Gillies
23	11,213	3,376	1963	Gillies
24	19,937	6,002	1971	Tuckerman
25	21,701	6,533	1978	Noll & Nickel
26	23,209	6,987	1979	Noll
27	44,497	13,395	1979	Nelson & Slowinski
28	86,243	25,962	1982	Slowinski
29	110,503	33,265	1988	Colquitt & Welsh
30	132,049	39,751	1983	Slowinski
31	216,091	65,050	1985	Slowinski
32	756,839	227,832	1992	Slowinski & Gage
33	859,433	258,716	1994	Slowinski & Gage
34	1,257,787	378,632	1996	Slowinski & Gage
35	1,398,269	420,921	1996	Joel Armengaud
36	2,976,221	895,932	1997	Gordon Spence
37	3,021,377	909,526	1998	Roland Clarkson
38	6,972,593	2,098,960	1999	Nayan Hajratwala
39	13,466,917	4,053,946	2001	Michael Cameron
40	20,996,011	6,320,430	2003	Michael Shafer

#	p	Digits in M_p	Discovery Date	Discoverer
41[*]	24,036,583	7,235,733	2004	Josh Findley
42[*]	25,964,951	7,816,230	2005	Martin Nowak
43[*]	30,402,457	9,152,052	2005	Curtis Cooper and Steven Boone
44[*]	32,582,657	9,808,358	2006	Curtis Cooper and Steven Boone
45[*]	37,156,667	11,185,272	2008	Hans-Michael Elvenich
46[*]	42,643,801	12,837,064	2009	Odd M. Strindmo
47[*]	43,112,609	12,978,189	2008	Edson Smith
48[*]	57,885,161	17,425,170	2013	Curtis Cooper

[*] It is not known whether any undiscovered Mersenne primes exist between the 41st and the 47th on this chart; the ranking is therefore provisional.

A.3   Perfect Numbers

A positive integer n is called a perfect number if it is equal to the sum of its positive factors (not including the number itself). Examples of perfect factors are 6 and 8 where

$$6 = 1 + 2 + 3$$
$$28 = 1 + 2 + 4 + 7 + 14$$

It is well know that if $2^p - 1$ is a prime, then $2^{p-1}(2^p - 1)$ is perfect. The first few perfect numbers are shown by the following table:

p	$2^{p-1}(2^p - 1)$	Perfect Number
2	$2^1(2^2 - 1)$	6
3	$2^2(2^3 - 1)$	28
5	$2^4(2^5 - 1)$	496
7	$2^6(2^7 - 1)$	8128
13	$2^{12}(2^{13} - 1)$	33550336
17	$2^{16}(2^{17} - 1)$	8589869056
19	$2^{18}(2^{19} - 1)$	137438691328

A.4   Amicable Numbers

Two numbers (m, n) are called amicable numbers if each number equals to the sum of the divisors of the other number (not including the number itself). Examples of perfect numbers are 220 and 284. where divisors of 220 are

$$1, 2, 4, 5, 10, 11, 20, 22, 44, 55, 110$$

The sum of these divisors is 284. However, the divisors of 284 are

$$1, 2, 4, 71, 142$$

The sum of these divisors is 220. Therefore, the pair $(220, 284)$ consists of two amicable numbers. The first few pairs of amicable numbers are shown by the following table:

Pair Number	Amicable Numbers	
1	220	284
2	1184	1210
3	2620	2924
4	5020	5564
5	6232	6368
6	10744	10856
7	12285	14595
8	17296	18416
9	63020	76084
10	66928	66992
11	67095	71145
12	69615	87633

A.5 Catalan Numbers

The Catalan numbers form a sequence of natural numbers, named after Eugne Charles Catalan (18141894), arise in a number of counting problems of combinatorics. The nth Catalan number C_n is given in terms of the binomial coefficients by

$$C_n = \frac{1}{n+1}\binom{2n}{n} = \frac{(2n)!}{(n+1)!n!}, n \geq 0$$

The Catalan numbers satisfy the following recurrence relation

$$C_{n+1} = \sum_{i=0}^{n} C_i C_{n-i}, C_0 = 1, n \geq 0$$

The first Catalan numbers can be found in the table below.

One of the properties of the Catalan numbers is given by the Hankel matrix. The Hankel matrix is a square matrix whose (i, j) entry is the Catalan number C_{i+j-2} has determinant 1, regardless of the value of n. For example, for $n = 4$, we obtain the 4×4 Hankel matrix

$$H = \begin{bmatrix} 1 & 1 & 2 & 5 \\ 1 & 2 & 5 & 14 \\ 2 & 5 & 14 & 42 \\ 5 & 14 & 42 & 132 \end{bmatrix}$$

Notice that $\det(H) = 1$.

The first Catalan numbers are shown by the following table:

n	Catalan Numbers
0	1
1	1
2	2
3	5
4	14
5	42
6	132
7	429
8	1430
9	4862
10	16796
11	58786
12	208012

Appendix B

Summation Formulas

B.1 Summation Formulas for Polynomials

1. $\displaystyle\sum_{k=1}^{n} k = \frac{n(n+1)}{2}$

2. $\displaystyle\sum_{k=1}^{n} k^2 = \frac{n(n+1)(2n+1)}{6}$

3. $\displaystyle\sum_{k=1}^{n} k^3 = \frac{n^2(n+1)^2}{4}$

4. $\displaystyle\sum_{k=1}^{n} k^4 = \frac{n(n+1)(2n+1)(3n^2+3n-1)}{30}$

5. $\displaystyle\sum_{k=1}^{n} k^5 = \frac{n^2(n+1)^2(2n^2+2n-1)}{12}$

6. $\displaystyle\sum_{k=1}^{n} k^6 = \frac{n(n+1)(2n+1)(3n^4+6n^3-3n+1)}{42}$

7. $\displaystyle\sum_{k=1}^{n} k^7 = \frac{n^2(n+1)^2(3n^4+6n^3-n^2-4n+2)}{24}$

8. $\displaystyle\sum_{k=1}^{n} k^8 = \frac{n(n+1)(2n+1)(5n^6+15n^5+5n^4-15n^3-n^2+9n-3)}{90}$

9. $\displaystyle\sum_{k=1}^{n} k^9 = \frac{n^2(n+1)^2(n^2+n-1)(2n^4+4n^3-n^2-3n+3)}{20}$

10. **Sum of Odd Numbers:**

$$1+3+5+7+\cdots+(2n-1) = n^2$$

11. **Sum of Even Numbers:**

$$2+4+6+8+\cdots+2n = n(n+1)$$

B.2 Summation Formulas for Geometric Series

(i)
$$1+2+2^2+2^3+\cdots+2^n = 2^{n+1}-1$$

(ii)
$$1+3+3^2+2^3+\cdots+3^n = \frac{3^{n+1}-1}{2}$$

B.3 Other Summations Formulas

(i)
$$1\cdot2+2\cdot3+3\cdot4+\cdots+n(n+1) = \frac{n(n+1)(n+2)}{3}$$

(ii)
$$1\cdot3+2\cdot4+3\cdot5+\cdots+n(n+2) = \frac{n(n+1)(2n+7)}{6}$$

(iii)
$$1\cdot2+3\cdot4+5\cdot6+\cdots+n(2n-1)(2n) = \frac{n(n+1)(4n-1)}{3}$$

(iv)
$$1^2+3^2+5^2+\cdots+(2n-1)^2 = \frac{n(4n^2-1)}{3}$$

(v)
$$1\cdot2\cdot3+2\cdot3\cdot4+\cdots+n(n+1)(n+2) = \frac{n(n+1)(n+2)(n+3)}{4}$$

Appendix C

Fibonacci and Lucas Numbers

C.1   First 30 Fibonacci Numbers

1	1	2	3	5	8
13	21	34	55	89	144
233	377	610	987	1597	2584
4181	6765	10946	17711	28657	46368
75025	121393	196418	317811	514229	832040

C.2   First 30 Lucas Numbers

1	3	4	7	11	18
29	47	76	123	199	322
521	843	1364	2207	3571	5778
9349	15127	24476	39603	64079	103682
167761	271443	439204	710647	1149851	1860498

C.3   Summation Formulas for Fibonacci Sequence

1. Sum of Fibonacci Numbers:

$$F_1 + F_2 + F_3 + F_4 + \cdots + F_n = F_{n+2} - 1$$

2. Sum of Fibonacci Numbers with Odd Subscripts:

$$F_1 + F_3 + F_5 + F_7 + \cdots + F_{2n-1} = F_{2n}$$

3. Sum of Fibonacci Numbers with Even Subscripts:

$$F_2 + F_4 + F_6 + F_8 + \cdots + F_{2n} = F_{2n+1} - 1$$

4. Sum of Squares of Fibonacci Numbers:

$$F_1^2 + F_2^2 + F_3^2 + F_4^2 + \cdots + F_n^2 = F_n F_{n+1}$$

C.4   Summation Formulas for Lucas Sequence

1. Sum of Lucas Numbers:

$$L_1 + L_2 + L_3 + L_4 + \cdots + L_n = L_{n+2} - 3$$

2. Sum of Lucas Numbers with Odd Subscripts:

$$L_1 + L_3 + L_5 + L_7 + \cdots + L_{2n-1} = L_{2n} - 2$$

3. Sum of Lucas Numbers with Even Subscripts:

$$L_2 + L_4 + L_6 + L_8 + \cdots + L_{2n} = L_{2n+1} - 1$$

4. Sum of Squares of Lucas Numbers:

$$L_1^2 + L_2^2 + L_3^2 + L_4^2 + \cdots + L_n^2 = L_n L_{n+1} - 2$$

ANSWERS

Exercises 1.3

1.(a) $1011_{(2)}$ (b) $10101_{(2)}$ (c) $1110001_{(2)}$ (d) $10101011_{(2)}$

3.(a) $101111_{(2)}$ (b) $111010_{(2)}$ (c) $1010101_{(2)}$ (d) $1011011_{(2)}$

5.(a) $x = 1$ (b) $x = 0$ (c) $x = 2$ (d) $x = 2$

7.(a) $101010_{(2)}$ (b) $10011010_{(2)}$

Exercises 1.4

1.(a) $347_{(8)}$ (b) $472_{(8)}$ (c) $1061_{(8)}$ (d) $1323_{(8)}$

3.(a) $77_{(8)}$ (b) $141_{(8)}$ (c) $157_{(8)}$ (d) $235_{(8)}$

5.(a) $y = 4$ (b) $y = 2$ (c) $x = 8$ (d) $x = 8$

7. (a) $x = 4$ (b) $x = 6$ (c) $x = 8$ (d) $x = 9$

Exercises 1.4.3

1.(a) $101100011_{(2)}$ (b) $11100101_{(2)}$ (c) $1111110_{(2)}$ (d) $10110001_{(2)}$

3.(a) $731_{(8)}$ (b) $553_{(8)}$ (c) $275_{(8)}$ (d) $132_{(8)}$

5.(a) $x = 4$ (b) $x = 110$ (c) $x = 3$ (d) $x = 101$

Exercises 1.5

1. (a) $28B_{(16)}$ (b) $CBA_{(16)}$ (c) $DEF_{(16)}$ (d) $CBA1_{(16)}$

3. (a) $91A_{(16)}$ (b) $9AB_{(16)}$ (c) $381_{(16)}$ (d) $A4B_{(16)}$

5. (a) True (b) True (c) False (d) False

7. (a) $1234_{(5)}$ (b) $1234_{(6)}$ (c) $1234_{(7)}$ (d) $1234_{(8)}$

9. $x = 7$ 11. $x = 6$ 13. $x = 5$ 15. $x = 8$

17. $x = 8$ 19. $x = 16$ 21. $x = 6$ 23. $x = 16$

25. $x = 10$ 27. $x = 4$

Exercises 1.5.4

1. (a) $1101001010_{(2)}$ (b) $110111100001_{(2)}$

3. (a) $D6C_{(16)}$ (b) $FC1_{(16)}$ (c) $356B_{(16)}$ (d) $81C_{(16)}$

5. (a) $x = B$ (b) $x = 1100$ (c) $x = A$ (d) $x = 1111$

7. (a) $E03_{(16)}$ (b) $1025_{(16)}$ (c) $18AB_{(16)}$ (d) $1CA7_{(16)}$

Exercises 2.1

1. Factors of 48 are: 1, 2, 3, 4, 6, 8, 12, 16, 24, 48

   Factors of 57 are:1, 3, 19, 57

3. The five primes that can be obtained are:

   2, 5, 17, 37, 101

5. For $x = 0, 1, \ldots\ldots, 9$ we find the primes:

   17, 19, 23, 29,37, 47, 59, 73, 89, 107

7. The primes between 50 and 100 are:

   53, 59, 61, 67, 71, 73, 79, 83, 89, 97

9. Substituting $p = 5, 7, 11, 13, 19$ in the formula gives:

27, 51, 123, 171, 291

11. For $k = 2$, we find the pair (17,19)

   For $k = 5$, we find the pair (41,43)

13. 5, 13, 17, 29, 37, 41

15. 4=2+2, 6=3+3, 8=3+5, 10=5+5, 12=5+7

17. $M = 2^{29} - 1 = 536870911 = 233 \times 1103 \times 2089$

19. Obtained primes are $5, 11, 83$

21. Obtained primes are $7, 1031, 25999$

Exercises 2.2

1. (a) 13   (b) 41   (c) 29

3. (a) 61   (b) 62   (c) 1

7. (a) 13   (b) 13   (c) 13

   (d) g.c.d(a, b) = g.c.d$(a + b, b)$ = g.c.d$(a - b, b)$

9. 47, $1363 = 47 \times 29$, $1739 = 47 \times 37$

11. 271, $16531 = 271 \times 61$, $18157 = 271 \times 67$

Exercises 3.1

1. (a) $1, 4, 7, 10, 13, \cdots$ (b) $2, 7, 12, 17, 22, \cdots$

3. (a) $x \equiv 2 \bmod 3$ (b) $x \equiv 1 \bmod 5$

5. 7   7. 0   9. 0   11. 1   13. 3   15. 9   17. 81   19. Yes

Exercises 3.2

1. (a) $x \equiv 2 \bmod 5$ (b) $x \equiv 2 \bmod 7$

3. (a) $x \equiv 5 \bmod 7$ (b) $x \equiv 10 \bmod 11$

5. (a) $x \equiv 2 \bmod 11$ (b) $x \equiv 3 \bmod 7$

7.(a) $x \equiv 2 \bmod 35$ (b) $x \equiv 1 \bmod 15$

9.(a) $x \equiv 1 \bmod 30$ (b) $x \equiv 2 \bmod 105$

11. $x \equiv 2 \bmod 231$

   yes, if we divide 22871 by 231, the remainder is 2

13. $x \equiv 5 \bmod 1001$

   no, if we divide 19019 by 1001, the remainder is 0

15. $x \equiv 158 \bmod 210$ 17.$x \equiv 147 \bmod 385$ 19. $x \equiv 31 \bmod 130$

21. $x \equiv 6 \bmod 77$ 23.$x \equiv 7 \bmod 143$ 25. $x \equiv 6 \bmod 1001$

27. $x \equiv 5 \bmod 1001$ 29.$x \equiv 7 \bmod 1683$ 30. $x \equiv 8 \bmod 2717$

Exercises 3.3

1. (a) $x = 4 + 3t$, $y = 1 - 5t$ (b) $x = 2 + 11t$, $y = 3 - 7t$

3. (a) $x = 2 + 9t$, $y = 1 - 7t$ (b) $x = 12 + 13t$, $y = -7 - 11t$

5. (a) $(0, 8)$, $(3, 1)$ (b) $(0, 3)$, $(12, -4)$

7. $5x + 3y = 22$

Exercises 4.2 (selected exercises)

3. (i) For $n = 1$, LHS=RHS=1

   (ii) Assume that the statement is true for $n = k$

   $1 + 7 + 13 + \cdots + (6k - 5) = k(3k - 2)$

   (iii) Show that it is true for $n = k + 1$

   Show that $1 + 7 + 13 + \cdots + (6k - 5) + (6k + 1) = (k + 1)(3k + 1)$

   LHS$= 1 + 7 + \cdots + (6k - 5) + (6k + 1)$

   $\quad = k(3k - 2) + (6k + 1) = (k + 1)(3k + 1)$=RHS

9.(i) For $n = 1$, LHS=RHS=3

   (ii) Assume that the statement is true for $n = k$

$3 + 3^2 + 3^3 + \cdots + 3^k = \frac{3(3^k-1)}{2}$

(iii) Show that it is true for $n = k+1$

Show that $3 + 3^2 + 3^3 + \cdots + 3^k + 3^{k+1} = \frac{3(3^{k+1}-1)}{2}$

LHS= $3 + 3^2 + 3^3 + \cdots + 3^k + 3^{k+1}$

$= \frac{3(3^k-1)}{2} + 3^{k+1}$

$= \frac{3^{k+1}-3+2\cdot3^{k+1}}{2} = \frac{3\cdot3^{k+1}-3}{2}$=RHS

11. (i) For $n = 1$, LHS=RHS=$\frac{1}{2}$

(ii) Assume that the statement is true for $n = k$

$\frac{1}{1\times2} + \frac{1}{2\times3} + \cdots + \frac{1}{k(k+1)} = \frac{k}{k+1}$

(iii) Show that it is true for $n = k+1$

Show that $\frac{1}{1\times2} + \frac{1}{2\times3} + \cdots + \frac{1}{k(k+1)} + \frac{1}{(k+1)(k+2)} = \frac{k+1}{k+2}$

LHS= $\frac{1}{1\times2} + \frac{1}{2\times3} + \cdots + \frac{1}{k(k+1)} + \frac{1}{(k+1)(k+2)}$

$= \frac{k}{k+1} + \frac{1}{(k+1)(k+2)} = \frac{k+1}{k+2}$=RHS

19. (i) For n=5, LHS $= 2^5 > 5^2 =$ RHS

(ii) Assume it is true for $n = k$. Assume that $2^k > k^2$

(iii) Show that it is true for $n = k+1$. i.e show that $2^{k+1} > (k+1)^2$

LHS $= 2^{k+1} = 2\cdot2^k > 2k^2 = k^2 + k^2 > k^2 + 2k + 1 = (k+1)^2$=RHS

Notice that we used Exercise 18, where $k^2 > 2k+1$

21. (ii) $9^k \equiv 1\bmod 8$

multiply both sides by 8, we obtain

$9^{k+1} \equiv 9\bmod 8$, or equivalently $9^{k+1} \equiv 1\bmod 8$

25. (ii) Assume it is true for $n = k$. This means that we assume

$1 \times 2 + 2 \times 3 + \cdots + k(k+1) = \frac{k(k+1)(k+2)}{3}$

(iii) Show it is true for $n = k+1$, i.e show that

$$1 \times 2 + 2 \times 3 + \cdots + k(k+1) + (k+1)(k+2) = \tfrac{(k+1)(k+2)(k+3)}{3}$$

LHS$= 1 \times 2 + 2 \times 3 + \cdots + k(k+1) + (k+1)(k+2)$

$$= \tfrac{k(k+1)(k+2)}{3} + (k+1)(k+2)$$

$$= \tfrac{(k+1)(k+2)(k+3)}{3} = \text{RHS}$$

33. (ii) Assume it is true for $n = k$. This means that we assume

$$1 \times 2 + 2 \times 2^2 + \cdots + k(2^k) = 2 + (k-1)2^{k+1}$$

(iii) Show it is true for $n = k + 1$, i.e show that

$$1 \times 2 + 2 \times 2^2 + \cdots + k(2^k) + (k+1)(2^{k+1}) = 2 + k(2^{k+2}$$

LHS$= 1 \times 2 + 2 \times 2^2 + \cdots + k(2^k) + (k+1)(2^{k+1})$

$$= 2 + (k-1)2^{k+1} + (k+1)(2^{k+1} = 2 + 2^{k+1}(2k) = 2 + k(2^{k+2}) = \text{RHS}$$

35. For $n = 1$, $(18 - 5)|13$.

    (ii) Assume it is true for $n = k$. This means that we assume

$$18^k - 5^k = 13b,\ b \text{ is an integer}$$

(iii) Show it is true for $n = k + 1$, i.e show that

$$18^{k+1} - 5^{k+1} = 13r,\ r \text{ is an integer}$$

LHS$= 18(18)^k - 5(5^k) = 18(5^k + 13b) - 5(5^k) = 13(5^k + b) = 13s$

43. It is true for $n = 1$

    (ii) Assume it is true for $n = k$. This means that we assume

$$2(7^k) + 3(5^k) - 5 = 24b,\ b \text{ is an integer}$$

(iii) Show it is true for $n = k + 1$, i.e show that

$$2(7^{k+1} + 3(5^{k+1} - 5 \text{ is divisible by 24}$$

    LHS$= 7 \times 2(7^k) + 15(5^k) - 5 = 7b \times 24 + 24 + 6(1 - 5^k)$

Each term is divisible by 24

45. It is clear that it it is true for $n = 1$

    (ii) Assume it is true for $n = k$. This means that we assume

$$2^{5k} + 30 = 31b,\ b \text{ is an integer}$$

(iii) Show it is true for $n = k + 1$, i.e show that

$$2^{5(k+1)} + 30 = 31r, \ r \text{ is an integer}$$

LHS$= 2^5(2^{5k} + 30 = 2^5(31b - 30) + 30 = 31b(2^5) - 31 \times 30.$

47. (ii) Assume it is true for $n = k$. This means that we assume

$$\frac{1}{4\times1^2-1} + \frac{1}{4\times2^2-1} + \frac{1}{4\times3^2-1} + \cdots + \frac{1}{4\times k^2-1} = \frac{k}{2k+1}$$

(iii) Show it is true for $n = k + 1$, i.e show that

$$\frac{1}{4\times1^2-1} + \frac{1}{4\times2^2-1} + \frac{1}{4\times3^2-1} + \cdots + \frac{1}{4\times(k+1)^2-1} = \frac{k+1}{2k+3}$$

LHS$= \frac{1}{4\times1^2-1} + \frac{1}{4\times2^2-1} + \frac{1}{4\times3^2-1} + \cdots + \frac{1}{4\times(k+1)^2-1}$

$$= \frac{k}{2k+1} + \frac{1}{4\times(k+1)^2-1} = \frac{k+1}{2k+3} = \text{RHS}$$

49. (ii) Assume it is true for $n = k$. This means that we assume

$$\left(1 + \tfrac{3}{1}\right) \times \left(1 + \tfrac{5}{4}\right) \times \left(1 + \tfrac{7}{9}\right) \times \cdots \times \left(1 + \tfrac{2k+1}{k^2}\right) = (k+1)^2 \text{ (iii) Show it}$$

is true for $n = k + 1$, i.e show that

$$\left(1 + \tfrac{3}{1}\right) \times \left(1 + \tfrac{5}{4}\right) \times \left(1 + \tfrac{7}{9}\right) \times \cdots \times \left(1 + \tfrac{2k+3}{(k+1)^2}\right) = (k+2)^2$$

LHS$= \left(1 + \tfrac{3}{1}\right) \times \left(1 + \tfrac{5}{4}\right) \times \left(1 + \tfrac{7}{9}\right) \times \cdots \times \left(1 + \tfrac{2k+3}{(k+1)^2}\right)$

$$= (k+1)^2 + \times \times \left(1 + \tfrac{2k+3}{(k+1)^2}\right) = (k+2)^2 = \text{RHS}$$

Exercises 4.3(selected exercises)

3. $S_1 = -6, S_2 = -5, S_3 = 3, S_4 = 18$

Using the finite differences method we find

$$
\begin{array}{ccccccc}
-6 & & -5 & & 3 & & 18 \\
& 1 & & 8 & & 15 & \\
& & 7 & & 7 & &
\end{array}
$$

$2a = 7, \Longrightarrow a = \tfrac{7}{2}$

$3a + b = 1, \Longrightarrow b = -\tfrac{19}{2}$

$a + b + c = -6 \Longrightarrow c = 0$

$p(n) = \tfrac{1}{2}n(7n - 19)$

5. $S_1 = \tfrac{1}{6}, S_2 = \tfrac{2}{11}, S_3 = \tfrac{3}{16}, S_4 = \tfrac{4}{21}$

Using the finite differences method we find for the numerator

$$1 \qquad 2 \qquad 3 \qquad 4$$

and for the denominator we obtain

$$6 \qquad 11 \qquad 16 \qquad 22$$

This means that $p(n) = n$, and $q(n) = 5n + 1$

Using the finite differences method we find

$$2 \qquad 13 \qquad 33 \qquad 62$$
$$\quad 11 \qquad 20 \qquad 29$$
$$\qquad 9 \qquad 9$$

$2a = 9, \Longrightarrow a = \frac{9}{2}$

$3a + b = 11, \Longrightarrow b = -\frac{5}{2}$

$a + b + c = 2 \Longrightarrow c = 0$

   Therefore, the formula is: $p(n) = \frac{1}{2}n(9n - 5)$

17. $S_1 = 8, S_2 = 26, S_3 = 56, S_4 = 100, S_5 = 169$

Therefore, the formula is: $p(n) = \frac{1}{3}n(n + 1)(n + 11)$

19. $S_0 = 0, D_1 = 6, D_2 = 18, D_3 = 18, D_4 = 6$

Therefore, the formula is: $f(n) = 4n + 18n(n - 1)/2$
$+ 18n(n - 1)(n - 2)/6 + 6n(n - 1)(n - 2)(n - 3)/24 = \frac{n(n+1)(n+2)(n+3)}{4}$

21. $S_0 = 0, D_1 = 8, D_2 = 28, D_3 = 32, D_4 = 12, f(n) = \frac{n(n+1)(n+2)(3n+5)}{6}$

Exercises 4.4

1. $\dfrac{1}{(5n - 4)(5n + 1)} = \dfrac{1}{5(5n - 4)} - \dfrac{1}{5(5n + 1)}$

$$
\begin{aligned}
LHS &= \frac{1}{5 \times 1} - \frac{1}{5 \times 6} + \frac{1}{5 \times 6} - - \frac{1}{3(3n + 1)} \\
&= \frac{1}{3 \times 1} - \frac{1}{5(5n + 1)} \\
&= \frac{n}{5n + 1} = RHS
\end{aligned}
$$

5. $\dfrac{1}{(3n-2)(3n+1)} = \dfrac{1}{3(3n-2)} - \dfrac{1}{3(3n+1)}$

$$
\begin{aligned}
LHS &= \frac{1}{3\times 1} - \frac{1}{3\times 4} + \frac{1}{3\times 4} - \frac{1}{3\times 7} + \cdots - \frac{1}{3(3n+1)} \\
&= \frac{1}{3\times 1} - \frac{1}{3(3n+1)} \\
&= \frac{n}{3n+1} = RHS
\end{aligned}
$$

7. $f(n) = \frac{n}{8n+1}$

9. $f(n) = \frac{n}{4(3n+4)}$

11. $f(n) = \frac{n}{5(4n+5)}$

Exercises 4.5

1. $x > 6, y > 8 \Longrightarrow x + y > 14 > 10$

3. $n = 2k+1, k = 0, 1, \cdots$
   Hence, $n^2 = 2r + 1$, where $r = 2k^2 + 2$

5. $n = 2k+1, k = 0, 1, \cdots$

   Hence, $n^2 + n + 2 = 2m$, where $m = 2k^2 + 3k + 2$

   7. $n = 2k+1, k = 0, 1, \cdots$

   Hence, $n^3 + n = 2s$, where $s = 4k^3 + 6k^2 + 4k + 1$

9. $n = 2k, k = 0, 1, \cdots$

   Hence, $n^3 - n = 2m$, where $m = 4k^3 - k$

11. $n = 2k+1, k = 0, 1, \cdots$, and $m = 2r + 1$, where $r = 0, 1, \cdots$

   Hence, $n^2 - m^2 = 2s$, where $s = 2k^2 + 2k - 2r^2 - 2r$

13. $x \equiv 2 \bmod 5$.

   Multiplying both sides by 4 gives

   $4x \equiv 3 \bmod 5$

15. Factoring by grouping gives

   $(x-1)(x^2+1) = 0$. Hence, $x = 1$ is the only real root.

17. $n = 2k$, $5(2k)^3 + 6(2k) - 11 = 2(20k^3 + 6k - 6) + 1$

19. $1^5 \equiv 1 \bmod 5$, $2^5 \equiv 2 \bmod 5$, $3^5 \equiv 3 \bmod 5$, $n^5 \equiv n \bmod 5$

21. $b = ka$, $c = ra$, $b + c = a(k + r)$

23. $a + b = 2k$, $b + c = 2r$, $a + c = 2(k + r - b)$

Exercises 4.6

1. Assume that $x + 6 \not< 10$. This means that $x + 6 \geq 10$. This gives the contradiction $x \geq 4$

3. Assume $\mid x + 3 \mid \not\leq 7$. This means $\mid x + 3 \mid > 7$. This gives the contradiction $\mid x \mid \neq 4$

9. Assume that r/s is a rational number. This gives $a/b \div s = m/n$, or $s = \frac{bm}{an}$, $bn \neq 0$. This means that s is a rational number.

10. Assume that $\frac{\sqrt{2}}{5} + 7 = \frac{a}{b}$ is a rational number. This gives $\sqrt{2} = \frac{5(a-7b)}{b}$. This means that $\sqrt{2}$ is a rational number. This is a contradiction.

11. Assume that $\frac{x}{x+1} \geq \frac{x-1}{x+2}$. This gives $1 \leq 0$. This is a contradiction.

Exercises 5.1

1. (a) $C = \{\text{red, black}\}$ (b)$D = \{2, 4, 6, 8, 10, 12\}$

(c) $A = \{0, 4\}$ (d) $B = \{3, 10, 17\}$ (e) $S = \{2, 4, 6\}$

3. (a) $A = \{a, e, i, o, u\}$, $n(A) = 5$ (b) $B = \{1, 12\}$, $n(B) = 2$

(c) $C = \{3\}$, $n(C) = 1$ (d) $D = \phi$, $n(D) = 0$

(e) $E = \{8, 21, 34, \cdots\}$, $n(E) = \infty$

5. $\phi = \{\}$. ϕ contains zero elements. However, $\{\phi\}$ contains one element, namely the Greek letter ϕ, therfore $\phi \neq \{\phi\}$

Exercises 5.2

1. (a) False. $\{1, 3\}$ is a set and not an element.

(b) True. Every element of $\{1, 3\}$ is an element of A.

(c) True.

(d) False. Every element of B is an element of A.

(e) True. Every element of $\{1, 3, 5\}$ is an element of B.

3. Subsets of K are: $\phi, \{a\}, \{b\}, \{c\}, \{a, b\}, \{a, c\}, \{b, c\}, \{a, b, c\}$

5. All sets are empty sets

Exercises 5.3

1.$(a) A^c = \{1, 3, 5, 7, 9\}$ $(b) \cup B = \{1, 2, 4, 6, 7, 8, 10\}$

$(c) A \cap B = \{4, 10\}$ $(d)(A \cup B)^c = \{3, 5, 9\}$

$(e)(A \cap B)^c = \{1, 2, 3, 5, 6, 7, 8, 9\}$

3.$(a) A^c = \{1, 3, 5\}$ $(b) A \cap B = \{6\}$

$(c)(A \cup B)^c = \{1, 5\}$ $(d)(A \cup B)^c \cap B = \phi$

$(e) A^c \cup B^c = \{1, 2, 3, 4, 5\}$

5.$(a) n(B^c) = 600$ $(b) n(B \cap S) = 120$

$(c) n(G \cup M) = 700$ $(d) n(M \cap S) = 0$

$(e) n(G \cap E) = 190$

7. (a) ϕ,   (b)$\{5\}$,   (c) $\{1, 2\}$,   (d) $\{5\}$,   (e)$\{3, 4, 5, 6, 7\}$

9. (a) ϕ (b)$\{1, 2, 3, 4, 5, 6, 7, 8, 10, 11\}$ (c) $\{7, 9, 11\}$ (d) $\{7, 9, 11\}$ (e)$2^3 = 8$

11. (a) U (b)ϕ (c) $\{3, 4\}$ (d) $\{1, 2, 3, 4, 5, 6\}$ (e)$2^6 = 64$

Exercises 5.4

1. We first find the intersection between the three sets $A, B,$ and C.

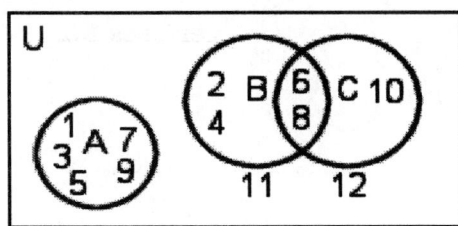

$A \cap B \cap C = \phi$

We then find the intersection between each pair, therefore we find

$B \cap C = \{6, 8\}, A \cap B = A \cap C = \phi$. Venn diagram is given by:

3. (a)$n(M \cap S \cap E) = 23, n(M \cap s) = 40, n(A \cap B) = 34, n(M \cap E) = 37$. Venn diagram is given by:

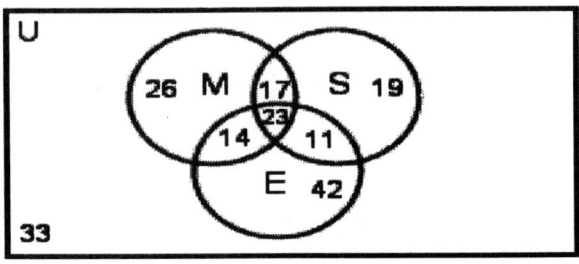

   (b) Number of students= 185   (c) 19   (d) 14   (e) 26

5. $A \cap B \cap C = \phi, A \cap B = \{4, 5, 6\}, B \cap C = \{8, 9\}, A \cap C = \phi$.

   Hence, Venn diagram is given by:

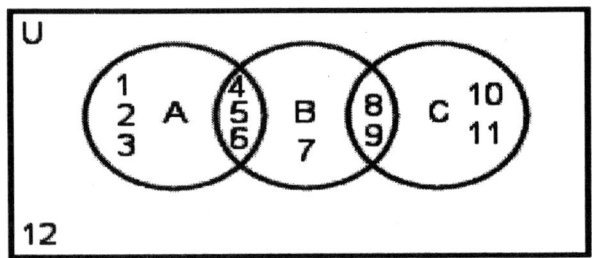

7. (a)$A \cap B \cap C = \phi, B \subset A, A \cap C = \{3\}$.

Hence, Venn diagram is given by:

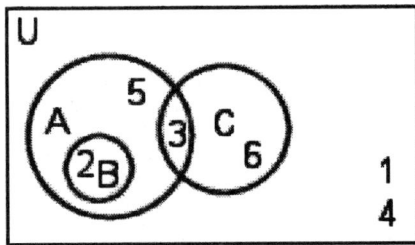

9. (a) $U = \{1, 2, 3, 4, 5, 6, 7, 8, 9, 10\}$

(b) $A \cap B \cap C = \phi$, (c) $A \cap B = \{1, 2\}$

(d) $(A \cup B)^c = \{8, 9, 10\}$

(e) $(A \cap B)^c = \{3, 4, 5, 6, 7, 8, 9, 10\}$

11. (a) The Venn diagram is given by the following figure.

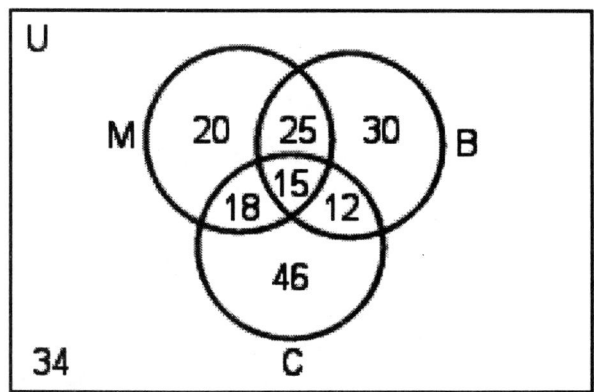

(b) 34   (c) 33   (d) 75   (e) 46

13. (b) 500   (c) 48   (d) 267   (e) 75

15. (b) 50   (c) 6   (d) 80   (e) 191

17. (b) 10   (c) 100   (d) 190   (e) 200

Exercises 6.1.1

1.

p	q	$\sim q$	$(p \vee \sim q)$
T	T	F	T
T	F	T	T
F	T	F	F
F	F	T	T

3.

p	q	$\sim p$	$\sim q$	$(\sim p \vee \sim q)$	$\sim (\sim p \vee \sim q)$
T	T	F	F	F	T
T	F	F	T	T	F
F	T	T	F	T	F
F	F	T	T	T	F

5. Notice here that three simple statements p, q, and r are given. Accordingly the number of rows should be $2^3 = 8$.

p	q	r	$(p \vee \sim q)$	$(p \vee \sim q) \wedge r$
T	T	T	T	T
T	T	F	T	F
T	F	T	T	T
T	F	F	T	F
F	T	T	F	F
F	T	F	F	F
F	F	T	T	T
F	F	F	T	F

7.

p	q	$p \wedge q$	$(p \wedge \sim q)$	$(p \wedge q) \wedge (p \wedge \sim q)$
T	T	T	F	F
T	F	F	T	F
F	T	F	F	F
F	F	F	F	F

9. The results of the last column are: T,T,T,F,T,T,T,T.

11. The results of the last column are: T,T,T,T,F,F,F,T

13. The results of the last column are: T,T,F,T,T,T,F,F

15. The results of the last column are: T,F,T,T,T,F,T,F

17. The truth value in this case is F as shown below.

p	q	r	$\sim p$	$\sim p \wedge r$	$(\sim p \wedge r) \wedge q$
F	T	F	T	F	F

19. The truth value in this case is T as shown below.

p	q	r	$q \vee r$	$\sim p$	$\sim p \wedge (q \vee r)$	$\sim p \wedge (q \vee r) \vee q$
F	T	F	T	T	T	T

Exercises 6.1.2

1.

p	q	$\sim q$	$(p \rightarrow \sim q)$
T	T	F	F
T	F	T	T
F	T	F	T
F	F	T	T

3.

p	q	$\sim p$	$(q \rightarrow \sim p)$	$\sim (q \rightarrow \sim p)$
T	T	F	F	T
T	F	F	T	F
F	T	T	T	F
F	F	T	T	F

5.

p	q	$\sim p$	$(\sim p \leftrightarrow q)$
T	T	F	F
T	F	F	T
F	T	T	T
F	F	T	F

7.

p	q	$p \wedge q$	$p \vee q$	$(p \wedge q) \leftrightarrow (p \vee q)$
T	T	T	T	T
T	F	F	T	F
F	T	F	T	F
F	F	F	F	T

9.

p	q	$p \wedge q$	$\sim p$	$(p \wedge q) \wedge (\sim p)$
T	T	T	F	F
T	F	F	F	F
F	T	F	T	F
F	F	F	T	F

11.

p	q	$\sim q$	$\sim q \rightarrow p$
T	T	F	T
T	F	T	T
F	T	F	T
F	F	T	F

The statement is neither a tautology nor a contradiction.

13.

p	q	$p \vee q$	$p \wedge q$	$(p \vee q) \leftrightarrow (p \wedge q)$
T	T	T	T	T
T	F	T	F	F
F	T	T	F	F
F	F	F	F	T

15. The statement in this exercise is a tautology statement as shown by the following truth table.

p	q	$p \vee q$	$p \rightarrow (p \vee q)$
T	T	T	T
T	F	T	T
F	T	T	T
F	F	F	T

17. The results of the last column are: F,T,T,T,T,T,T,T

19. The results of the last column are: T,F,F,T,F,F,F,F

Exercises 6.2

1. The switching circuit is given below.

A	B	B'	$A \vee B'$	$B \wedge (A \vee B')$
0	0	1	1	0
0	1	0	0	0
1	0	1	1	0
1	1	0	1	1

3.

A	B	A'	$A' \vee B$	$A \wedge B$	$(A \wedge B) \wedge (A' \vee B)$
0	0	1	1	0	0
0	1	1	1	0	0
1	0	0	0	0	0
1	1	0	1	1	1

5. The results of the last column are: 1,1,1,1

7. The results of the last column are: 0,1,1,0

Exercises 6.3.1

1.

x	y	x'	$x + y$	$(x + y) + x'$
0	0	1	0	1
0	1	1	1	1
1	0	0	1	1
1	1	0	1	1

3.

x	y	x'	$x' + y$	$x(x' + y)$	$x(x' + y) + x$
0	0	1	1	0	0
0	1	1	1	0	0
1	0	0	0	0	1
1	1	0	1	1	1

5. The results of the last column are given by: $0, 0, 1, 0, 1, 1, 1, 0$

7. The results of the last column are given by: $0, 0, 0, 1, 0, 1, 1, 1$

9. The results of the last column are given by: $0, 1, 1, 1, 1, 1, 1, 1$

Exercises 6.3.2

1. $E = 1 + 1 + 1 = 1$

3. $E = (1 + y)(1 + y') = 1 + (y + y') + y \cdot y' = 1$

5. $E = 0$

7. $E = x(y + y') = x$

9. $E = x(y + y') + x'(z + z') = x + x' = 1$

11. $E = xy'(z + z') + x'y'(z + z') = y'(x + x') = y'$

13. $E = xz(y + y') + xz'(y + y') = x(z + z') = x$

15. $E = xz(x + x') + xy'(z + z') = xz + xy'$

17. $E = y'z + x'z$

19. $E == z' + xy'$

21. $E = y + y'z'$ or $E = z' + yz$

23. $E = 1$

27. $E = y + y'z' + x'y'$, $E = y + y'z' + x'z$
$E = z' + yz + x'z$, or $E = x' + xy + xz'$

29. $E = x + x'z$, or $E = z + xz'$

Exercises 7.2

1. Number of choices $= 3 \cdot 2 = 6$

3. Number of meals $= 3 \cdot 4 \cdot 2 = 24$

5. Number of selections $= 3 \cdot 2 \cdot 4 = 24$

7. Number of plates $= 26 \cdot 26 \cdot 26 \cdot 10 \cdot 9 \cdot 8 = 12654720$

9. Number of outcomes $= 2^8 = 256$

11. Number of outcomes $= 2^5 \cdot 6^4 = 41472$

13. Number of plates $= 24 \cdot 25 \cdot 25 \cdot 9 \cdot 10 \cdot 10 = 13500000$

15. Number of meals $= 3 \cdot 4 \cdot 3 \cdot 5 = 180$

17. Number of odd four-digit numbers $= 9 \cdot 10 \cdot 10 \cdot 5 = 4500$

Exercises 7.3

1.   244   3. 2   5. $\lceil \frac{169}{26} \rceil = 7$

7. $\lceil \frac{n}{5} \rceil = 2$, $n = 6$ 9. $\lceil \frac{9}{2} \rceil = 5$ 11. $\lceil \frac{162}{150} \rceil = 2$

13. $\lceil \frac{n}{7} \rceil = 3$, $n = 15$ 15. 5 possible remainders $0 \cdots 4$, then $\lceil \frac{n}{5} \rceil = 2$, $n = 6$

Exercises 7.4

1. (a) $4P3 = 24$ (b) $7P3 = 210$ (c) $9P4 = 3024$ (d) $8P0 = 1$

3. (a) 7   (b) 168   (c)2   (d) 3

5. $8C5 = 56$ 7. -44 11. $n = 8$

13. $n = 10$ 15. $n = 6$ 17. $n = 10$

19. For $n = 6, 6! = 720, 6^3 = 216$, It is true for $n = 6$
Assume it is true for $n = k$, i.e assume that $k! > k^3$
Show it is true for $n = k + 1$, i.e show $(k + 1)! > (k + 1)^3$

$$
\begin{aligned}
(k + 1)! \ &= \ (k + 1) \cdot k! \\
&> \ (k + 1)k^3 \text{ by using the induction assumption} \\
&> \ (k + 1) \cdot (k + 1)^2 \text{ since } k^3 > (k + 1)^2. \text{ This means that} \\
(k + 1)! \ &> \ (k + 1)^3.
\end{aligned}
$$

This means that the statement holds for $n = k + 1$

21. For $n = 4, 4! = 24, 2^4 = 16$, It is true for $n = 4$
Assume it is true for $n = k$, i.e assume that $k! > 2^k$
Show it is true for $n = k + 1$, i.e show $(k + 1)! > 2^{k+1}$

$$
\begin{aligned}
(k + 1)! \ &= \ (k + 1) \cdot k! \\
&> \ (k + 1)2^k \text{ by using the induction assumption} \\
&> \ 2 \cdot 2^k \text{ since } k + 1 > 2. \text{ This means that} \\
(k + 1)! \ &> \ 2^{k+1}.
\end{aligned}
$$

This means that the statement holds for $n = k + 1$

23. For $n = 1, 1! = 1, (1 + 1)! - 1 = 1$, It is true for $n = 1$
Assume it is true for $n = k$, i.e assume that
$1(1! + 2(2!) + \cdots + k(k!) = (k + 1)! - 1$
Show it is true for $n = k + 1$, i.e show that
$1(1! + 2(2!) + \cdots + k(k!) + (k + 1)(k + 1)! = (k + 2)! - 1$

$$
\begin{aligned}
LHS &= 1(1! + 2(2!) + \cdots + k(k!) + (k + 1)(k + 1)! \\
&= (k + 1)! - 1 + (k + 1)(k + 1)! \text{ by using the induction assumption} \\
&= (k + 2)(k + 1)! - 1 \\
&= (k + 2)! - 1
\end{aligned}
$$

This means that the statement holds for $n = k + 1$

Exercises 7.5

1.(a) 28   (b) 120   (c) 15   (d) 1

3. 16 5. 2^6 7. $n = 10$ 9. $n = 8$

11. $n = 7$ 13. $n = 6$ 15. $n = 8$ 17. $n = 7$

Exercises 7.6

1. $(x + y)^6 = x^6 + 6x^5 y + 15x^4 y^2 + 20x^3 y^3 + 15x^2 y^4 + 6xy^5 + y^6$

3. $(3x + y)^5 = 243x^5 + 405x^4 y + 270x^3 y^2 + 90x^2 y^3 + 15xy^4 + y^5$

5. $(3x + 2y)^5 = 243x^5 + 810x^4 y + 1080x^3 y^2 + 720x^2 y^3 + 240xy^4 + 32y^5$

7. $(x^2 + y)^5 = x^{10} + 5x^8 y + 10x^6 y^2 + 10x^4 y^3 + 5x^2 y^4 + y^5$

9. $(x + y)^{16} = x^{16} + 16x^{15} y + 120x^{14} y^2 + 560x^{13} y^3 + \cdots$

11. $(x - 2y)^{36} = x^{36} - 72x^{35} y + 2520x^{34} y^2 - 57120x^{33} y^3 + \cdots$

13. $\binom{10}{6} = 210$

15. $n = 4 + 8 = 12$, hence, coefficient $= \binom{12}{8} = 495$

17. $9 + k = 14 \implies k = 5$, hence, coefficient $= \binom{14}{5} = 2002$

19. $r + 2 = n, \binom{n}{2} = 28 \implies n = 8, r = 6$

21. $(x+1)^6 = x^6 + 6x^5 + 15x^4 + 20x^3 + 15x^2 + 6x + 1$

23. (b) $(1+6)^n = 7^n$

25. $\frac{4^n}{4^n} = 1$

Exercises 8.1

1. (a) $P(E) = \frac{1}{2}$ (b) $P(A) = \frac{1}{8}$ (c) $P(B) = 0$

3. (a) $P(\text{green}) = \frac{7}{11}$ (b) $P(\text{blue}) = \frac{4}{11}$ (c) $P(\text{red}) = 0$

5. (a) $\frac{7}{16}$ (b) $\frac{1}{4}$ (c) 0

7. (a) $\frac{103}{200}$ (b) $\frac{79}{400}$ (c) $\frac{7}{20}$ (d) $\frac{181}{400}$

9. (a) $\frac{1}{12}$ (b) $\frac{1}{6}$ (c) $\frac{7}{36}$

11. (a) $\frac{1}{6}$ (b) $\frac{7}{12}$ (c) $\frac{5}{18}$

13. (a) $\frac{1}{2}$ (b) $\frac{1}{2}$ (c) $\frac{1}{2}$

15. (a) $\frac{1}{2}$ (b) $\frac{1}{4}$ (c) $\frac{3}{26}$ (d) $\frac{1}{26}$ (e) $\frac{1}{52}$

17. $\frac{1}{36}$ 19. $\frac{n(E)}{112} = \frac{1}{8} \implies n(E) = 14$

21. (a) P(even) $= \frac{1}{2}$, (b) P(ends in 6) $= \frac{1}{10}$, (c) P(divisible by 6) $= \frac{4}{25}$

23. $P(E) = \frac{32}{225}$ 25. $P(E) = \frac{25}{126}$

Exercises 8.2

1. $P(A^c) = 0.42$ 3. $P(A \cup B) = 0.7$ 5. $P(A \cup B) = 0.69$

7. $P(A \cup B) = \frac{5}{12}$ 9. $P(A \cap B) = 0.15$ 11. $P(A \cap B) = 0$

13. $P(B^c) = 0.4$ 15. (a) $\frac{1}{4}$ (b) 1 (c) $\frac{1}{2}$ (d) $\frac{1}{2}$

17. (a) $\frac{3}{4}$ (b) $\frac{1}{8}$ (c) $\frac{7}{8}$ (d) $\frac{7}{8}$

19. (a) $\frac{11}{18}$ (b) $\frac{1}{9}$ (c) $\frac{11}{18}$ (d) $\frac{1}{6}$ (e) $\frac{4}{9}$

21. (a) $\frac{2}{3}$ (b) $\frac{1}{18}$ (c) $\frac{5}{9}$ (d) $\frac{1}{36}$ (e) $\frac{17}{36}$

23. (a) $\frac{3}{5}$ (b) $\frac{7}{40}$ (c) $\frac{41}{60}$ (d) $\frac{13}{60}$ (e) $\frac{17}{24}$

Exercises 8.3

1.(a) $\frac{1}{6}$ (b)$\frac{1}{3}$ 3.(a)$\frac{1}{4}$ (b)$\frac{1}{2}$ 5.(a) $\frac{2}{11}$ (b)$\frac{1}{3}$ 7.(a) 0, (b) 0

9. (a) $\frac{74}{100}$ (b) $\frac{67}{100}$ (c) $\frac{22}{100}$ (d) $\frac{15}{26}$ (e)$\frac{11}{47}$

11. (a) 0.184 (b) 0.631 (c) 0.731 (d) $\frac{131}{318}$ (e)$\frac{129}{269}$

13. (a) $\frac{11}{18}$ (b) $\frac{1}{9}$ (c)$\frac{11}{18}$ (d)$\frac{3}{5}$ (e)$\frac{1}{3}$

15. (a) $\frac{2}{3}$ (b) $\frac{1}{18}$ (c)$\frac{5}{9}$ (d)$\frac{1}{6}$ (e)$\frac{1}{6}$

17. (a) $\frac{3}{5}$ (b) $\frac{7}{40}$ (c)$\frac{41}{60}$ (d)$\frac{13}{24}$ (e)$\frac{21}{46}$

19. (a) $\frac{11}{15}$ (b) $\frac{13}{60}$ (c)$\frac{7}{10}$ (d)$\frac{1}{3}$ (e)$\frac{1}{7}$

Exercises 8.4

1. (a) P(RG)+P(GR)= $\frac{21}{50}$, (b) P(RR)+P(GG)= $\frac{29}{50}$

   (c) P(RG)+P(GR)+P(GG)= $\frac{51}{100}$

3. (a) P(different colors) = $\frac{31}{50}$, (b) P(same color) = $\frac{19}{50}$, (c) P(1R red) = $\frac{1}{2}$

5. (a) P(same color) = $\frac{7}{25}$, (b) P(two red) = $\frac{54}{125}$, (c) P(two green) = $\frac{36}{125}$

7. (a) P(A) = $\frac{9}{25}$

Exercises 9.1

1. $f(-1) = 0,\quad f(0) = -1,\quad f(1) = 4$

3. $f(-3) = 4,\quad f(0) = 1,\quad f(3) = 2$

5. $g(0) = 1,\quad g(3) = 2,\quad g(8) = 3$

7. $h(0) = \sqrt{3},\quad h(-1) = 2,\quad h(3) = 0$

9. $f(0) = 1,\quad f(-2) = 0,\quad f(3) = \frac{5}{11}$

11. $f(0) = \frac{1}{6},\quad f(1) = 1,\quad f(-1) = 0$

13. $f(-1) = 0,\quad f(0) = -3,\quad f(2) = 1$

15. 1 17. $k = 2$ 19. $k = 1$

21. $k = 1, \pm 2$ 23. $k = -5, 2$ 25. $a = 1, b = 2$

27. $a = 4, b = 5$ 29. $a = -3, b = -5$ 30. $a = 2, b = 3$

Exercises 9.2

1. $D = \{x \mid x \in R\}$ 3. $D = \{x \mid x \in R, x \neq -1\}$

5. $D = \{x \mid x \in R\}$ 7. $D = \{x \mid x \in R, x \neq -6, 0\}$

9. $D = \{x \mid x \in R, x \neq -2, 2\}$ 11. $D = \{x \mid x \in R, x \neq -7, 1\}$

13. $D = \{x \mid -\infty < x \leq 4\}$

15. $D = \{x \mid 6 \leq x < 14 \text{ or } 14 < x < \infty\}$

17. $D = \{x \mid x \in R, x \neq 7, -1$

19. $D = \{x \mid x \leq 3, x \neq 1$

21. $D = \{x \mid x \in R\}$

23. $D = \{x \mid -4 \leq x < -1 \text{ or } -1 < x \leq 4\}$

Exercises 9.3

1. 1   3. 0   5. ∞ 7. ∞ 9. $-\frac{1}{2}$

11. 1   13. $3x$ 15. x^2 17. 0   19. -4

Exercises 9.4

1. $O(1)$ 3. $O(n^2)$ 5. $O(n^2)$ 7. $O(n\log_2 n)$ 9. $O(n\log_2 n)$

11. $O(n^4)$ 13. $O(1)$ 15. $O(n)$ 17. $O(n^3)$ 19. $O(n!)$

21. $O(n!)$ 23. $O(n^2)$ 25. $O(n^3)$ 27. 2   29. 3

31. -8 33. $O(1)$ 35. $O(n)$ 37. $O(n^2 \log n)$ 39. $O(n^3)$

Exercises 10.1

1. $a_{n+1} = 3a_n - 2, a_0 = 2$

3. $a_{n+2} = 7a_{n+1} - 12a_n, a_0 = 2,\quad a_1 = 7$

5. $a_n = 6^n$ & 7. $a_n = -2(3)^n$

9. $a_n = 2 + 5(6)^n$ 11. $a_n = 2(5)^n + 7^n$

13. $a_n = 4^n + n(4^n)$ 15. $a_n = 2(3)^n + n(3^n)$

17. $4, 18, 60, 186, \cdots$ 19. $-3, -9, -54, -486, \cdots$

21. $1, 3, 6, 15, 33, \cdots$ 23. $1, 3, 4, 17, 21, \cdots$

25. $2, 3, 6, 18, 108, \cdots$ 27. $1, 2, 4, 5, 7, 8, \cdots$

Exercises 10.2

1. $4181, 17711$

3. $317811, 1346269$

5. $832040, 3524578$

7. 10 9. 4

11. $2, 5, 7, 12, 19, 31, 50, 81, 131, 212, \cdots, \frac{212}{131} \approx 1.618$

13. $6, 28, 34, 62, 96, 158, 254, 412, 666, 1078, \cdots$

   where $\frac{X_{10}}{X_9} = \frac{1078}{666} = 1.618$, the golden ratio

15. $1 + 1 + 2 + 3 + 5 + 8 + 13 + 21 = 54 = F_{10} - 1$

17. $F_7^2 + F_8^2 = 13^2 + 21^2 = 610 = F_{15}$

19. It is true for $n = 1$. Assume that it is true for $n = k$. This means that

$$F_1 + F_2 + \cdots + F_{2k-1} = F_{2k}$$

We should show that it is true for $n = k + 1$, i.e we should show that

$$F_1 + F_2 + \cdots + F_{2k-1} + F_{2k+1} = F_{2k+2}$$

$$\text{LHS} = (F_1 + F_2 + \cdots + F_{2k-1}) + F_{2k+1} = F_{2k} + F_{2k+1} = F_{2k+2}$$

Exercises 10.3

1. $843, 3571$

3. $439204, 1149851$

5. 10 7. 2

9. It is true for $n = 2$. Assume it is true for $n = k$, i.e assume

$$L_k^2 - L_{k-1}L_{k+1} = 5(-1)^k$$

We should show that it is true for $n = k + 1$, i.e we should show that

$$L_{k+1}^2 - L_k L_{k+2} = 5(-1)^{k+1}$$

$$
\begin{aligned}
\text{LHS} = L_{k+1}^2 - L_k L_{k+2} &= L_{k+1}^2 - (L_{k+1} - L_{k-1})(L_k + L_{k+1}) \\
&= -(L_k L_{k+1} - L_k L_{k-1} - L_{k-1}L_{k+1}) \\
&= -\left(L_k L_{k+1} - L_k L_{k-1} - L_k^2 + 5(-1)^k\right) \\
&= -(L_k L_{k+1} - L_k(L_{k-1} + L_k)) + 5(-1)^{k+1} \\
&= -(L_k L_{k+1} - L_k L_{k+1}) + 5(-1)^{k+1} \\
&= 5(-1)^{k+1}
\end{aligned}
$$

Therefore it is true for all $n \geq 2$.

11. It is true for $n = 2$. Assume that it is true for $n = k$, i.e.

$$F_{k+1} + F_{k-1} = L_k$$

To show it is true for $n = k + 1$, i.e we should show that

$$F_{k+2} + F_k = L_{k+1}$$

$$
\begin{aligned}
F_{k+2} + F_k &= F_{k+1} + F_k + F_{k-1} + F_{k-2} \\
&= (F_{k+1} + F_{k-1}) + (F_k + F_{k-2}) \\
&= L_k + L_{k-1} \\
&= L_{k+1}
\end{aligned}
$$

13. Notice that $\alpha + \beta = 1, \alpha - \beta = \sqrt{5}, \frac{1}{\alpha}^2 = \beta^2, \frac{1}{\beta^2} = \alpha^2$.

$$
\begin{aligned}
L_{n+2} - L_{n-2} &= (\alpha^{n+2} + \beta^{n+2}) - (\alpha^{n-2} + \beta^{n-2}) \\
&= \alpha^n(\alpha^2 - \alpha^{-2}) - \beta^n(\beta^{-2} - \beta^2) \\
&= \alpha^n(\alpha^2 - \beta^2) - \beta^n(\alpha^2 - \beta^2) \\
&= (\alpha^n - \beta^n)(\alpha^2 - \beta^2) \\
&= (\alpha^n - \beta^n)(\alpha - \beta)(\alpha + \beta) \\
&= \sqrt{5}(\alpha^n - \beta^n) = 5F_n
\end{aligned}
$$

15. $\dfrac{1}{\sqrt{5}}$

17. $\alpha^{2n} + \beta^{2n}$

19. $4\alpha^n \beta^n$

21. $r = 103$ 23. $r = 152$ 25. $r = 14$ 27. $r = 16$

Exercises 11.1

1. (a) dimension is 3×4, hence $A_{3 \times 4}$

   (b) $a_{23} = 1, a_{32} = 3, a_{34} = 6, a_{14} = 0$

   (c) $j = 4$

3. $c_{11} = 1, c_{12} = 4, c_{13} = 9$

   $c_{21} = 4, c_{22} = 7, c_{23} = 12$

   Therefore, $C_{2 \times 3} = \begin{bmatrix} 1 & 4 & 9 \\ 4 & 7 & 12 \end{bmatrix}$

5. $D_{3 \times 3} = \begin{bmatrix} 1 & 2 & 3 \\ 2 & 4 & 6 \\ 3 & 6 & 9 \end{bmatrix}$

7. $A_{3 \times 3} = \begin{bmatrix} 3 & 2 & 2 \\ 2 & 4 & 2 \\ 2 & 2 & 5 \end{bmatrix}$

9. $A_{1 \times 3} = \begin{bmatrix} 0 & -1 & -2 \end{bmatrix}$

11. $A_{3 \times 3} = \begin{bmatrix} 1 & 1 & 1 \\ 0 & 1 & 1 \\ 0 & 0 & 1 \end{bmatrix}$

$$13.\ A_{3\times3} = \begin{bmatrix} 1 & 2 & 3 \\ 0 & 4 & 6 \\ 0 & 0 & 9 \end{bmatrix}$$

Exercises 11.2

$$1.(a)\ A + B = \begin{bmatrix} 0 & 3 \\ 6 & 1 \end{bmatrix} \qquad (b)\ B - C = \begin{bmatrix} -4 & 0 \\ 1 & 2 \end{bmatrix}$$

$$(c)\ 2A + 3B = \begin{bmatrix} -1 & 7 \\ 15 & 4 \end{bmatrix} \qquad (d)\ A - B - C = \begin{bmatrix} -1 & 0 \\ -2 & -3 \end{bmatrix}$$

$$3.(a)\ 2A + \frac{1}{3}B = \begin{bmatrix} 3 & 6 \\ 4 & 5 \\ 8 & 6 \end{bmatrix} \qquad (b)\ A - \frac{1}{3}B = \begin{bmatrix} 0 & 0 \\ 2 & 4 \\ 1 & 6 \end{bmatrix}$$

$$(c)\ A^T + B^T = \begin{bmatrix} 4 & 2 & 9 \\ 8 & 0 & -2 \end{bmatrix} \qquad (d)\ A^T - B^T = \begin{bmatrix} -2 & 2 & -3 \\ -4 & 6 & 10 \end{bmatrix}$$

7. $x = 3, y = 4, u = 5, v = -5$ 9. $a = 5, b = 6, c = 7, d = 8$

11. $a = 5, b = 6, c = 7, d = 8$ 13. $x = 4, y = 6, u = 3, v = 5$

15. $x = 1, 4, y = 2, 1/2, u = 3, 8, v = 4, 3/2$

17. $x = 1, y = 2, u = 3, v = 4$

Exercises 11.3

1.(a) $A_{2\times5}B_{5\times3} = K_{2\times3}$ (b) $C_{3\times4}D_{4\times6} = M_{3\times6}$

(c) $D_{4\times6}C_{3\times4}$ is undefined   (d) $A_{2\times5}A_{2\times5}$ is undefined

(e) $A_{2\times5}A^T_{5\times2} = N_{2\times2}$

3. $AB = \begin{bmatrix} 13 & 12 \\ 9 & 1 \end{bmatrix}$ 5. $AB = \begin{bmatrix} 4 & 2 & 1 \\ 0 & -2 & 1 \\ 3 & 2 & -2 \end{bmatrix}$

7. $A^T A = \begin{bmatrix} 14 & 1 \\ 1 & 6 \end{bmatrix}$

9. $AB = \begin{bmatrix} -4 & -7 \\ 7 & 12 \end{bmatrix}$, $\qquad BA = \begin{bmatrix} -1 & 5 \\ -2 & 9 \end{bmatrix}$

11. $x = 1, y = 3, u = 3, v = -2$ 13. $a = 3, b = 3, c = 4, d = 6$

15. $x = 1, y = 2, u = 3, v = 4$ 17. $x = 1, y = 2, u = 5, v = 6$

19. $x = \pm 4, y = 1, u = \pm 4, v = 3, 11$ 21. $x = 2.5, y = 2$

23. $x = 2, y = 3$ 25. $x = 1, y = 4, u = 3, v = 2$

27. $AA^T = A^T A = I_2$

29. $AA^T = \begin{bmatrix} \frac{2}{a} & 0 & 0 \\ 0 & \frac{2}{a} & 0 \\ 0 & 0 & \frac{2}{a} \end{bmatrix}$

31. $x = 5, y = 6, u = 9, v = 10$

Exercises 11.4

1. $x = 1, y = 3$,   3. $x = 3, y = 4$,   5. $x = 8, y = 9$

7. $x = 1, y = 2, z = 3$ 9. $x = 3, y = 2, z = 1$ 11. $x = 3, y = 3, z = 2$

13. $x = 5, y = 1, z = -2$ 15. $x = -1, y = 2, z = 3$

17. $x = 4, y = -2, z = 3$ 19. $x = 1, y = 3, z = 1$

21. $x = 1, y = 2, z = 3$ 22. $x = 1, y = 2, z = 3$

Exercises 12.1

1.(a) $\det(A) = 39$ (b) $\det(B) = 18$ (c) $\det(C) = -25$

3.(a) $\det(A) = 6$ (b) $\det(B) = -18$ (c) $\det(C) = 6$

5.(a) $\det(A) = 8$ (b) $\det(B) = 16$ (c) $\det(C) = 8$

7. $x = 1$ 9. $x = 2, -1$ 11. $x = \pm 3$

13. $x = \pm 2$ 15. $x = -1, 0, 3$ 17. $n!$

19. $x = 1, 2$

Exercises 12.2

1.(a) 5   (b) -15 (c) -135
(d) -5 (e) 150   (f) 150

3. (i) 6   (ii) 6   (iii) 96   (iv) 12

5. $|A| = \pm 1$ 7. $|A| = 5$ 9. $x = a$

Exercises 12.3

1.(a) $A^{-1} = \begin{bmatrix} 3 & -4 \\ -5 & 7 \end{bmatrix}$ (b) $B^{-1} = \begin{bmatrix} \frac{5}{3} & -\frac{4}{3} \\ -1 & 1 \end{bmatrix}$

(c) C^{-1} does not exist because $\det(C) = 0$

3. $A^{-1} = \begin{bmatrix} 2 & -3 \\ -3 & 5 \end{bmatrix}$, but $\begin{bmatrix} 2 & -3 \\ -3 & 5 \end{bmatrix}^{-1} = \begin{bmatrix} 5 & 3 \\ 3 & 2 \end{bmatrix} = A$

5. $x = -1, 2$ 7. $x = 0, \pm 2$

9. $x = 0, \pm 3$ 11. $x = 1, y = -1$

13. $x = 3, y = 2$ 15. $x = 1, y = 2, u = 2, v = 3$

17. $x = 4, y = 3, u = 2, v = 6$ 19. $x = 4, y = 12, u = 8, v = 16$

21. $x = 7, y = 5, u = 2, v = 1$

23. $k \neq 6, x = 1, y = 3$ 25. $x = 1, y = 2, z = 1, u = 1$

27. $x = 1, y = 2, z = 3, u = 4, v = 5$ 28. $x = 1, y = -5, z = 7, u = 1, v = 25$

Exercises 12.4

The exercises can be easily solved by studying Examples 1–3.

Exercises 13.1

1.(a) $R = \{(1,3), (1,5), (1,7), (3,5), (3,7), (5,7)\}$

  (b) $R = \{(1,5), (3,3), (5,1), (5,7), (7,5)\}$

  (c) $R = \{(1,1), (3,3), (5,5), (7,7)\}$

  (d) $R = \{(3,1), (5,1), (5,3), (7,1), (7,3), (7,5)\}$

3. $R = \{(1,1), (2,3), (2,4), (3,2), (4,4)\}$

5. $R = \{(1,1), (1,2), (3,3), (3,5), (4,5), (5,2)\}$

7.

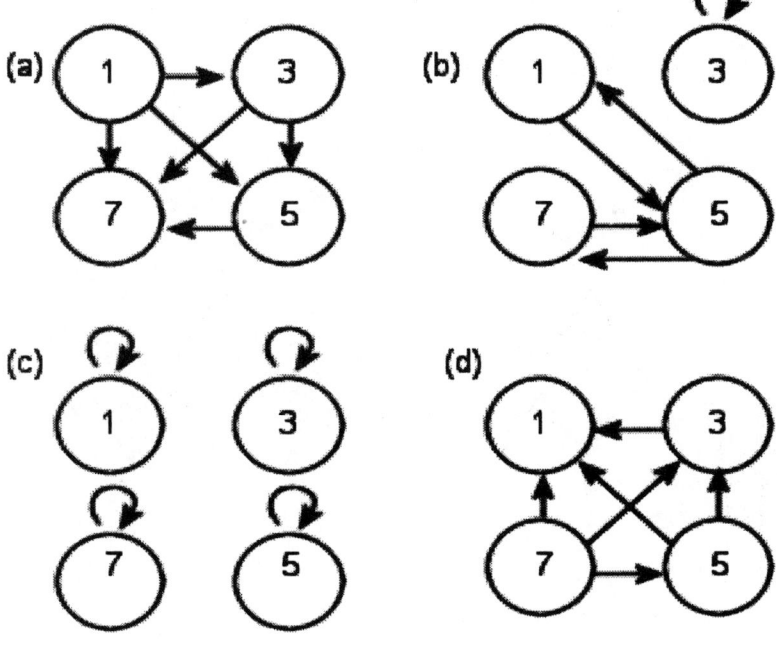

Exercises 13.2

1. The binary tree is given below.

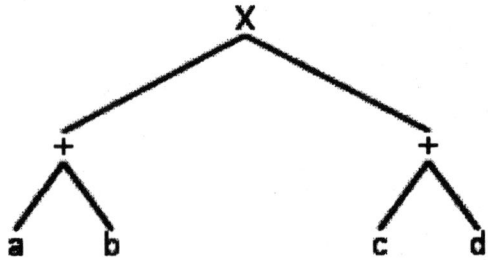

3. The binary tree is given below.

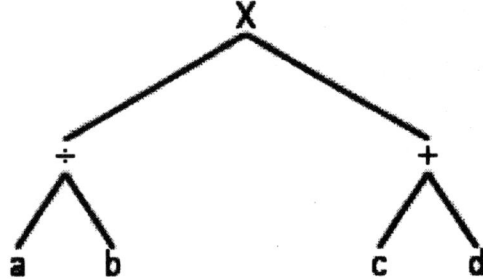

5. The binary tree is given below.

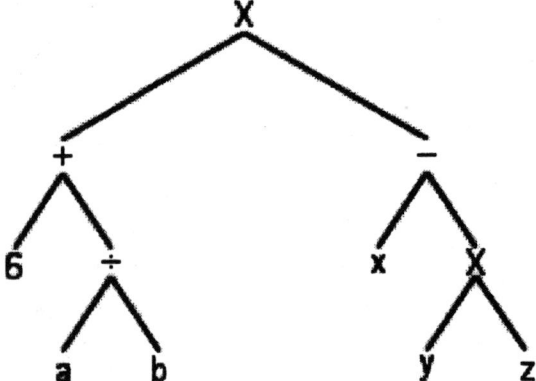

7. The binary tree is given below.

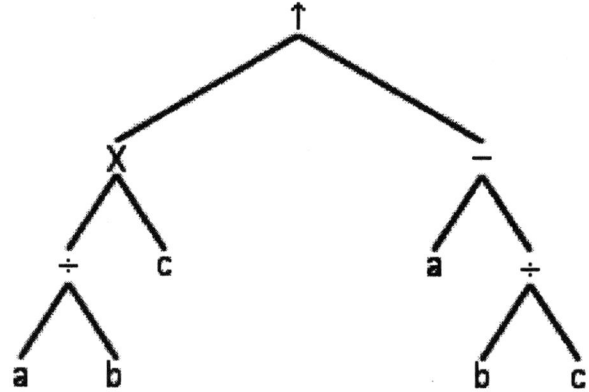

9. $(6 \div (3 \times 4)) + (6 - 4)$

11. $(4 \times (3 + 2)) - (x \div (u + v))$

13. $3 \times ((x + y) + (u - v))$ 15. $(3 \uparrow 2) \times (8 \div (2 \times 2))$

Exercises 13.3

1. $(x \uparrow y) + (a - b)$

3. $(2 \uparrow 3) + (6 \div 2) = 11$

5. $xy - (x + y)$

7. $((4 \div 2) + 2) \times ((4 + 3) \uparrow 2) = 196$

9.

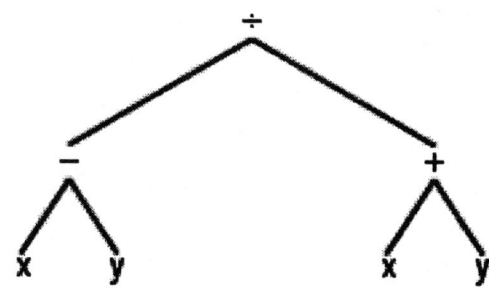

Polish notation is: $\div - xy + xy$

11. Polish notation is: $\times + \div 621 \uparrow +432$

13. $xy - xy + \div$

15. $62 \div 1 + 43 + 2 \uparrow \times$

17. 19

19. 169

21. 5

23. -2

Bibliography

[1] R. N. Aufman, J.S.Lockwood, R.D. Nation, and D.K. Clegg, *Mathematical Excursions*, Houghton Mifflin Company, Boston, (2004).

[2] N.L. Biggs, *Discrete Mathematics*, Oxford, London, (2002).

[3] M. L. Bittinger, *Logic, Proof, and Sets*, Addison-Wesley, New York, (1982).

[4] D. J. Booth, *Foundation Discrete Mathematics for Computing*, Chapman & Hall, London, (1995).

[5] S. S. Epp, *Discrete Mathematics with Applications*, Brooks/Cole, New York, (1995).

[6] T. Feil and J. Krone, *Essential Discrete mathematics for Computer Science*, Prentice Hall,new jersey, (2003).

[7] R. P. Grimaldi, *Discrete and Combinatorial Mathematics*, Addison-Wesley, New York, (1994).

[8] V.E. Hogatt Jr. and I.D.Ruggles, A Primer for the Fibonacci Numbers–Part V, The Fibonacci Quarterly, 2(1), (1964), 59–65.

[9] V.E. Hogatt Jr. and I.D.Ruggles, A Primer for the Fibonacci Sequence–Part V, The Fibonacci Quarterly, 2(1), (1964), 59–65.

[10] R. Johnsonbaugh, *Discrete Mathematics*, Prentice Hall, New Jersey, (1997).

[11] R. Johnsonbaugh, *Algorithms*, Prentice Hall, New Jersey, (2004).

[12] B. Kolman, R. C. Busby, and S. Ross, *Discrete Mathematical Structures*, Prentice Hall, New Jersey, (1996).

[13] B. Kolman, *Linear Algebra with Applications*, Prentice Hall, New Jersey, (1997).

[14] T. Koshy, *Fibonacci and Lucas Numbers with Applications*, Wiley, New York, (2001).

[15] T. Koshy, *Discrete mathematics with Applications*, Elsevier Academic Press, (2004).

[16] A.V.Levitin, *Introduction to the Design and Analysis of Algorithms*, Addison-Wesley, (2003).

[17] J. C. Molluzzo, and F. Buckley, *A First Course in Discrete Mathematics*, Wadsworth, New York, (1986).

[18] A.S. Posamentier and I. Lehmann, *The Fabulous Fibonacci Numbers*, Prometheus Books, New York (2007).

[19] K. Rosen, *Elementary Number Theory and its Applications*, Addison Wesley Longman, New York, (2000).

[20] R. J. Trudeau, *Introduction to Graph Theory*, Dover, New York, (1993).

[21] S. Venit, and W. Bishop, *Elementary Linear Algebra*, PWS, Boston, (1996).

[22] N. N. Vorobiev, *Fibonacci Numbers*, Birkhauser, Boston, (2002).

[23] W.D.Wallis, *A Beginner's Guide to discrete mathematics*, Birkhauser, Boston, (2003).

[24] M. A. Weiss, *Algorithm, Data Structure, and Problem Solving with C++*, Addison-Wesley, New York, (1996).

[25] D. B. West, *Introduction to Graph Theory*, Prentice Hall, New Jersey, (1996).

[26] M.J.Zerger, Mathematics Teacher, 89(1)3, 26.

[27] M.J. Zerger, The Golden State–Illinois, Journal of Recreational mathematics, 24(1) 24–26.

Index